Mercenaries

Edited by
Abdel-Fatau Musah and J. 'Kayode Fayemi

Foreword by Lord Avebury

Mercenaries
An African Security Dilemma

Pluto Press
LONDON

First published 2000 by Pluto Press
345 Archway Road, London N6 5AA

British Library Cataloguing in Publication Data
A catalogue record for this book is available from
the British Library

ISBN 9780745314716 pbk
    0745314716
Library of Congress Cataloging in Publication Data
Mercenaries : an African security dilemma / edited by Abdel-Fatau
Musah and J. 'Kayode Fayemi ; foreword by Lord Avebury.
       p.   cm.
    ISBN 0–7453–1471–7
    1. African—History—1960–   2. Mercenary troops—African. I. Musah,
Abdel-Fatau. II. Fayemi, 'Kayode.
    DT30.2.M47 2000
    960.3'2—dc21   99–34771
                                                                   CIP

Designed and produced for Pluto Press by
Chase Production Services, Chadlington, OX7 3LN
Typeset from disk by Stanford DTP Services, Northampton
Printed on Demand by CPI Antony Rowe, Eastbourne

Contents

About the Centre for Democracy and Development   vii
Foreword by Lord Avebury   viii
Acknowledgements   xi
Abbreviations and Acronyms   xii

Introduction   1

1. Africa in Search of Security: Mercenaries and Conflicts –
   An Overview   13
   Abdel-Fatau Musah and J. 'Kayode Fayemi

2. Private Military Companies and African Security 1990–98   43
   Kevin A. O'Brien

3. A Country Under Siege: State Decay and Corporate Military
   Intervention in Sierra Leone   76
   Abdel-Fatau Musah

4. The Hand of War: Mercenaries in the Former Zaire 1996–97   117
   Khareen Pech

5. Mining for Serious Trouble: Jean-Raymond Boulle and his
   Corporate Empire Project   155
   Johan Peleman

6. Mercenaries, Human Rights and Legality   169
   Alex Vines

7. The OAU Convention for the Elimination of Mercenarism
   and Civil Conflicts   198
   Kofi Oteng Kufuor

8. Understanding the African Security Crisis   210
   Eboe Hutchful

9. Arresting the Tide of Mercenaries: Prospects for Regional
   Control   233
   'Funmi Olonisakin

   Conclusion   257
   Appendices   265
   I   Mercenaries: Africa's Experience, 1950s–1990   265
   II   Convention for the Elimination of Mercenarism in Africa 275
   III OAU Resolution on the Activities of Mercenaries (1967) 281
   IV   OAU Declaration on the Activities of Mercenaries in
        Africa (1971)   283
   V   OAU Convention for the Elimination of Mercenaries in
        Africa (1972)   286
   VI   The Report by the UN Special Rapporteur on the Use of
        Mercenaries, 1998   289

   Notes on Contributors   321
   Index   323

About the Centre for Democracy and Development

The Centre for Democracy and Development (CDD) is a non-profit, non-governmental, independent research, information and training institution dedicated to policy-oriented scholarship on questions of democratic development and peace-building in the West African sub-region, concentrating on the Gambia, Ghana, Liberia, Nigeria and Sierra Leone. The Centre has offices in London, UK, Accra, Ghana and Lagos, Nigeria. Its work is grounded in an appreciation of the practical problems that have inhibited democratic reform in the region, and in the need to strengthen civil society institutions and government bodies in their work. The Centre aims to provide accurate information on and rigorous analysis of democratic developments in the region. It also plans to offer strategic training to people interested in fostering democratic development in the region.

Programme on Conflict Management, Regional Security & Peacebuilding

In seeking to develop an institutional framework for analysing conflict and its links with democracy and development, this programme recognises structural stability as a foundation for sustainable development. By doing this, we acknowledge the importance of a holistic security agenda in which the management of violence is subject to civilian, democratic control. In addition to our work on PMCs we are also actively involved in civil-military relations and peacebuilding training, light weapons research and developing a monitoring mechanism for conflict prediction in West Africa.

Foreword

Lord Avebury

There has been a pressing need for this study, which examines the connections between African conflicts, the extraction of minerals, and the use of private military companies (PMCs). Armed opposition groups exist in Angola, Algeria, Burundi, Central African Republic, Chad, Democratic Republic of Congo, Republic of Congo, Egypt, Ethiopia, Guinea Bissau, Liberia, Rwanda, Senegal, Sierra Leone, Somalia, Sudan, Uganda and Western Sahara. In some cases, the fighting arises from unresolved aspirations towards self-determination, while in others, religious and ethnic differences are at issue. In a significant proportion of these conflicts, however, the ownership and control of valuable resources is a factor, making the outcomes interesting to multinationals, and enabling the state participants to consider using PMCs to help secure victory over their opponents. Johan Peleman traces the astonishing deeds of Jean-Raymond Boulle and his America Mineral Fields, whose tentacles reached into several African countries, with the help of good connections and military assistance.

Despite the adoption of Codes of Conduct for arms sales by the EU, the OSCE and the UN, large quantities of arms and ammunition are still pouring into Africa, cutting into the resources available for development and undermining the continent's ability to catch up with the rest of the world. Some of this arsenal is funded by the award of prospective mineral licences, as in the notorious transaction for the supply of $10 million military equipment to Sierra Leone arranged with the Thai businessman Rakesh Saxena, described by Abdel-Fatau Musah.

In spite of the weakness of international law in regulating the activities of PMCs, a general consensus has evolved that non-state foreign involvement in Africa's conflicts should not be allowed. Recognising the trend, the PMCs are increasingly trying to diversify their activities from pure combat operations to incorporate the provision of military advice, the procurement of arms and the supply of passive security services. For public relations purposes and because of existing legislation against mercenary activity, they usually advertise the latter activity and are cagey about the former. They claim to work only for governments and not for armed oppositions, and they say they operate strictly within the bounds of domestic and international law. Kevin O'Brien says that IRIS, a group formed in June 1998, works for UNITA and is thus an exception to the rule, while Alex Vines claims that there is now an increasing number of mercenary operations conducted with non-state parties. Obviously, UNITA has had a great deal of foreign assistance in using the proceeds of illegal diamond sales to fund large-scale acquisition of military equipment and the boundary between PMCs and arms brokers is not sharply defined. Regardless of any doubts PMCs may harbour working for non-state groups, they have not always acted in the interests of the people, as the example of PMC support for the dictator Mobutu in the former Zaire, dealt with by Khareen Pech, clearly demonstrates.

At the same time South Africa, with its experience of one particular firm, enacted legislation to control the activities of PMCs based on her territory. The obvious flaw with that approach is that this business can be run from anywhere in the world, and individual African countries have no power to regulate PMCs headquartered in the UK, for instance.

International agreements so far hold out no hope either. The OAU Convention for the Elimination of Mercenarism, analysed by Kofi Oteng Kufuor, relies on the definition of a mercenary from the UN Convention, and since they both deal only with foreigners who take a direct part in hostilities, they have no effect on the indirect but nevertheless key impact of the PMCs on several African conflicts. The difficulty of enforcement is possibly the reason why so few countries have ratified the UN Convention. Nor do these Conventions have anything to say about the related operations of arms brokers, and the financial role played by resource extraction multinationals.

At the bottom end of the scale, the PMCs may shade off into Private Security Companies (PSCs), flourishing particularly where, as Eboe Hutchful emphasises, the coercive power of the state has declined. These enterprises are seen as legitimate but have their own problems. When multi-million dollar capital works are undertaken in connection with resource development, the owners need to protect their assets from sabotage, and either the state provides that service, or the owner has to engage a contractor for the purpose. If the armed forces and police of the state are seen as incapable of doing the job, there will be a strong argument for engaging private security guards, who may be more effective, but are obviously less accountable.

There are no easy answers to a complex and difficult problem, as perhaps the fact that some PMCs are themselves calling for regulation demonstrates. Africa will need to establish and refine its own sub-regional and continental conflict resolution mechanisms, to reduce the plague of wars afflicting her peoples; licencing of mineral extraction will have to become transparent so that Africa's wealth is not misapplied to the funding of civil wars, and the arms brokers must be put out of business. J. 'Kayode Fayemi argues that small arms proliferation poses a serious challenge to the demilitarisation agenda in Africa, and here, the enlarged European Union could help, not only by enforcement of its new, stricter code on conventional arms exports, but also by persuading the rest of the OSCE, the source of illicit weapons in many of Africa's conflicts, to adopt the same rules.

Whether those strategies could make mercenaries permanently unnecessary, to use Funmi Olonisakin's vivid phrase, is doubtful. Conflicts are never going to be eliminated entirely from Africa, and as the Legg Report says, '[PMCs] are on the scene and likely to stay on it'. However, the international community can go a long way in drastically reducing the proliferation of PMCs by helping to build the capacity of autonomous conflict prevention mechanisms in the developing world, even while western states adopt proactive national legislation to curtail PMC activity coordinated from their territories. The government has promised a Green Paper on the subject in the next twelve months, and this admirable collection by the Centre for Democracy and Development sets the scene for an intelligent discussion of the issues which must be faced in a timely and comprehensive manner.

Lord Eric Avebury

Acknowledgements

This book has had a long gestation period and many people have contributed to its final emergence. We would like to thank first, the contributors – seasoned researchers and activists – whose deep insights and incisive analyses of the mercenary phenomenon in Africa are brought vividly to the reader in this book. Besides the authors, many others have played significant roles in the course of producing this book. In acknowledgement of that, we wish to express our profound gratitude to Liberal peer and CDD patron, Lord Eric Avebury, for his authoritative grasp of the issues, constant encouragement to us and agreement to write the foreword. We also wish to thank Bjorn Warkalla of the University of Berlin who helped with useful editorial work including significant updating of the appendices, during his internship at CDD and Emmanuel Kwesi Aning of the Centre for Development Research in Denmark for providing hitherto unpublished primary documents. Mohammed Faal helped greatly in gathering material and providing additional editorial support just as Charlotte Jenner's administrative support was crucial in seeing through the project. Finally, we wish to register our appreciation to Roger van Zwanenberg and the editorial staff at Pluto Press for taking a keen interest in the project right from its inception. That *Mercenaries: An African Security Dilemma* is in the public domain is due partly to the unwavering encouragement they gave throughout the book's production.

Notwithstanding the numerous assistance received from various quarters, primary responsibility for the final product lies with the editors and individual authors.

Editors
June 1999

Abbreviations and Acronyms

ACRF	African Crisis Response Force
ACRI	African Crisis Response Initiative
ADFL	*Alliance des forces democratiques pour la libération du Congo* (Alliance of Democratic Forces for the Liberation of Congo)
AFRC	Armed Forces Revolutionary Council
AHSG	Assembly of Heads of State and Government
AMF	America Mineral Fields
ANC	African National Congress
ANC	Armée Nationale Congolaise
APC	All People's Congress
APR	*Armée Patriotique Rwandaise*
ARMSCOR	Armaments Corporation of South Africa
ASAS	Association of Southern African States
BBC	British Broadcasting Corporation
BCL	Bougainville Copper Ltd
BE	Branch Energy
BICC	Bonn International Center for Conversion
BMATT	British Military Advisory and Training Team
BP	British Petroleum
BRA	Bougainville Resistance Army
CAMA	Campaign Against Mercenaries in Africa
CAMEO	Canadian Association for Mine and Explosive Ordnance
CAR	Central African Republic
CCB	Civil Co-operation Bureau
CDF	Civil Defence Force
CEMA	Convention for the Elimination of Mercenaries in Africa
CEO	Chief Executive Officer
CFAO	*Compagnie Française de l'Afrique de l'Ouest*
CIA	Central Intelligence Agency
CNDD	*Conseil National pour la Défense de la Démocratie*
CPP	Convention People's Party
DCC	Directorate of Covert Collection

DGSE	*Direction Générale de la Sécurité Extérieure*
DPKO	Department of Peacekeeping Operations in the UN
DRC	Democratic Republic of the Congo
DSC	Defence Systems Colombia
DSL	Defence Systems Ltd
DST	*Direction de la Surveillance du Territoire*
DTAP	Democracy Transition Assistance Program
EC	European Community
ECOMOG	Economic Community of West African States Monitoring Group
ECOWAS	Economic Community of West African States
ELN	Castroite National Liberation Army
EO	Executive Outcomes (Pty) Ltd or Executive Outcomes CC or Executive Outcomes Ltd
EU	European Union
FAA	*Forcas Armadas Angolanas* (Armed Forces of Angola)
FALA	*Forqas Arinadas de Libertaqdo de Angola* (UNITA's armed wing)
FAR	*Forces Armées Rwandaises*
FAZ	*Forces Armées Zaïroises* (Zairean Armed Forces)
FBC	Fourah Bay College
FCO	*Foreign and Commonwealth Office*
FLEC	Front for the Liberation of Cabinda
FLS	Front Line States
FNLA	*Frente Nacional de Libertaqdo de Angola* (Front for the National Liberation of Angola)
GNP	Gross National Product
GPS	Global Positioning System
GSG	Gurkha Security Guards Ltd
GSPR	*Garde Spéciale du Président de la République*
GURN	Government of Unity and National Reconciliation
HRW/A	Human Rights Watch (Africa)
HSE	Health, Safety and the Environment
ICI	International Chartered Incorporated
ICRC	International Committee of the Red Cross
IDAS	International Defence and Security Ltd
IDDA	International Danger and Disaster Assistance
IFOR	Implementation Force
IGAD	Inter-Governmental Authority on Development
IMF	International Monetary Fund
ISC	International Security Consultants
ISDSMC	Inter-State Defence and Ministerial Committee
JPMC	Joint Political-Military Commission
KLA	Kosovo Liberation Army
KMS	Keeny-Meeny Services

LLDC	Least of the Less Developed Countries
MAP	Mass Awareness and Participation
MHC	Mercenary-Hiring Company
MK	*Umkhonto we Sizwe* (Spear of the Nation)
MMD	Movement for Multiparty Democracy
MNF	Multinational Force
MOJA	Movement for Justice in Africa
MPLA	*Movimento Popular de Libertaqdo de Angola* (Popular Movement for the Liberation of Angola)
MPRI	Military Professional Resources Incorporated
MSP	Military Stabilisation Program
MTS	Military Technical Services
NATAG	Nigerian Technical Assistance Group
NATO	North Atlantic Treaty Organisation
NCDD	National Council for the Defence of Democracy
NCO	Non-commissioned Officer
NEC	National Electoral Commission
NGO	Non-governmental Organisation
NIS	National Intelligence Service
NPFL	National Patriotic Front of Liberia
NPRC	National Provisional Ruling Council
OAS	Organisation Armée Secrète
OAU	Organisation of African Unity
ONUC	*Opérations des Nations Unies au Congo*
ONUMOZ	United Nations Operation in Mozambique
OSCE	Organisation for Security and Co-operation in Europe
PANAFU	Pan-African Union
PESC	Public enterprise security company
PMC	Private Military Company
PNG	Papua New Guinea
PONAL	Colombian National Police
RENAMO	National Resistance Movement of Mozambique
RPF	Rwandan Patriotic Front
RSLMF	Republic of Sierra Leone Military Forces
RUF	Revolutionary United Front
SAAF	South African Air Force
SADC	Southern African Development Community
SADCC	Southern African Development Co-ordination Conference
SADF	South African Defence Force (before April 1994)
SANDF	South African National Defence Force (after April 1994)
SAP	South African Police
SARM	*Service d'Action et de Renseignement Militaire*
SAS	Special Air Service
SAS Ltd	Security Advisory Services Ltd
SCOA	*Société Commerciale Ouest Africaine*

SDB	Department of State Security
SDM	*Sociedada de Desenvolvimento Mineiro de Angola* (Mining Development Society of Angola)
SFOR	Stabilisation Force
SGS	Special Gurkha Services Ltd
SI	Sandline International
SIS	Secret Intelligence Service
SLPP	Sierra Leone People's Party
SMC	Standing Mediation Committee
SOF	Soldier of Fortune
SOFA	Status of Forces Agreement
SPLA	Sudanese People's Liberation Army
SPS	Special Project Service Ltd
SRC	Strategic Resources Corporation
SRL	Sierra Rutile Ltd
SSD	State Security Department (Sierra Leone)
TeleServices	*Tele Service Sociedade de TelecomunicaVoes, Seguranqa e Servicos*
ULIMO	United Liberation Movement for Democracy in Liberia
UN	United Nations
UNAMIR	United Nations Assistance Mission for Rwanda
UNAVEM	United Nations Angola Verification Mission
UNDP	United Nations Development Programme
UNHCR	United Nations High Commissioner for Refugees
UNITA	*Unido Nacional para a Independencia Total de Angola* (National Union for the Total Independence of Angola)
UNITAF	Unified Task Force (Somalia)
UNOMIL	United Nations Observer Mission in Liberia
UNOSOM	United Nations Operation in Somalia
UNPROFOR	United Nations Protection Force
UXO	Unexploded Ordnance
WHO	World Health Organisation
WNLF	West Nile Liberation Front

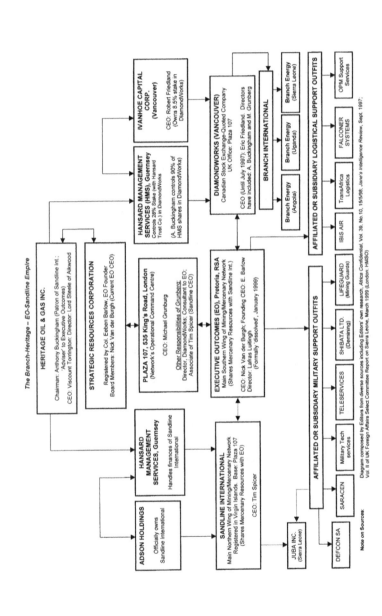

The Branch-Heritage – EO-Sandline Empire

ADSON HOLDINGS
Officially owns
Sandine International

HANSARD MANAGEMENT SERVICES, Guernsey
Handles finances of Sandine International

HERITAGE OIL & GAS INC.
Chairman: Anthony Buckingham (Patron of Sandine Int.;
Adviser' to Executive Outcomes)
CEO: Viscount Torrington: Director: Lord Steele of Aikwood

STRATEGIC RESOURCES CORPORATION
Registered by Col. Eeben Barlow, EO Founder
Board Members: Nick Van der Burgh (Current EO CEO)

PLAZA 107, 535 King's Road, London
(Network's Operational Command Centre)

CEO: Michael Grunburg

Other Responsibilities of Grunberg:
Director, DiamondWorks; Consultant to EO:
Associate of Tim Spicer (Sandine CEO)

SANDLINE INTERNATIONAL
Main Northern Wing of Mining/Mercenary Network
Registered in Virgin Islands. Base: Plaza 107
(Shares Mercenary Resources with EO)

CEO: Tim Spicer

EXECUTIVE OUTCOMES (EO), Pretoria, RSA
Main Southern Wing of Mining/Mercenary Network
(Shares Mercenary Resources with Sandine Int.)

CEO: Nick Van der Burgh; Founding CEO: E. Barlow
Director: Lafras Lutfingh
(Formally 'dissolved', January 1999)

IVANHOE CAPITAL CORP.
(Vancouver)

CEO: Robert Friedland
(Owns 8.5% stake in DiamondWorks)

HANSARD MANAGEMENT SERVICES (HMS), Guernsey
Controls 26% Stake (with Hansard Trust Co.) in DiamondWorks

(A. Buckingham owns 90% of HMS shares in DiamondWorks)

DIAMONDWORKS (VANCOUVER)
Canadian Stock Exchange-Quoted Company
UK Office: Plaza 107

CEO (Until July 1997): Eric Friedland. Directors
have included: A. Buckingham and M. Grunberg

BRANCH INTERNATIONAL

Branch Energy
(Angola)

Branch Energy
(Uganda)

Branch Energy
(Sierra Leone)

AFFILIATED OR SUBSIDIARY MILITARY SUPPORT OUTFITS

DEFCON SA

JUBA INC.
(Sierra Leone)

SARACEN

Military Tech
Services

TELESERVICES

SHIBATA LTD.
(Denning)

LIFEGUARD
(Mining Guards)

AFFILIATED OR SUBSIDIARY LOGISTICAL SUPPORT OUTFITS

IBIS AIR

TransAfrica
Logistics

FALCONER
SYSTEMS

OPM Support
Services

Note on Sources: Diagram composed by Editors from diverse sources including Editors' own research. *Africa Confidential*, Vol. 39, No. 10, 15/5/98. *Jane's Intelligence Review*, Sept. 1997;
Vol. II of UK Foreign Affairs Select Committee Report on Sierra Leone, March 1999 (London: HMSO)

Anatomy of Interlocking Mining and Mercenary Network

Introduction

The widespread deployment of mercenary forces in Africa's internal conflicts in the 1990s underlines an increasing acceptance by the international community of a profession until now considered pariah. This growing influence of corporate armies is occurring at a particular point in history worth explaining. Mercenary forces were used widely by the colonising powers in the 1870s and subsequent years when they first invaded the continent. They are being used again now, as the continent's main resources are being exploited on the cheap by transnational extracting companies. Are mercenaries becoming the shock forces of corporate recolonisation, or, as some contend, a 'necessary evil' in out-of-area conflict management in the post-Cold War arena?

Whichever way it is looked at, mercenary intervention has defined and introduced a new dimension to violent internal conflicts and the scramble for resources in peripheral states since the mid-1980s. States which have experienced mercenary intervention include Angola, the Democratic Republic of Congo (formerly Zaire), Congo-Brazzaville, Liberia, Senegal, Sudan and Sierra Leone in Africa; the Balkan states in Europe and Papua New Guinea in the Indian Ocean. 'Private Security', a qualitatively new form of mercenary activity, has thus become prominent on the agenda of global peace discourse since the end of the Cold War. Even as superpower-inspired proxy wars in the world's peripheral states are being replaced by regionalised wars within national borders, old-fashioned mercenaries of the 1970s, mostly contracted by foreign powers, are being transformed into corporate and increasingly autonomous mercenary armies to take advantage of the new security dilemma. Private military companies (PMCs) are thus assuming an important role in the balance of power both within individual states and in international security as a whole.

Many reasons have been advanced to explain this new phenomenon. Some argue that the mounting financial and human costs incurred by multilateral intervention in seemingly intractable civil wars have brought about intervention fatigue among states that have traditionally supported these efforts. Another reason is that the proliferation of low-intensity conflicts around the world has stretched multilateral peacekeeping resources to the limit; supplementary sources of intervention have therefore become a necessity, hence the hiring of these corporate dogs of war.

There are also those who see an increasingly blurred line in cases when private armies are invited into conflict zones by incumbent governments. They argue that since most of the conflicts which have involved private security arrangements in recent times have been primarily internal, any effort by a *de jure* government to protect regime security should be considered a legitimate cause for intervention. Finally, supporters of private military companies see multilateral peacekeeping as too costly, slow, incompetent and ineffective. This, they claim, is because it is premised on principles of neutrality, consent of warring factions and intervening forces – factors that put too many bureaucratic hurdles in the way of effective intervention.

In the light of the above, some scholars see mercenary outfits and private armies as the peace-enforcer *par excellence* in the post-Cold War era, particularly in weak, resource-laden states that have been stripped of superpower protection amidst increasing threats of collapse from within. Some mercenary-watchers see private militaries as the best equipped to deliver in post-Cold War conflict situations and are eager to attribute impressive developmental and humanitarian qualities to them:

> Private forces can start up and deploy faster than multinational forces, and may carry less political baggage, especially concerning casualties ... They have clearer chains of command, more readily compatible military equipment and training, and greater experience of working together than do *ad hoc* multinational forces. [In addition] they may be financially less expensive than other foreign forces.[1]

Others, in pointing out the readiness to take sides in internal conflicts as an advantage PMCs have over multilateral forces, insist that 'The growth of [private] military companies is a private sector response in part related to the shortcomings of negotiations in resolving [conflicts].'[2]

What is really influencing the upsurge of mercenarism in the 1990s? There are four interconnected reasons, which the following chapters explain to a greater or lesser extent.

1. The consequences of the defeat of the American ground forces in Vietnam have persuaded a generation and more of American and Western generals that the use of Western and particularly American ground forces in foreign conflicts is a mistake, as body bags arrive back home with all the attendant domestic publicity. The increas-

ingly negative public opinion within traditional interventionist states about deployment of troops in out-of-area conflicts has thus strengthened intervention reluctance.

2. Western powers have, over the last 150 years, been willing to intervene militarily anywhere in the world whenever it suited their interests, without domestic hindrance or influence. The methods of that intervention were seriously curtailed by the Cold War, due to the threat posed by the rival claim of the Warsaw Pact states to global strategic space. Since the fall of the Berlin Wall, however, the era of 'intervention' has returned, so long as that intervention can be suitably dressed up for public consumption, and can be done without loss of men and materials. The deployment of mercenaries makes probable such a scenario.

3. We live in an era of neoliberalism. The main thrust of monopolistic finance capital is directed at the continuous search for markets and the usurpation of the traditional domain of the state. The neoconservative philosophy that sees the state as overburdened and incompetent in matters of political economy has led to pressures on governments to download. Consequently, private enterprise has become the ultimate lever to pry concessions from, and roll back, the state. The Weberian notion is that the modern state emerged because of its successful claim to monopoly of the legitimate use of physical force in the enforcement of its order. Today, however, the increasing encroachment of sub-state and transnational groups on the traditional zones of state prerogative – citizen's welfare, the prisons, general security, intelligence, and defence of territorial integrity of states – means that the maintenance of sovereignty is slipping out of the grip of governments. This is becoming even more evident in the peripheral states. Thus, according to Nossal,

> engaging the private sector to provide public security efficiently, effectively, and relatively cheaply is the logical outgrowth of a neoconservative theology that preaches that the private sector will do a more competent job building, owning, and profiting from what used to be the 'public works' in many countries: highways, bridges, prisons, garbage disposal, waterworks, electricity grids.[3]

Instruments of the market, such as COCOM, the now defunct Co-ordinating Committee on Multilateral Exports, were key in bringing Stalinist economies of the East to their knees, forcing the pace of capitalist restoration and formally ending the Cold War. The end of the Cold War, in turn, has brought serious repercussions for the relations of states to questions of security. No more, it seems, is there a need to maintain bloated armies in the East and West – hence the massive downsizing of the militaries in the North, and the creation of a huge labour pool of potential mercenaries. Nor are there any compelling reasons for the East or the West to continue guaranteeing the security of client peripheral states, whose value to these powers did not go beyond that of pawns in now obsolete ideological and strategic battles. The absence of superpower protection and the increasing sub-state threats from within have made African and other weak states most vulnerable to explosive conflicts.

Meanwhile, IMF/World Bank insistence on austerity has cut budgets not only for social security, but also for the militaries in these countries. The result is increased internal threats and inadequate resources to contain them. Endemic poverty, corrupt political leadership, economic, ethnic and religious strife have led sometimes to virtual collapse of state structures, creating what is now termed a 'security vacuum in Africa'. Mercenary organisations, whether the classical 'vagabond' archetype or the refined PMC variant, are staking a compelling claim to this vacuum, taking advantage of the lethargy and lack of political will within the international community to intervene directly in peripheral conflicts, against a background of increased marginalisation of African concerns in multilateral fora.

4. Africa has been a rich continent in terms of natural resources. Western nations have always been apprehensive about owning and using these resources for the benefit of their own people. An environment unstable to Western investment, particularly in the extraction industry, poses a threat to the unfettered exploitation of these resources. However, the growing partnerships between Western transnationals and private mercenary companies ensure that those mercenaries are paid with part of these resources, thus allowing the transnationals to maintain their ownership and exploitation of resources within the Western fold.

The continent of Africa has been the most cruelly exploited of any part of the world for over 300 years. The superficial independence of the 1960s has led to a growing impoverishment of the region. This impoverishment has now come to a head, as the diverse groups within the region fight among themselves for the crumbs of this precarious system. The continent is in serious turmoil, and while the Western powers wish to maintain their hold on its resources, they do not want to lose their own skin in the process. The hope is that mercenary forces will keep Africa safe for multilateral banks and companies in the millennium.

Preying on the vulnerability of kleptocratic regimes, corporate armies are repackaging violence in pseudo-market frills, with their eyes firmly set on creating safe havens around enclaves that are rich in natural resources. As Harding noted, 'what they [the mercenary outfits] have done so successfully to date, is to interpret political instability in Africa as a market issue, and they [have] positioned themselves perfectly in that market'.[4]

Who are these private armies? What are their real objectives? What distinguishes them from earlier mercenaries? What are the interrelationships between them and Western/transnational institutions? Unchecked, what would be the eventual impact of their activity on the security landscape of Africa? How is international and national legislation keeping pace with the rapidly changing face of mercenarism against the backdrop of a changing global security environment? These are some of the questions this book attempts to grapple with.

Contributors to the book have, in one way or the other, come mentally and physically in contact with the phenomenon of private peace-enforcement in Africa. They include seasoned researchers in peace studies and activists from the non-governmental sector specialising in security, complex emergencies and humanitarian law. Several of them have had direct experience with mercenary activities in the African theatre of war, both as victims and monitors. Their incisive collective analyses of the conflicts unfolding in Africa, and the role of freelance armies, have been harnessed as part of the efforts by the Centre for Democracy and Development and its partners to contribute to the understanding of the security conundrum in post-Cold War Africa, with a view to offering viable policy options for its management.

Brief Overview of the African Encounter with Mercenarism

Mercenarism – the practice of foreign professional soldiers freelancing their labour and skills to a party in conflict for fees higher and above those of soldiers of the state in conflict – is as old as conflict itself. Mercenary activity triggered the First Punic War as far back as 264 BC. Earlier, in 334 BC, Greek mercenaries were used by Persia in the war against Alexander the Great. Examples of mercenary practice were prevalent during the Roman Empire, and even today, in the Vatican, the special Swiss Guard is still used for the protection of the Papacy.

Closer to our era, the use of mercenaries by major European colonial powers in the attempt to maintain control of overseas territories is well documented. Invariably, such special troops have been hired to carry out assignments that regular armies would not be called upon to perform, because they are either morally repugnant and/or illegal under assumed or laid down rules of engagement in armed conflict under international law. These assignments included sabotage operations in rival empires, crushing rebellions, assassinations and scorched earth operations against rebellious subjects.

In Africa, the genesis of mercenary activity in the 1960s coincided with the peoples' struggles to assert their right to self-determination and independence. Mercenarism was characterised by indisciplined individuals or small groups loosely organised in bands. Nossal called them 'vagabond mercenaries'.[5] Contract killers such as 'Mad' Mike Hoare, Jacques 'Black Jack' Schramme, Callan and Bob Denard (real name Gilbert Bourgeaud) have become household names for all the wrong reasons: they were hired by former colonial powers and other external interests to undermine the rights of the peoples of Algeria (1956), the Congo (1960s), the Comoros Islands (1970s–90s), Benin (1970s) to choose their leaders and assert their political independence, or to destabilise sovereign states (Guinea, Angola, Seychelles, Mozambique). They assassinated Heads of State of sovereign nations (Lumumba in the Congo, Ahmed Abdallah on the Comoros Islands), and staged the violent overthrow of, or attempted coups against, governments.

In the aftermath of the Cold War and with the exacerbation and pro-liferation of internal conflicts in Africa, clones of Bob Denard such as Eeben Barlow, Tim Spicer, Nick van der Berg, now with a corporate identity, are squeezing into the security vacuum in Africa created jointly by the hasty exit of the superpowers and the implosion of states. Internal conflicts that always ran concurrently with inter-state wars but were suppressed by Cold War priorities, have intensified, proliferated, and threatened to tear states apart. Above all, they have forced their way on to the public agenda. Unfortunately, the huge interest generated by these conflicts at the level of rhetoric has not been matched by a vigorous practical international response. At best, attempts by the international community to respond to these crises have been half-hearted and inadequate, as was demonstrated in the UN peacekeeping efforts in Angola and Somalia.[6] At times, the response, such as that to the conflicts in Somalia and Rwanda, has been an outright disaster.

Mercenary organisations and their business associates have taken advantage of the inertia within the international community. They have presented themselves as the only cost-effective and efficacious alternative. Apologists for mercenaries use the same excuse to proffer intellectual rationalisation for, and urge international acceptance of, the work of mercenaries. Those opposed to private intervention in wars have tried to expose the hollowness of corporate mercenary propaganda behind the atrocities and exploitation committed daily by these soldiers of fortune in Africa, and campaign for the international community to live up to its post-Cold War responsibility as the guarantor of global security. This book is an attempt to positively influence this debate. It does so by subjecting private military activity in Africa to surgical scrutiny. It unravels the dialectical interconnections between mercenarism, increased impoverishment and unending conflicts in Africa. Beyond the corporate mercenaries themselves, the book throws a critical searchlight on the forces that aid and abet this activity either by hiring, promoting or condoning it. Finally, the book proposes sustainable conflict management policy options which, in the view of the authors, are capable of resolving the African security conundrum without recourse to mercenaries.

Structure of Book

The chapters have been carefully arranged to cover the entire phenomenon of private intervention in conflicts in a free-flowing, logical and intelligible manner. The first third of the book provides a comprehensive overview of the African experience of mercenarism. It follows the evolution of the mercenary trade on the continent, linking historical epochs and strategic goals to the nature and forms of corresponding mercenary activity. The next third undertakes a case study of contemporary mercenary intervention in internal conflicts. It unravels the convoluted international networks underpinning the mercenary business: the interconnections between foreign powers, their intelligence networks, mining transnationals, financiers, African states and their leaders/warlords. Central to this discourse are the post-Cold War rivalries among foreign and regional powers in Africa, as well as the fierce inter-mercenary struggles for control over the African conflict market. Besides the analyses of the political economy of state inversion and foreign or private military intervention, the case studies take up related themes such as the spread and use of light weapons and of inhumane and illegal weapons systems. They bring into focus the illegal trade in drugs and minerals, the upsurge in banditry and child warfare and the issue of human rights and international law.

The concluding section examines three important issues related to corporate conflict intervention. First, it subjects existing legislation on mercenaries to a stringent critique, offering practical suggestions as to how it could be updated and, more importantly, effectively implemented by leading actors, including the ordinary potential and real victims of mercenary activity in Africa. This is followed by an in-depth analysis of the security crises that have faced Africa since independence and the outcomes of various attempts by states, policy-makers and sub-state groups to manage them. Finally, this section discusses existing multilateral and regional conflict management mechanisms in Africa and how they could be transformed to keep pace with the fluid nature of post-Cold War conflicts and to exclude mercenaries.

Several of the chapters discuss similar issues or themes in varying degrees of detail. They also profile some of the better known mercenary outfits. At first sight, this seems repetitive and therefore a structural weakness in the book. However, given the rapid mutations, relabelling and ramifications that characterise the modern mercenary industry, and the lengths to which mercenary groups can go in the attempt to hide their corporate identities, the reader has a better opportunity of solving the mercenary jigsaw puzzle by piecing together the complementary inputs supplied by different contributors.

Synopsis

In Chapter 1, 'Kayode Fayemi traces the dialectical relationships between the nature and dynamics of conflicts that have unfolded in Africa since the anti-colonial revolution, with the structural and motivational transformation of mercenaries to adapt to each historical change. He analyses the genesis of mercenary intervention in Africa, linking it to the peoples' struggles for self-determination and against encroachment by external forces on their wealth. With the dependency syndrome inherent in Africa's political economy, Fayemi sees any external intervention in African conflicts that 'does not seek to address the root causes of these conflicts' only as an inflammatory factor to internal violence and, consequently, a mortal danger to Africa. He places corporate mercenary intervention, which is principally motivated by potential super-profits, squarely in this category.

Fayemi believes that to construct a viable security paradigm for Africa there is a need to understand the nature and dynamics not only of the military dimension but, more importantly, of the many and diverse threats that cross-fertilise and reinforce one another into complex emergencies. In the same vein, he takes issue with those who see the mercenary dilemma solely in technical terms, without regard to the economic, social, political and humanitarian consequences of private military intervention in African conflicts, and condemns their pseudo-intellectual rationalisation of such intervention.

In Chapter 2, Kevin O'Brien undertakes a comprehensive study of PMCs in Africa since 1990. In greater detail than perhaps any other researcher has achieved to date, O'Brien traces the role and extent of private military intervention involvement in Africa, queries the factors that underpinned the dramatic entry of PMCs into the African security scene and examines the clients for whom the PMCs work. He does this by constructing a typology of the interventions in terms of the number of military companies involved and their interrelationships in each particular African conflict since 1991. His aim is to paint a much clearer picture of the trends in African conflict resolution over the past decade. The chapter is a revelation, exposing the often underestimated extent of mercenary infiltration and size of the mercenary-business network active on the African continent.

In Chapter 3 (A Country Under Siege), Abdel-Fatau Musah takes up the case study on the civil war in Sierra Leone with special emphasis on private and foreign peace-enforcement. This chapter examines the role of mercenary outfits, particularly Executive Outcomes (EO) and Sandline International (SI), and other foreign forces in the Sierra Leone conflict within the context of an integrated and comprehensive security paradigm. To achieve this goal, Musah analyses the Sierra Leone crisis and the concomitant foreign intervention by querying two interrelated factors: firstly, the role of the internal political economy, and also the social and generational contradictions, in the process of state inversion and explosion of violence. Secondly, the influence of the market-driven global political economy on these internal processes. Finally, he asks whether the drama unfolding in Sierra Leone can be classified as an assertion of sovereignty or a capitulation to corporate recolonisation.

Khareen Pech continues the theme of state inversion and foreign military intervention in Chapter 4, The Hand of War, as she looks at the role of soldiers of fortune in the last days of Mobutu Sese Seko's kleptocracy in Zaire. The chapter throws light on the frantic efforts of the Mobutu clan to hire mercenaries in a desperate, last-ditch but doomed attempt to stop the corrupt regime falling to Kabila's rebel forces in 1997. For the first time, we see the introduction of Serb mercenaries, fresh from their ethnic cleansing genocidal campaigns in the Balkans, to the conflict arena in Africa. Through Pech's narrative of the Congo drama, the reader once more sees the frenetic jostling among mercenary forces – EO, French, Belgian and Serb groups – for lucrative financial and mineral contracts from a failed regime on the brink of extinction. We are also familiarised with the web of interconnections between foreign intelligence outfits and mercenaries and with the conflicts of interest and lack of morality that underpin the mercenary business. Pech introduces yet another underlying theme to mercenarism – the phenomenon of a regionalised war within a single state – as she describes the role of Rwanda, Uganda and Angola in the fall of Mobutu.

But it was not the besieged Mobutu regime alone that hired the services of mercenary outfits. Kabila's rebel alliance was also a target of mercenary forces. This is the subject of the fifth chapter, Mining for Serious Trouble, in which Johan Peleman discusses the frenetic efforts of mining financiers to secure lucrative deals with Kabila and his alliance even before the warring general had captured power. This chapter, like the preceding one, exposes the links between foreign powers and shady businessmen behind the deployment of mercenary forces in African conflicts. Beyond that, Peleman follows the transformation of a major mining financier – Jean-Raymond Boulle – from petty investor into major player in the mining and mercenary business in Africa. This chapter once again throws light on the fierce struggles between rival mining companies, backed by opposing mercenary outfits, for control of the African conflict market.

In the next chapter, Mercenaries, Human Rights and Legality, Alex Vines casts serious doubts on the oft-trumpeted success stories of mercenary outfits such as EO and Sandline in internal conflicts. He analyses the involvement of mercenaries in conflict areas as far apart as Papua New Guinea and Angola, making use of his own observations and competent reports such as the UN Special Rapporteur's account of mercenaries. He concludes that mercenary outfits have no perceivable role in a conflict management that seeks to root out the underlying causes of war. On the contrary, mercenaries exacerbate conflict situations around the globe, and he documents a catalogue of human rights abuses by mercenaries. Central to Vines's thinking is the future of mercenarism. He distinguishes private military companies from private security companies and concludes that while the days of the former are numbered, there is a role for the latter in medium-term conflict management as long as they do not slide into mercenary activities. He accuses PMCs of many sins – the erosion of national self-determination, the lack of transparency, their training in psychological warfare against civilians and their use of combatants with track records of human rights abuse. He acknowledges that, at times, genuine mining companies and other groups may require the services of security guards in hostile environments, but calls on these groups to implement contractual and procedural structures to ensure respect for human rights in their security arrangements.

Kofi Oteng Kufuor follows up Vines's call for strong regulatory mechanisms to control mercenarism by subjecting *The OAU Convention for the Elimination of Mercenaries and Civil Conflicts* to close scrutiny in Chapter 7. He probes the issue of whether the OAU Convention actually prohibits incumbent governments in Africa from relying on the services of mercenaries for the purpose of suppressing rebel uprisings. He examines the various approaches that could be used to interpret the Convention and the resulting uncertainty about whether its provisions bar governments from recruiting soldiers of fortune. The chapter concludes by suggesting means by which the uncertainty could be cleared up.

Chapter 8 – Understanding the African Security Crisis – is a masterly treatment of the concept of security by Eboe Hutchful. He discusses the underlying philosophical understandings of 'security' and probes the interconnections between security and such concepts as 'the legitimate use of force', 'the modern state' and sovereignty. He analyses the dramatic changes that have occurred in the appreciation of 'security' in Africa in the last one and a half decades. Hutchful traces the transformation of security into a 'racket' and links this development with the emergence of strongmen, warlords and mercenaries, creating what he terms 'the ruler's dilemma' in Africa. He looks at the exacerbating influences that have been exerted on this dilemma post-Cold War by the pervasive powers of liberalisation and the global military-industrial complex. He concludes by proposing policies and structures that are capable of pulling the continent back from the insecurity precipice. Among his proposals are social and cultural demilitarisation, a paradigm shift from security as a military concept to its appreciation as a comprehensive entity based on structural stability – that is, on sustainable development aimed at eradicating poverty and ensuring good governance, inclusive politics and the respect of basic rights. He laments the jettisoning of the idea of an African security umbrella at the inception of the OAU in 1963 and calls for a return to this idea through the empowering of regional security frameworks.

Funmi Olonisakin picks up the theme of regional security mechanisms in the concluding chapter, Arresting the Tide of Mercenaries: Prospects for Regional Control. She explores the potential available in sub-regional organisations to develop mechanisms for conflict management that would render mercenary intervention in conflicts both impossible and irrelevant. Along the way, she takes issue with advocates of regulating mercenary activity by querying the basis of their position. Olonisakin does not gloss over the deficiencies of existing sub-regional security structures such as ECOMOG and SADC. Among their limitations she lists rivalries between states, the hesitation to speak up on gross human rights abuses in neighbouring states, and poverty that limits capacity-building for efficient conflict intervention. But she does not suggest that the baby be thrown out with the bath water. In evolving sub-regional structures to contain internal conflicts in Africa, Olonisakin sees the need for a critical role for responsible and accountable regional hegemonies, such as Nigeria and South Africa, as well as a rethink of the link between instability in a particular state and wider regional security.

All told, this ground-breaking book raises an alternative voice in the private security debate, and it is hoped that the arguments advanced by its contributors will introduce a much needed balance in the discourse about internal conflicts and conflict management tools in Africa and, in particular, have an influence on mercenary activities in the African security landscape.

Introduction Notes

1. Herbert M. Howe, 'Private security forces and African stability: the case of Executive Outcomes', *The Journal of Modern African Studies*, No. 36, February 1998, pp. 308–9.
2. David Shearer, 'Exploring the Limits of Consent: Conflict Resolution in Sierra Leone', *Millennium: Journal of International Studies*, No. 3, Vol. 26, 1997, p. 859; and also private communication with author.
3. Kim Richard Nossal, 'Roland Goes Corporate: Mercenaries and Transnational Security Corporations in the Post-Cold War Era', *Civil Wars*, Vol. 1, No. 1 (Spring 1998), p. 31.

1

Africa in Search of Security: Mercenaries and Conflicts – An Overview

Abdel-Fatau Musah and J. 'Kayode Fayemi

The time has come to begin a policy of constructive engagement with military companies. Doing so may create possibilities for these companies to complement international and regional peace-keeping efforts ... There is a role for military companies, particularly where they are already involved, in building and sustaining stability ... What is required is a more pragmatic approach that assesses the effectiveness of – and engages with – private armies.

D. Shearer, *Private Armies and Military Intervention*

[Executive Outcomes'] experience has demonstrated that established private groups enjoy potential advantages over State-centric militaries. Private forces can start up and deploy faster than multinational (and perhaps) national forces and may carry less political baggage ... Additionally, they have a clearer chain of command, more readily compatible military equipment and training, and greater experience of working together than do ad hoc multinational forces.

Herb Howe, *Private Security Forces and African Stability:*
the Case of Executive Outcomes

Private Military Companies (PMCs) are not 'arms dealers', but are more packaged services providers ... The annual cost of implementing and managing the concepts of PMCs is likely to be far less than the cost of even one UN peacekeeping force and, if PMCs continue to take root, the likelihood of their use in the place of an 'official' UN sponsored deployment will continue to increase ...

Sandline International Promotional Brochure

With the seemingly endless nature of internal conflicts in parts of Africa, a growing trend has emerged among scholars and policy-makers to establish a causal link between the cessation of conflicts, failure (or absence) of international action and the rise of the 'good mercenaries' or 'private security forces'. The involvement of Executive Outcomes in Angola first drew attention to this phenomenon in post-Cold War Africa, even though it has existed in various forms throughout the 1970s and 1980s. A recent investigation into the activities of Sandline International in Sierra Leone by the UK Parliament has brought into sharper focus the role of private armies in African conflicts.[1]

In search of security and quick-fix solutions to problems of internal conflicts, international security scholars of the realist school have tended to emphasise the coercive power of the 'established' and 'legitimate' State and its agents (this time a private security force) as the primary source of stability and order. Just as the realist conception of international politics helped to preserve the status quo in the Cold War era by dominating the policy-makers' agenda with concepts like 'stability', 'order' and 'balance of power', and was used to explain and justify all manner of Cold War era gung-ho diplomacy, the new school of conflict resolution decries innovative peacekeeping methods in the form of regional peacekeeping mechanisms. It criticises their supposedly ineffective even-handedness, highlights the shortcomings of negotiated settlements and promotes the virtues of military force in the resolution of conflict.[2] By reading violent sub-national challenges to the supreme authority of the nation-state as irrational, and expressing disgust at the methods used by some belligerent parties and the need to 'do something' about these 'barbaric acts', the potent identity-based factors and wider perceptions of economic and social injustice which combine to fuel these conflicts conveniently escape serious attention. In the same vein, the idea of accountable governance and democracy as conflict management tools without recourse to violence is quickly dismissed as 'utopian', 'too imprecise' and 'idealistic'. Even so, these writers find ready audiences in foreign exchequers and multilateral agencies where isolationist policy-makers are always looking for escape routes and a justification for the abdication from international responsibility and obligations.[3] The words of Stanley Hoffman, written 22 years ago, about how 'realist principles came to define the terms of policy relevance and a steady stream of realist scholars found themselves not merely in the corridors but also the kitchens of power'[4] still ring true.

Viewed historically though, it is difficult not to be shocked by the realist revisionism that dominates much of the mainstream contemporary scholarship on this subject,[5] which betrays a bias towards conflict management through military force, rather than towards transforming the existing order whose malaise has been at the root of the conflict. The idea that contemporary identity-based conflicts will respond to the available tools of conflict management reminiscent of the Cold War era is one that they hold dear. Words like 'utopian', 'wishful thinking' and 'idealist' are therefore common in their descriptions of alternative visions of collective security and conflict management that call for a shift in focus from the immediate to the long-term, a reorientation of perspectives from the superficial symptoms to root causes and a concentration on process, rather than event. Although they agree that the nation-state no longer holds a monopoly over coercive instruments, they remain nostalgic about its primacy and expect unquestioning sub-national loyalties to the state even as it is caught between the paradox of transnational and sub-national loyalties in a world without borders. This is of course a major problem in Africa where boundaries of a state rarely match boundaries of a 'nation' and borders bear little congruence with the ethnic distribution of their component units. In effect, any challenge to the state's supreme authority can only elicit a perpetual condition of anarchy, the solution of which resides in a one size, fit-all conflict management service package and all you have to do is 'Dial an army'.

These mainstream 'end justifies the means' explanations and/or justifications of the use of private armies as guarantors of post-Cold War stability in Africa are increasingly assuming hegemonic dimensions in diplomatic circles, and pose an inherent danger for the future of conflict management on the continent. There is a reluctance on the part of the international community to discuss the sub-state and transnational phenomenon, and the whole structure of international relations still rests on the state as the basic unit of interaction.[6] Alternative visions of a new security order that might take into account the changing nature of internal conflict and the place of the nation-state are blocked from mainstream discourse, thus offering little hope of a comprehensive understanding of conflict management in situations of deep-rooted conflicts. According to Mackinlay,

> There is an inflexibility of language, habit and *modus operandi*, which stifles the consideration of the real nature of the problem. Heads of fragile and failing States actively collude in the stifling process for they stand to lose international aid and trade if the *de facto* secessionist regimes within the States are somehow recognised by global organisations. There is an overwhelming sense of delusion and hypocrisy about the inviolable nature of State borders ... for less obvious reasons the international community goes along with the fiction, continuing to deal with States on the same basis as before, unwilling to see that beyond the capital city several competing autonomies have sprung up in what were previously administrative no-go areas.[7]

There is a need to challenge the organic interconnection between the security imperative and the political logic responsible for the rise of mercenaries and the permanence of conflict in Africa, so that we can better understand how the post-Cold War pressures of sub-nationalism and globalisation have exacerbated the crisis of governance in the nation-state.

This chapter offers an overview of mercenary involvement in post-Cold War conflicts in Africa. We seek to achieve a greater understanding of the mercenary–instability complex by examining:

- the linkages between the rise in internal conflicts and the proliferation of mercenary activities in the 1990s;
- the differences and similarities in the methods adopted by Cold War mercenaries and contemporary counterparts;
- the convoluted network between private armies and business interests;
- the linkages between mercenaries, regime security and poverty, to the connection between mercenary activities and arms proliferation.

In exploring solutions to the upsurge of mercenaries on the continent, the chapter argues for a redefinition of the legal and political understanding of the term *mercenaries* and calls for new international legislation. It also argues for comprehensive solutions to the root causes of conflict and the strengthening of regional mechanisms for conflict management in Africa.

For the purpose of this chapter, 'mercenary' is the term used to describe the practice of professional soldiers freelancing in external circumstances away from their countries of origin or residence. Although we have used the term 'private armies', 'privatised peacekeepers' and 'private security forces' interchangeably, we believe it is more appropriate to use 'mercenary' for conceptual clarity. Hence, even when it is not used to describe organised or freelance activities that fit into the definition given below, the underlying meaning is the same.

What is Mercenarism?

Mercenarism – the practice of professional soldiers freelancing their labour and skills to a party in foreign conflicts for fees higher and above those of native counterparts – is as old as conflict itself. Examples can be found in Ancient Greece, Rome, and subsequent empires. The use of mercenaries by the major European colonial powers in their attempts to maintain control over their overseas territories is well documented. Invariably, such special troops have been hired to carry out assignments that regular armies would not be called upon to perform because they are considered morally repugnant and illegal under customary or laid down rules of war. These assignments include sabotage operations in rival empires, crushing rebellions, assassinations and scorched earth operations against rebellious subjects.

Since the late eighteenth century, mercenaries have, for the most part, been involved in African conflicts as extensions of the mercantile expeditions that pioneered the colonisation of African territories as well as in the suppression of African aspirations for self-determination and political independence. Many served as advance guards in the colonial scramble for Africa's trade and territories. Trading concerns like the United African Company – operating in Anglophone Africa – and Compagnie Française de l'Afrique de l'Ouest (CFAO) and Société Commerciale Ouest Africaine (SCOA) had their own security forces responsible for forcefully paving the way for the expropriation of resources. Since World War II, mercenaries have gained some prominence, some would say notoriety, for their involvement with governments and anti-government groups in Africa.

The 1960s and 1970s will, however, remain the golden age of mercenaries in Africa and of their impact on stability in the continent. This was a period in which their activities generated serious attention from the Organisation of African Unity (OAU). Mercenaries such as 'Mad' Mike Hoare, Jacques Schramme and Bob Denard were hired by former colonial powers and other external interests to undermine the rights of the peoples of the Congo (1960s), the Comoros Islands (1970s–1990s), Benin (1970s) to choose their leaders and assert their political independence, or to destabilise sovereign states (Angola, Mozambique). They assassinated heads of state of sovereign nations (Patrice Lumumba of Congo, Ahmed Abdallah of Comoros Islands), and staged the violent overthrow of (or attempted coups against) governments.

Given Africa's long history as a proxy zone of superpower rivalry and Cold War arms proliferation, the continent was expected to produce evidence of the much desired peace dividend and arms reduction in a supposedly reconciled world.[8] In one respect at least this has proved to be the case – inter-state border conflicts that used to be a problem of Africa's international relations as a result of the continent's history of artificial borders were significantly reduced. However, in the 1990s Africa became the battlefield for internal conflicts of even greater ferocity, most of them resulting from the struggle for power and control of local state resources. Simultaneously, the downsizing of the military that occurred in the post-Cold War era and post-apartheid South Africa ensured that regular officers and soldiers sought employment in irregular operations where belligerent parties were in conflict.

In no time, a lot of these soldiers found themselves in wars in Angola, Mozambique, Sierra Leone, Sudan, Liberia, Congo (Brazzaville), Congo (Kinshasa) and the Great Lakes region. Corporate clones of Jacques Schramme and Bob Denard such as Colonels Eeben Barlow and Tim Spicer squeezed into the security vacuum created jointly by the formal exit of the superpowers from Africa's war zones, the winding down of inter-state wars and the increase in deep-rooted, identity-based internal conflicts.

Although they are presented as stabilising agents for Africa's development, there can be little doubt that these corporate organisations are still largely motivated by the pursuit of private gain. The vigour shown by today's mercenaries, however, has its antecedents in the activities of corporate mercenaries who provided security services for old-style monarchs and dictators in the Middle East and Africa in the post-colonial struggles of the 1960s and the post-independence political rivalries of the 1970s. Colonel David Stirling, founder of the original Special Air Service (SAS), formed WatchGuard International which provided among other services: military surveys and advice, Head of State Security, Training of Close Escort Units and Training of Special Forces to combat insurgency and 'guerrilla warfare'. Although he operated largely in Oman, Saudi Arabia, Yemen and the rest of the Middle East, WatchGuard also had businesses in East and Central Africa, protecting the Lonhro mines and copperbelt and organising Head of State Security for President Kenneth Kaunda. Another security outfit known as UNISON, strongly backed by a retired Commander-in-Chief of the Allied Forces in Northern Europe, Sir General Walter Walker, was set up in 1973.

For all the transformation in language, diversification of customer base, and the non-ideological, professional outlook of contemporary mercenaries, they still retain a dubious relationship with, and dangerous liaisons to, governments and, in some cases, act as fronts for government inspired operations – especially in the search for markets for what has become a fiercely nationalistic arms industry. Since many of the active recruiters are recently retired high-ranking officers, who are part of the 'old boys network' acting as advisers to foreign governments friendly to their own governments, there is a sense in which they are seen and encouraged to fulfil their government's foreign policy objectives. For example, BDM based in McLean, Virginia, has on its board ex-CIA operatives like Frank Carlucci – who became the United States' Secretary of Defense in the Reagan administration, whilst Military Professional Resources Inc (MPRI),[9] based in Alexandria, Virginia, has ex-Generals like Carl Vuono on its board and a database of more than 2,000 retired generals, admirals and officers available for contract work. In the United Kingdom, from where the bulk of mercenary outfits operate these days, Defence Systems Limited (DSL) is run by Major-General Stephen Carr-Smith, an ex-SAS officer with close links to the British military establishment. DSL is sometimes regarded in the business as Britain's officially backed 'mercenaries', while Pentagon officials have been known to describe Vinnell Corporation as 'our own little mercenary army in Vietnam'.

Just as today's corporate mercenary outfits try to convince potential clients of their access to governments and at the same time stress their independence, WatchGuard International, according to Colonel Stirling, is 'a profit-making, private commercial enterprise which moreover works for foreign governments' (*Sunday Times*, 18 January 1970). In a similar manner, General Walker told the *Sunday Times* of 29 July 1974 that: 'Unison is an entirely non-militant organisation. Its members would act only if there was a break up of law and order ... .' Compare this to an extract from Sandline International's Managing Director, Colonel Tim Spicer, whose testimony to the Foreign Affairs Select Committee in response to a question whether he was acting as a front for British policy in Sierra Leone, was, 'I am confident that what we did was consistent with United Kingdom's foreign policy and law ... although I cannot comment for the Foreign Office ... I do not work for the Foreign Office.'[10]

Although the British government has consistently denied working with Sandline International in its alleged mercenary operation in Sierra Leone, there can be little doubt that unofficial links continue to dominate the 'old boy network' relationship between serving officers, diplomats and retired diplomats. Yet there is evidence to suggest that even the lone soldier of fortune thought to retain no links with government officials also believes that his actions were consistent with his government's foreign policies.

Political Context of Mercenary Activities in Africa during the Cold War

As well as being a proxy territory for the Cold War superpower rivalry, Africa was also the battleground for a peculiar form of fiercely nationalistic competition over territories that pitted the British against the French, the French against the Americans, the Belgians and the Portuguese, and all of them against the Soviets. World War II saw the conscription of several thousand colonial subjects from Africa, the Americas and Asia to fight alongside the armies of their colonial masters. The combat experience of these conscripts became useful for intellectuals and politicians from the colonial dependencies for agitational purposes and they proved crucial in launching the decolonisation revolution. On the one hand, nationalist leaders distrusted the native members of the colonial armies, regarding them largely as agents of the status quo. On the other hand, Africans in colonial armies had become empowered by their association with white soldiers during the world war. As one historian put it, 'Africans had fought alongside white men, killed white men, seen brave Africans and white cowards, slept with white women, met white soldiers who treated them as equals, or who were like themselves, hardly educated ... .'

The response by the colonialists was to use newly discharged soldiers from the metropolitan states to crush, sabotage, frustrate or delay the aspirations for self-determination. For example, after World War II, Colonel Stirling of WatchGuard International helped to set up Capricorn Africa – a society of elitist whites and blacks intended to persuade nationalist blacks in Tanganyika (now Tanzania), Kenya, Rhodesia (independent Zimbabwe) and Nyasaland (Malawi and Zambia) to reject the concept of majority rule in favour of 'responsible', elitist, white-weighted-majority rule.[11] The French embarked on the same scheme during the Algerian War of Independence, through the presence of the underground Organisation Armée Secrète (OAS) contingent from France (some of whom, like Christian Tavernier, later played key roles in Congo and Angola). Thus, in Africa, the genesis of mercenary activity in the immediate aftermath of World War II coincided with the peoples' struggles to assert their rights to self-determination and sovereignty.

The problem, however became acute during the crisis in the Congo in 1961, when the then leader of the Katanga region and Prime Minister of Congo, Moise Tshombe recruited 500 mercenaries led by the notorious French 'dog of war', Christian Tavernier, the Belgian Jacques Schramme and the South African Colonel 'Mad' Mike Hoare. Although the United Nations, which had a peacekeeping force in Congo at the time, was unequivocal in its condemnation of the presence of mercenaries,[12] the demands were never heeded by the supplying countries like France and Belgium, whose determination to maintain a presence in Congo outweighed respect for international obligations.[13] Indeed, in many cases the leading mercenaries of the time were not merely individual soldiers of fortune. They were also linked to the security and intelligence services of their countries as foreign military advisers and provided their governments with the opportunity to dissociate official policy from the actions of the soldiers. For example, a contingent of French mercenaries involved on the side of the secessionists in the Nigerian civil war was believed to have been given official cover by Jacques Foccart – the late African Affairs Chief at the Elysée, whilst pilots from Egypt were engaged on the Nigerian side in charge of the Czech L29 Delphins in 1969.[14]

In several resolutions, the Organisation of African Unity maintained a principled opposition to the presence of mercenaries.[15] By 1967, the Assembly of Heads of State and Government urged 'all States of the world to enact laws declaring the recruitment and training of mercenaries in their territories a punishable crime and deterring their citizens from enlisting as mercenaries'.[16] In 1970, in response to the invasion of Guinea and attempted overthrow of the government of Sekou Toure, the OAU decided to put in place a convention for the elimination of mercenaries in Africa, the first such convention to be drafted in the world.[17] In 1971, the OAU adopted a declaration on mercenaries in Africa in consideration of 'the great threat which the activities of mercenaries represents to the independence, sovereignty, territorial integrity and the harmonious development of Member States of the OAU'.[18] This was presented in the form of an Expert Committee Report to the 19th Ordinary Session of the Council of Ministers meeting in Rabat, Morocco, in June 1972. This report laid the groundwork for the eventual draft Convention for the elimination of mercenaries in Africa. The Convention defines a mercenary as:

> anyone who, not a national of the State against which his actions are directed, is employed, enrols, or links himself willingly to a person, group or organisation whose aim is:
> a) to overthrow by force of arms or by any other means the government of that Member State of the OAU
> b) to undermine the independence, territorial integrity or normal working of the institutions of the said State
> c) to block by any means the activities of any liberation movement recognised by the OAU.

African states also spearheaded the international campaign leading to the adoption of several resolutions condemning the use of mercenaries and to Article 47 of the Geneva Convention, which outlaws the use of mercenaries. The International Convention against the Use, Financing and Training of Mercenaries has been signed and ratified by 21 countries.

Notwithstanding these various attempts at continental and universal condemnation of mercenaries in the early 1970s, there was a significant growth in their presence in African conflicts. By the mid-1970s, it had become a major issue in the anti-colonial struggle in Angola, which was then a major proxy zone of superpower rivalry on the continent. Still smarting from the forced withdrawal from Vietnam, no fewer than 200 Americans arrived at San Salvador in Northern Angola in 1975, with the implicit backing of the Central Intelligence Agency.

Having formed the ruling government after gaining independence from Portugal in 1975, the Popular Movement for the Liberation of Angola (MPLA), led by Dr Agostinho Neto, became the target of other liberation movements that competed equally for power with support of the Americans. These included the Zaire-based and US-backed Front for the National Liberation of Angola (FNLA), headed by Holden Roberto, who received a lot of assistance from the US military, and the National Union for the Total Independence of Angola (UNITA), operating in Southern Angola under the leadership of Dr Jonas Savimbi. In their mutually exclusive agendas to stop the left wing Marxist-Leninist MPLA government, they were backed by other western countries such as Britain, from where an advance party of mercenaries arrived to take charge of Roberto's FNLA troops. While the motives of these demobilised ex-servicemen were completely non-ideological, they exhibited a strange bloodthirstiness which Holden Roberto had come to rely on. They had served in specialised forces such as the Special Air Service (SAS), or in British colonial forces in the Middle East or Africa protecting local potentates whose sole objective was the repression of freedom struggles. Some of them, including the notorious 'Colonel' Callan Georgiou, 'Captain' Charlie, Gustavo Marcelo Grillo, Daniel Gearhart, John Nammock and Gary Martin Acker were captured and interrogated by the MPLA government and brought before an international tribunal. There is a well-documented account of mercenary activity in Angola during the anti-colonial struggle. More importantly, the trial of mercenaries in Angola provided extensive evidence of the tacit or actual alliance between governments and mercenaries for the first time.[19]

Given the humiliation of American and British-inspired mercenaries in Angola, one would have thought the world had witnessed the demise of freelance soldiers in internal conflicts. But no sooner had media attention on Angola cooled than the 'dogs of war' were back in Benin in 1977, trying to overthrow the Marxist regime of Matthew Kerekou. In the same year, a contingent led by Colonels Mike Hoare and Christian Tavernier returned to the Katanga region of Zaire to help President Mobutu Sese Seko crush a rebellion. In 1978, a coup d'état led by ex-French legionnaire and mercenary, Bob Denard, infamous for his political assassination attempts and subversion, succeeded in overthrowing the government of the Comoros Islands. Hoare and Denard went into Seychelles in 1981 to overthrow the 'unfriendly and ideologically opposite' government of President Albert Rene, but their attempt failed.

Arguably, the agenda of the freelance soldiers often coincided with the policies of their home governments, although money was usually the primary motive. Since many were recently demobilised or retired officers well-connected within military, diplomatic and political circles, they almost always had quasi-governmental backing for their operations. Unlike unscrupulous terrorists like Colonel Callan, their activities were often approved by sections of Western intelligence outfits, if not by governments, with enough room for plausible denial when things went awry. In all of the Cold War cases, it is difficult to imagine the absence of complicity on the part of governments engaged in the protection of spheres of influence, territorial aggrandisement or mineral control, and many mercenaries were willing to act as agents in these complex missions. The post-Cold War mercenary industry built on this arrangement and developed a wider, more sophisticated network.

The Mercenary Agenda in Contemporary Africa

'Unlike the renegade dogs of war that tore through Africa in the 1960s ... the new wave of military privateers consider themselves legitimate businessmen. You won't spot them, though; they are more likely to wear tailored suits than combat fatigues ... .'[20]

Although there have been consistent and often successful attempts to distinguish the new mercenaries from the lawless 'guns for hire gangs' who ran riot in Africa 30 years ago, the post-Cold War mercenary industry has benefited from the tactics and strategies of the sophisticated wing of the 1960s mercenaries. The international political arena within which they operate has also changed somewhat from the proxy war zones of inter-state border conflicts to a world in which the internal dimensions of conflicts have become the most pronounced. While the financial motive is still central to the new mercenaries, those of the 1960s were generally less concerned about their image. In fact, many of them thrived on ideological and romantic notions of their importance.

In the post-Cold War era, mercenary activity has changed in terms of organisational structure, links with business, target clients and manner of 'reward' for services. Renegades like Colonel Callan and others involved in the Angolan disaster have been replaced by sharp-suited and media-friendly corporate players like Eeben Barlow of Executive Outcomes (EO), one of the best known mercenary outfits in the industry today, and Tim Spicer of Sandline International, based in London. In their public relations-speak, they are not 'mercenary outfits' or 'arms dealers' but 'packaged services providers' who render 'professional military assistance' to 'established' and/or 'legitimate' governments.[21] Unlike earlier mercenaries, their activities are diversified to a high degree. Their customers range from sovereign governments to international agencies, foreign embassies to corporate entities – usually involved in oil, timber logging, exploration and mineral prospecting. In many instances, the same directors and executives run both these corporate bodies and the mercenary outfits. Tony Buckingham, a Director of Sandline International, established Branch Energy, which later merged with Carson Gold to form DiamondWorks, a Canadian quoted company.

Remuneration for contracts implemented also comes in different forms – not just cash – mostly in the form of trade deals whereby security services and executive protection are offered in exchange for long-term diamond or oil concessions. In many cases, it appears that this is often preferred to cash deals, as it allows mercenary outfits to maintain a strategic stranglehold on their client state.

It is no longer a secret that a major objective of the new mercenary business is access to natural and mineral resources. Confident that most of the groups in desperate need of their services are not in a position to pay in cash, 'security firms' demand payments in the form of mining concessions and oil contracts. Executive Outcomes' contract with the government of Angola is believed to have included a diamond concession awarded to their subsidiary, Branch Energy. In total, the contract is said to have been worth US$40 million.[22] The same is true of the contract with the Strasser-led NPRC government in Sierra Leone, which close observers believed to have been worth US$50 million in cash and mining concessions.[23]

Ironically, according to the UNDP Human Development Index, these two countries sit at the bottom of the list of the world's poorest countries. Yet the very resources that they ought to have utilised in alleviating poverty and reducing conflict are now mortgaged on needless and avoidable wars. According to Enrique Bernales Ballesteros, the UN Special Rapporteur on Mercenaries, firms like EO and Sandline begin to exploit the concessions received by associating with established mineral companies or setting up a convoluted network of associated companies which engage in 'legitimate' business. In this manner, they acquire a significant, if not hegemonic presence in the economic life of the country in which they are operating.[24] In the case studies contained in this book on Sierra Leone, Angola and Zaire,indicate a clear and consistent correlation between the activities of the mercenary outfits and the rising fortunes of mineral prospecting and distribution corporations in these war-torn countries. Yet this association is regularly denied by leading figures of the mercenary organisations like Sandline and EO. In the recent investigation conducted by the Foreign Affairs Committee of the British House of Commons on the role of Sandline in Sierra Leone, the following conversation on ownership took place between Colonel Spicer and Mr Mackinlay, MP:

Mackinlay: I understand Mr Buckingham is Chairman of Heritage Oil which own Sandline, is that correct?
Spicer: I am not quite sure what Mr Buckingham's position in Heritage is. He is very closely associated with Heritage, but Heritage does not own Sandline.
Mackinlay: Does Plaza 107 own Sandline?
Spicer: No.
Mackinlay: Who owns Sandline?
Spicer: It is owned by a group of investors and they are incorporated in a holding company.
Mackinlay: In the Papua New Guinea inquiry, there was a diagram produced showing the relationships between various companies of Sandline, i.e. Heritage Oil, LifeGuard, Branch Energy, Executive Outcomes, IBIS Air. Would it be possible to produce a diagram of the relationship between these key companies for the Committee?
Spicer: It would be possible to produce a diagram that refers to these companies, but what I would point out is that Sandline is a completely separate company and it is not owned by any of the other companies referred to. It may have contractual relations with some of the organ-

isations referred to, but I do not have the benefit of having that diagram in front of me.[25]

Although Colonel Spicer did his best to deflect the question about the association of this set of companies, the same Tim Spicer described Tony Buckingham as chairman and CEO of Sandline during the Bougainville investigations in Papua New Guinea.[26] In a meeting with Michael Grunberg in his capacity as representative of DiamondWorks in London, the activities of EO and Sandline International and their associated companies came up for scrutiny. Grunberg heads Plaza 107,[27] a holding company for Branch Energy (BE) – the defunct mineral-exploiting company operating in Angola, Uganda and Sierra Leone. Whilst in existence, EO was the major shareholder in BE, with 60 per cent of BE Angola, 40 per cent of BE Uganda and 40 per cent of BE Sierra Leone. BE merged with Carson Gold to form DiamondWorks in 1996[28] which was how Michael Grunberg inherited his new role. Interestingly, it strikes pertinent observers as curious that Michael Grunberg, who had written to CAMA (Campaign Against Mercenaries in Africa), was arrested alongside Tim Spicer by HM Customs and Excise in connection with the sanctions-busting arms shipment to Sierra Leone where, incidentally, DiamondWorks also has major diamond concessions.

This structural difference is crucial to the understanding of the contemporary nature of mercenary activity. The new mercenaries have perfected their public relations and legal jargon so as to divert prying eyes and confuse the trail of the web of intrigues and deals surrounding the supposedly altruistic notions of saving 'legitimate' governments from 'warlords' and 'barbaric thugs'. These amateurishly disguised tracks of their corporate web also help explain the inextricable linkage of mercenaries to the issues of instability, light weapons proliferation and protracted conflict in their regions of operation.

There is a sense in which the current militarisation in Africa must be seen as a function of a dominant cartel comprised of arms manufacturers, mineral exploiters, corporate mercenaries and Africa's authoritarian governments and warlords, all of whom believe that interest must be defined as power which by its very nature must be authoritarian in order that unbridled profiteering can be pursued with military dispatch. The activities of private peacekeepers, the problem of light weapons proliferation and the linkage to resource exploitation in troubled West African States is but one critical example of the operations of this power elite.

In exploring the causes and potential cures of conflict in Africa, the United Nation's Secretary-General, Kofi Annan, recently referred to 'interests external to Africa', who, 'in the competition for oil and other precious resources in Africa continue to play a large and sometimes decisive role, both in suppressing conflict and sustaining it'.[29] The

Secretary-General also referred to the role of 'international arms merchants in African conflicts', and 'how access to resources by warring parties ... has highlighted the impact that international business interests can have on the success or failure of peace efforts'.[30]

This has wider implications both for the demilitarisation agenda and for democratic consolidation on the continent. The rise of corporate mercenaries and the linkages to proliferation of small arms and to mineral conglomerates pose a mortal danger to democracy in the region. Ironically, in a globalised world in which public interest in international peacekeeping has waned considerably,[31] the security vacuum created is now filled by unregulated private military armies often linked to international business interests intent on exploiting resources in countries gripped by conflict rather than on democracy and development.

Beyond these often undeniable links to governments lies the convoluted network with armaments producers in a fiercely competitive and nationalistic arms market. By their very nature, the conflicts mercenaries get embroiled in are primarily internal ones. More importantly, in most cases they are not guerrilla wars fought over political ideologies. In spite of these changes in the post-Cold War era, the notion of African security has remained inclusive and parochial, and Africa's national security elites have become more interested in the management of violence and primacy of power politics. The challenge to this perspective often comes from those who believe that the existing order needs transformation, not management, and the refusal to reach a middle ground has frequently resulted in internal conflicts. The weapons used for these wars are the more affordable, small, mobile and user-friendly light arms, since there are no mentors to provide the precision-guided missiles or stealth bombers available in the Cold War years through foreign military assistance programmes or as part of defence pacts signed between metropolitan centres and their satellite states.

The level of light weapons and small arms proliferation continues to pose a serious challenge to the demilitarisation agenda in Africa. This has been the subject of extensive debate in multilateral circles in particular, at the level of the UN, OAU, EU and ECOWAS in recent times. For example, a panel of government experts appointed by the UN's Secretary-General identified uncontrolled availability of small arms and light weapons as both a causal and exacerbating factor in Africa's conflicts. According to the panel, not only did the weapons contribute to 'fuelling conflicts but also [to] exacerbating violence and criminality'.[32]

When the recently ousted president of Congo, Henri Lissouba, came into power through popular elections, he knew he had to do something about the fragile peace in Congo-Brazzaville.[33] In no time, he found help via retired Israeli agents who not only offered to train his own loyalists but included a US$10 million worth of arms delivery in the package deal.[34] In a similar deal that fell through, Sandline International transported weaponry and men worth US$38 million to Papua New Guinea in March 1997. The head of Sandline, Colonel Tim Spicer, and his band of South African mercenaries were arrested and later deported and the Prime Minister, Sir Julius Chan, was forced to resign as a result.

Not only did the incidents confirm the linkages between arms suppliers and mercenaries, they also reveal the extent to which collapsing states have become dependent on non-state external actors for their survival. In the past, ex-colonial powers such as France and Britain would have stepped in with the necessary weapons or support lines in defence of incumbent regimes, but this is no longer the case. Even France, regarded as the most paternalistic of all ex-colonial powers, only sent troops to the former Zaire in the recent crisis to 'protect and/or evacuate' French citizens there. By understanding the way in which light weapons flow through and to regions of conflict, policy options and opportunities for control can be provided. It seems entirely reasonable to surmise that, given the strong linkages between mercenaries and regime security, any reduction in the involvement of mercenaries in African conflicts may help to control arms proliferation on the continent and reduce the mortgaging of the resources needed for economic development in these beleaguered states.

All these points expose as utopian the predilections of scholars who see the use of today's mercenaries as the effective antidote for insecurity in zones of complex emergencies, but pay little or no attention to the macroeconomic and political implications of their role in terms of subversion of the very state sovereignty the mercenaries claim to protect. The scholarly position is further undermined by fluid loyalties, which have resulted in intra-mercenary rivalries all over the continent. Writers like Herb Howe have argued that 'EO's permanence, as against the ad hoc nature of other mercenary groups and the need to keep open the possibility of future employment, have undoubtedly bolstered allegiance and decreased the chances of unprofessional behaviour.'

Yet old habits die hard. In spite of the appearance of permanence, there is evidence to suggest that combatants have shifted loyalty among belligerent parties. Just like EO's shift from UNITA to MPLA in Angola, recent analysis of 'hired guns' in Sierra Leone and the DRC reveal an exchange of loyalties in line with competition for the control of mineral resources. The competition has become fiercer, with the Jean-Raymond Boulle group pitted against the erstwhile EO, linked to Branch Energy and DiamondWorks. Now that EO no longer exists as a formal company, it is not difficult to predict a more diffused and ultimately bleak future for these competing armed groups. Yet, rather than condemn outright the involvement of these hired combatants, policy-makers and scholars decry the use of mercenaries on the side of the warlords whilst ignoring their use by governments in zones of conflict.

What Future for Mercenarism?

While it is in the interest of the new mercenaries that the world remains in a perpetual state of instability, their 'solutions' are often short-term. They articulate stability in terms of regime security intended to shore up the existing power distribution, not its transformation for the good of all. They also substitute charity for justice. The way in which Sierra Leone was plunged into another orgy of instability remains a pertinent example of why such intervention is only a stopgap measure. Although EO executives trumpet claims of their company's resounding success in Sierra Leone, and several scholars have helped promote that view, it was always predictable that the peace would be tenuous because the dialectics of hostility, suspicion, and violence were barely addressed by the standard conflict management package prescribed by groups like EO for all kinds of conflicts.

It was hardly surprising, therefore, that what was driven underground rose with a greater ferocity to consume the 14-month-old democratic government of Tejan Kabbah, throwing the country into yet another period of violence. Colonel Eeben Barlow was later to argue that had EO's contract not been terminated by the civilian regime, the elected president would not have found himself in this situation. In fact, Michael Grunberg, a Plaza 107 executive, recently boasted to representatives of CAMA that they (EO) predicted President Kabbah's overthrow within 100 days of the termination of the contract with Executive Outcomes. According to him, he was overthrown on the 95th day.[35]

Even if one were to concede that the actions of Executive Outcomes helped to achieve a level of tenuous peace in Sierra Leone, coercive power remained the defining characteristic of their search for stability. This clearly underestimates the transformed nature of internal conflicts and the shrinking dimensions of state power. It is this underestimation of the local patterns of conflicts that lead to often distorted and incomplete threat perceptions and security assessments and undermines the prospects of peace settlements that are not based on negotiations. Whilst it makes good copy for Western news media to dismiss the agitation of those who have been victims of state collapse as 'warlord politics', it is still important to understand the sub-state phenomenon that fuels structural instability, however true that may be. Mercenary outfits are not only incapable of undertaking such a comprehensive venture, it also goes against the grain of their gung-ho tactics.

In spite of the problems associated with the activities of the new mercenaries, do they have a future in a complex world of seemingly endless internal conflicts? There can be no doubt that a market will always exist for people who jump to the rescue of beleaguered potentates and dictators, but there is also no doubt that they will continue to prove ineffective in the long run. The question should be asked differently: should there be a future for mercenaries?

To the extent that national security is predicated on the ability to protect state sovereignty, preserve territorial integrity and maintain regime security, sub-national interests which become de-legitimised in their cultural, ethnic, environmental and economic dimensions will continue to challenge the primacy of the state. In the effort to help shore up the status quo, the ways the ruling elite affected the power dynamics were not taken into account. Also, the differing components of the nation-state and its centrifugal contradictions were underrated or simply interpreted as threats to regime security. Thus, the conception of homogeneity between regime security and state survival as exemplified by most mercenaries and their proponents tended to obscure the contradictions within the state, so much that any challenge to the idea of the state is often interpreted as a direct challenge to the legitimacy of the government in power, as we have seen in the Biafran war, the Eritrean War of Independence and the ongoing wars in Southern Sudan, Sierra Leone and the Democratic Republic of the Congo. It is analytically useless in the resolution or management of conflicts to demonise local players whose tactics appear abominable and abhorrent. Often these internal actors first had to threaten regime security in order to receive the attention of decision-makers and the international community.

Yet, if mercenaries are not to have a future, the international community must take account of the changing patterns of international security in the light of the new nature of internal conflict, and reflect this in the conflict management mechanisms they put forward. International legislation can but reflect the mindset of the political society it represents (which includes civil society in its current form). International legal instruments should emphasise the need to understand the interplay between systemic and local factors in order to accurately predict the nature of the emerging threats and plan their management. Without this necessary change in mindset of both the African security elite and the international policy-makers who influence them, all international instruments aimed at curbing the mercenary trade will remain ineffective. Yet this can only happen if there is a challenge to the dominant discourse in mainstream scholarship, which gives the impression that 'the only alternative is now private security forces'.

Ultimately, comprehensive solutions to the root causes of conflict must be explored by recognising the essential linkages between underdevelopment, instability and the presence of mercenary operations in a region. To this end, there is a need to critically assess what the new forms of private military activities mean for African security.

Re-conceptualising the Security Environment

To re-conceptualise the African security environment, we need to outline what is missing in the current conception. In the realist (read Western) conception of security, the security of units below the level of the state has rarely, if ever, been an important issue. In the words of Hans Morgenthau, 'there are inherent supra-sectional loyalties which transcend particularistic interests of individuals and groups within such societies, thus containing domestic conflict'. According to him, 'protection of the nation against destruction from without and disruption from within is the overriding concern of *all citizens*'.[36]

In Morgenthau's world-view, then, individual security is synonymous with national security, and political action is only meaningful if it seeks to promote the inviolability of the nation-state. Yet we know that sometimes – and increasingly – even the upholding of existing state boundaries is not considered a core value by important segments of a country's population. This is often the case because there is a yawning gap between the core values of the ruling regime and those cherished by large segments of the population. In such a situation of deep-rooted communal cleavages, there can be very few common values or interests on a national level, apart from those of the ruling class.

To capture the changing nature of security, we must avoid the static approach of structural realism to international conflicts. This emphasises technical and external factors, separates external from internal causes of dissent, and the economic from the political. More consistent attention ought to be paid by security scholars and policy-makers to the political context within which state policies are defined. Since mercenaries thrive in conditions deemed suitable for the modern state as an arena of continuing anarchy, this emphasis on the technical aspects of security and the abdication of responsibility by regional and international institutions allows the survival of privatised peacekeepers and a false comparison between private armies and regional security mechanisms. Instead, a historical perspective which explains the nature of the state civil society relations and how state power relates to key economic and social forces in society within a global setting, would offer a better framework for analysis and help determine the root causes of conflict and mechanisms for conflict management and transformation.

The implications of the global situation for Africa thus bear emphasis, especially in the shaping of decision-makers' world-views and threat analyses. Because of the inability of the ruling elite to relate 'defence needs' to 'economic means', the necessary connection between technology, strategy and economics is often lost. In effect, this encouraged misperceptions, since the ruling elite saw only what they believed to be the country's most enduring threats. It also led to the neglect of important situational factors, which ought to have made the analysis of perceived threats more accurate. For example, the concentration on espionage, sabotage and subversion by insecure and often undemocratic regimes appeared to have followed on the heels of the inability to solve mounting economic and other non-military aspects of security problems. The suggestion here is not that African governments do not face objective security threats, but that their deferral of them is often a mental defence mechanism to divert attention from criticism by people excluded from the political process, and the situation in Sierra Leone partly stems from this.

While a strong correlation remains between imperialism and some patterns of instability in Africa, the main thrust of our argument is that the ruling elite's perception of national security threats has largely been coloured by the Cold War environment, not the situational factors in their local environment, nor the predispositions or personalities of the ruling elite themselves. Hence, the late Mobutu Sese Seko, for example, managed a repressive but collapsing state for 30 years on the myth that he was invincible, and that he *was* the state, without whom the country would disintegrate. Even though this mindset has produced costly negative results, the notion of the strong ruler is still prevalent, and several African dictators continue to take advantage of it as long as the international community refuses to embrace the need to promote a broader set of competing autonomies within the context of the state and toward the promotion of good governance and democratisation.

On the other hand, since the notion of an authoritarian state is fast losing its appeal among its victims, internal conflicts almost always result from it, spearheaded by those who have been denied equality of opportunity and representation in the development of their communities. It is the security vacuum created in such conflicts that the new mercenaries seek to fill. By presenting their involvement as a humanitarian gesture in a world that has grown increasingly disinterested in African affairs, mercenaries posing as security outfits are able to argue, somewhat persuasively, that they are offering services that others are either incapable or unwilling to provide. This seems logical if one takes into consideration the increasing reluctance of the international community to assist with conflict resolution in Africa.

Consequently, several of the mushrooming mercenary cum security outfits now provide services to international agencies providing disaster relief in East and Central Africa, to embassies in Congo and Sierra Leone and to governments and multinational firms operating in the region. It is even rumoured that the OAU and the UNHCR may be endorsing the work of these outfits by offering contracts to establish and train a permanent peacekeeping force and carry out protection contracts in war zones respectively.[37]

Few would disagree with the short-term humanitarian solutions that some of these outfits describe in their glossy corporate brochures. Yet it seems undeniable that they are substituting charity for justice, and that their activities encourage the world community to shirk its responsibilities and obligations. Even worse, their presence encourages the perception that stability and peace can only be achieved within the context of prestige and power politics, concentrating on superficial security rather than on a humanistic development security policy based on a national consensus. The danger is only temporarily assuaged in the 'winner takes all' scenarios so appealing to ruling elites. In fact, we have argued elsewhere that the reason behind the failure of the UN peacekeeping mission in Somalia was the subordination of a political solution to the military pursuit of a key player and the unclear nature of the mission's mandate, not the decision to intervene itself.[38]

In the absence of the political will to address deep-rooted conflicts in a comprehensive way at the international level, regional mechanisms still offer the best hope for conflict management and preventive diplomacy in Africa and globally, in spite of their evident problems.

Caught between the extremes of supra-nationalism, as represented by globalisation, and the challenges of sub-nationalism that have been exacerbated by the politicisation of ethnicity, regionalism offers the best cure for the weakened nation-state in Africa. Indeed, it would appear that any prospect of sustained demilitarisation and democratisation in Africa must build on the tender fabric of regionalism if it is to have any chance of success. Given the declining external security threats and the need to curb the rising tide of internal strife, promoting a professional peace-building mechanism within the global framework of preventive diplomacy would seem critical in the region. The last decade has witnessed the strengthening of regional autonomy in Africa, especially in its ability to manage conflict. Although ECOMOG (the Economic Community of West African States Monitoring Group), is seen in several circles as a standard feature of Nigeria's sub-imperialist agenda, there is now a strong perception of it as a potential mechanism for an effective conflict management model, not only in Africa, but in the rest of the world.[39] Indeed, the ECOMOG ideal has now become the template for NATO's action in trying to stop genocide in Kosovo. Yet regional autonomy can be influenced by national and sub-national factors. It is also susceptible to superpower influence and control, which may be opposed to the goals of demilitarisation and democratic consolidation, especially if the latter do not offer the required stability for capitalist development.

For example, Nigeria's consistent commitment to regional peacekeeping may have arisen from unresolved tensions at home. Ironically, the concentration on the pacification of sub-regional threats resulted in a simultaneous neglect of internal threats. Although the recognition of internal threats as the most serious may question the necessity of a standing army, we still support a standing peacekeeping army within the region whose role is clear and measurable, but whose size reflects the identified needs of the states involved in regional peacekeeping.

In rethinking regionalism, we must go beyond the pro forma creation of a peacekeeping force that remains technical in form and content. For regionalism to be an effective antidote to globalisation and ethnicisation it must permeate the nation-state in a far deeper manner. Otherwise, if the current non-state challenges to the nation-state in Africa are a measure of what to expect in future, then the prospects for demilitarisation are slim, possibly non-existent. It is for this reason that acknowledging the need for a multi-dimensional understanding of security without redefining the concept of sovereignty undermines the creation of a comprehensive security agenda. In arguing for a reorientation of the concept of sovereignty in the sub-region which de-emphasises artificial colonial boundaries, the motive is not territorial revisionism. Instead, we are revisiting the territorial state where artificial boundaries have legitimised arrested development in several states that are largely juridical entities with little or no control beyond their capitals.

Translated into a sustainable security agenda, it is safe to argue in favour of a West-Central Africa security and development mechanism, but one that is properly structured, rather than a victim of ad hoc decisions such as ECOMOG. If a structured mechanism is available and deployable at short notice, it should be possible to convince small states like Sierra Leone and Gambia that the protection of their territorial integrity does not necessarily depend on a standing army, if there is a standing peacekeeping command to which they too can contribute soldiers.

A systemic change of the type we are suggesting requires extensive work. A good place to start might be:

- the harmonisation of all conflict management mechanisms to create a single mechanism empowered to undertake peacekeeping and perhaps peace enforcement missions
- developing a peacekeeping model with an accountable command, control and information system
- developing the necessary linkages between security, democracy and development in the regional integration process
- conceptualising an architecture of conflict management for twenty-first-century Africa in which the military plays a less significant role.[40]

Within the context of regional security mechanisms, it should be easier to reduce the relevance of mercenary outfits in Africa. If states and sub-national entities have confidence in the impartiality of regional mechanisms to broker peace in a fair-minded and even-handed manner, the demand for private security forces would decrease significantly. In

West Africa, for example, it has been argued that harmonising the various conflict mechanisms is the key to building confidence. Not only would this improve the operational effectiveness of the present arrangement, it would also minimise the 'undercurrents of fear of a neo-hegemonic leadership of a harmonised structure by a country like Nigeria'.[41] One could add to this the likelihood of developing appropriate and enforceable legal frameworks within the context of sub-regional security arrangements against the influx of mercenaries, while at the same time protecting those who have taken to hiring mercenaries when they had no choice. In Southern Africa, training and operational activities of the SADC peace operations are already being harmonised, and this is one area the reanimated East African Community is also interested in.

While effective regional mechanisms remain the most long-lasting method of curbing the activity of mercenaries, they still need to be backed up by an effective legal framework. The recent involvement of NATO in Kosovo further promotes the efficacy of regional mechanisms, provided they are subsumed under the authority of the UN.

Improving the Legal Frameworks of Mercenarism: Between Regulation and Redundancy

In the short-term, there is a need to put in place an appropriate legal framework in the light of changes that have occurred in the international environment. Until now, the Organisation of African Unity has prided itself on having probably the strongest convention on the elimination of mercenaries. It would appear that this convention has now become a Cold War relic in urgent need of review in order to deal with today's realities. Although political will is required to implement any new or amended convention, improving the legal framework might be the most appropriate start. The OAU has always seen the presence of mercenaries on the African continent as unacceptable. Between 1961 and 1977, the organisation passed several resolutions against the scourge of mercenaries.[42] In July 1977, the Assembly of Heads of State and Governments approved the Convention for the Elimination of Mercenaries in Africa (CEMA).

Looked at from the perspective of present realities, the first problem lies with the convention's definition of a mercenary. A mercenary is defined as:

anyone who, not a national of the State against which his actions are directed, is employed, enrols, or links himself willingly to a person, group or organisation whose aim is:

- to overthrow by force of arms or by any other means the government of that Member State of the OAU
- to undermine the independence, territorial integrity or normal working of the institutions of the said State
- to block by any means the activities of any liberation movement recognised by the OAU.

Whatever informed this definition of a mercenary, it was never foreseen that established governments would recruit mercenaries or fighters for the sole purpose of maintaining regime security, such as we have witnessed in Angola, Sierra Leone and Congo (Kinshasa). Beyond its statist conception, the definition provokes more questions that it answers in its additional clauses, perhaps unintentionally. For example, when it states that a mercenary is 'one motivated to take part in hostilities essentially by the desire for private gain and in fact is promised by or on behalf of a party to the conflict material compensation', it extends the boundaries of what are traditionally known as mercenaries, especially since it is arguable that 'private gains' do not come necessarily in the form of money. Up till now, mercenaries have been hired professional soldiers who fight for any group without regard for political interests or issues. As we argued in our review of mercenary activities of the 1960s and 1970s, they do not always have pecuniary motives. Some retain a strong sense of loyalty to ideological positions, which sometimes becomes the primary reason for their involvement.

This becomes even more crucial in the light of recent experience since the Cold War. The mujahedin soldiers from Afghanistan who assist the National Islamic Front government in the Sudanese civil war are probably doing so as soldiers of faith, but they have also been promised material compensation by a party to the conflict, in this case Iran. Although the definition quite correctly excludes soldiers who may be acting officially as authorised peacekeepers in third party conflicts, it considers as a mercenary someone who 'is not sent by a State other than a party to the conflict on official mission as a member of the armed forces of the said State'. Stripped bare of obfuscation, this definition would not exclude as 'mercenary' the involvement of Che Guevara and his Cuban colleagues in the Congolese conflicts, although the material gain still needed to have been proven and Che and his group could very well have argued that they were involved with a group trying to liberate Congo. Even so, it is possible to argue that the material gain this time, for a revolutionary of Che Guevara's clout, would have been the successful spread of the Cuban brand of socialism and the influence that comes with that.[43]

In the same vein, the now confirmed involvement of Rwandese and Ugandan troops in the ousting of Mobutu Sese Seko in Zaire and the subsequent war against Laurent Kabila could technically fall within the OAU's definition of mercenarism, since those involved were not there as members of the armed forces on a declared, official mission, although their countries would have benefited vicariously from the overthrow of a ruler perceived as a regional threat. However, the official involvement of Cuban forces in Angola in the 1970s and 1980s by invitation of the Angolan government excludes such a force being described as a mercenary involvement. These are grey areas, which can be manipulated by promoters of corporate mercenaries as well as their opponents to suit their side of the argument. There is therefore a need for more clearly defined parameters.

With regard to the recent conflict in Sierra Leone and the involvement of Executive Outcomes, it would still be difficult to enforce the clauses of CEMA, since the ousted government of Sierra Leone could technically argue that EO did not in fact 'take part in the hostilities' since it was contracted in an 'advisory and training' capacity – the well-worn euphemism for private armies on the battle front in local conflicts. These examples clearly indicate the weakness of the legal framework based on this inadequate and confusing definition of a mercenary.

Much more serious in the context of the mercenary convention are the questions provoked about the sovereignty of the nation-state and the principle of non-interference enshrined in the OAU charter in the context of the Cold War rivalry at the time of independence. For example, if it was acceptable for Nigeria and Biafra to engage the services of mercenaries or foreign forces for private gain in the 1967 civil war, then the Sierra Leonean authorities could have easily argued with little fear of contradiction that it reserved the right to choose whatever means it considered appropriate to restore peace and stability to the country, including the use of Executive Outcomes. After all, Nigeria was able to block any external interference at the level of the OAU in the civil war, claiming it was an internal problem.

Although the OAU has recently reviewed its conflict resolution mechanism so as to reflect the changing nature of conflict, disputes over the interpretation of the convention are to be resolved according to the principles of the OAU and UN charters.[44] Since it has its antecedents in the OAU definition, even the UN definition suffers from the same flaws as the CEMA definition. It defines a mercenary, for example, as a person:

who is specifically recruited locally or abroad in order to fight in an armed conflict; does in fact take a direct part in the hostilities; is motivated to take part in the hostilities essentially by the desire for private gain and in fact is promised by or on behalf of a party to the conflict nor a resident of a territory controlled by a party to the conflict; is not a member of the armed forces of a party to the conflict; and, is not sent by a State other than a party to the conflict on official mission as a member of the armed forces of the said State.[45]

This, of course, is inadequate if the convention is to have any teeth to combat the renewed upsurge in the activities of privatised mercenaries in the region. To give the convention teeth, however, would require recognition of the inextricably intertwined nature of conflicts, which traverses current approaches in peacekeeping and peacemaking towards peace creation as a means of addressing security problems in a comprehensive manner.

Taken to its logical conclusion, this may well lead us to an alternative view of security and stability which expands the vision of Africa's founding fathers – a comprehensive review of the OAU's charter. The experience of Africa in the last 40 years leads one to conclude that the state-centrism and the principles of non-interference which are at the core of the charter should no longer be seen as the dominant defining characteristics of the continent, especially given the tumultuous changes that have happened in the world. Given the loss of their monopoly over coercive power, it is time to diminish the self-sustaining politics of anarchy required of state-centrism and seek a redefinition of the sovereignty concept based on community consensus. This redefinition ought to de-emphasise the post-colonial boundaries which have been constant sources of territorial vulnerability and internal feuds. The perception of state-centrism offered by realists undoubtedly blurs many social processes within the international and domestic environments, such as the interaction of classes and the effect of ethnic and other social divisions, and prevents any serious discussion of these issues.

Evidence from more sophisticated views of security confirms that it is the interaction of social forces – political and ethnic groups divided along and between artificial boundaries – rather than international anarchy or expansionist motives on the part of governments that provide the most fundamental explanation of conflict in African countries. Until the internal and inter-state dimensions of security assume equal place with the international dimension of security in national security literature, the imbalance between the international and the domestic environments and between individual, national and international perspectives of security, will continue to be a source of controversy in the modern African nation-state.

In a continent where ruling elites desperately seek to stick to their old, ineffective ways even as other continents draw lessons from the changing nature of international relations and world conflicts, the only lesson that a significant section of marginalised Africans has learned from the experiences of Eritrea, South Africa, Rwanda, Namibia and now Congo is that *force* is the *only* means of achieving change; not dialogue, not international or continental pressure. This societal response to insecurity has made it easier for people at the receiving end of this challenge to accept the idea of 'private military companies' helping to 'restore peace' to the ever warring countries of Africa, no matter how dubious such claims are.

That is why it will be unfortunate if security in Africa is left to such an interpretation, which tends to limit peace and stability from achieving their emancipatory and humanistic ends. This would mean that nothing has been learned and the continent will remain in a Cold War time warp. This is an inevitable outcome of an international community which is not much interested in genuine African security, and a world too eager to cede its responsibilities to private armies.

For now, orthodox discourse on this subject dismisses any peace agenda that offers an alternative vision as naive and unrealistic. This is, of course, not surprising since realist scholars since the days of Morgenthau have always regarded state-centrism as the centrepiece of security and allowed no avenue for transformatory politics. Contemporary scholars of this school of thought continue to pursue the logic of anarchy, in their lopsided emphasis on the coercive form of power while ignoring its consensual dimension. We must challenge the underlying thesis of the realist discourse that seeks to ideologise peace-building through 'the end justifies the means' approaches to international security, in a bid to promote the hopelessness of collective security approaches and reify status quo views of power politics.

Conclusion

A commitment to a realistic international legal framework by countries at the continental level is the immediate step that needs to be taken to address this serious problem. The recently passed Foreign Military Assistance Law in South Africa, which outlaws South African citizens from participating in mercenary activities in a foreign country, sets a good example for what needs to be done in national parliaments. Although it would be naive to expect a sudden disappearance of mercenary and security outfits as a result of legislation, mercenaries must be made aware of the stiff penalties that await them should they be caught violating the laws of the land. The fact that Executive Outcomes has now formally withdrawn from the mercenary trade, even if it is still involved through its associated companies, is a victory for long-term stability and security in Africa.

Besides, there is a desperate need to bring this matter to public attention in the same way that the campaign against landmines has forced governments and public institutions to deal with a scourge they had pretended did not exist, because of pressure from the arms industry. Advocacy is therefore the key to ensuring that national legislation and international action rally round the need to make mercenaries accountable to the constituencies they serve and bring their activities in peace support operations under the regulation of regional peacekeeping mechanisms. The newly established International Criminal Court and the African Peoples and Human Rights Court should be empowered to deal with cases related to violation of international law by mercenaries. Nevertheless, the legal framework and the campaign to end mercenarism can only deal with the problem at the level of legislation. In a sense, the choice is not between regulation and redundancy. Regulation has its role within the ambits of national and international law, but the much touted success of the corporate mercenary is, in a way, an outcome of the inadequacy of the current conception of security and political action, based as it is on the questionable state-centric position. The ultimate solution still lies in a complete overhaul of the power politics complex through efforts to achieve community consensus. We are in no doubt that the most unyielding security problem on the African continent is that of identity and poverty-driven conflicts. Until the underlying causes of insecurity, which are deeply embedded in the retrenchment of the state in response to globalisation, and the challenge to it by local non-state actors, are addressed and resolved, mercenary activities can, at best, be band-aids; they cannot bring about a sustainable peacebuilding process.

Notes Chapter 1

1. See Foreign Affairs Committee, Sierra Leone (Vols I & II), *Report and Proceedings of the Committee* (The Stationery Office, 1999), 3 February 1999.
2. See Herb Howe, 'Private security forces and African stability: the case of Executive Outcomes', *Journal of Modern African Studies*, Vol.36, No.2, 1998 and David Shearer, 'Exploring the Limits of Consent: Conflict Resolution in Sierra Leone', *Millennium: Journal of International Studies*, Vol.26, No.3, pp. 845–60 for versions of this view.
3. Not unexpectedly, Robert Kaplan's famous 1994 *Atlantic Monthly* essay on 'The Coming Anarchy' became a must-read for every 'wannabe' US ambassador posted to Africa and for State Department desk officers dealing with Africa. Former UN Secretary-General, Boutros Boutros-Ghali also convened a special seminar led by Kaplan to discuss his Conradian diagnoses of the African predicament. In the same manner, David Shearer's IISS Adelphi paper on Private Armies is recommended effortlessly by Foreign Office diplomats and corporate mercenaries alike. Indeed, one of the first people to recommend the paper to this writer was Michael Grunberg of Sandline International and associated companies.
4. Stanley Hoffman, 'An American Social Science: International Relations', *Daedalus*, Vol.106, No.3, Summer 1977.
5. For our purpose, this refers mainly, but not exclusively, to the literature published in premier international security and African affairs journals such as *International Security, Survival, Millennium, Foreign Policy*, and *Journal of Modern African Studies*.
6. See John Mackinlay, 'War Lords', *RUSI Journal*, Vol.143, No. 2, April 1998, p. 26.
7. Ibid., p. 26.

8. See Marek Thee, *Whatever Happened to the Peace Dividend?* (Spokesman, 1991).

9. It may be recalled that the same Carlucci who was the CIA field officer in Congo played a key role in the elimination of Patrice Lumumba and the installation of Mobutu Sese Seko.

10. Minutes of Evidence taken, Foreign Affairs Select Committee Report, Vol.II, p. 138.

11. See Wilfred Burdett, *The Whores of War: Mercenaries Today* (Longman, 1977).

12. See UNDoc.5/4741 of 1961. The UN even launched 'Operation Rumpunch' to expel mercenaries from the Congo in August 1961.

13. Belgium pointedly refused to enact any laws banning their citizens from involvement, claiming they have no control over what their citizens do elsewhere, even though other western countries like France and Britain stamped their citizens' passports with a caveat, 'Not valid for the Congo.'

14. See J. John Stremlau, *The International Politics of the Nigerian Civil War* (Yale University Press, 1977). Also see Olusegun Obasanjo, *My Command: A Personal Account of the Nigerian Civil War* (Heineman Press, 1980).

15. OAU Doc.ECM/Res.55 (III), Addis Ababa, 10 September 1964.

16. AHG/Res.49 (iv), September 1967.

17. See Appendix I.

18. See the OAU's Declaration on the Activities of Mercenaries in Africa. Adopted at the meeting of Heads of Government on 23 June 1971.

19. See for example, Burdett, *The Whores of War*, which includes extensive accounts of the trial of mercenaries in Angola.

20. David Barzagan, 'Going private: a users' guide to the capital's [London's] freelance armies', *Guardian* (London), 8 September 1997.

21. Sandline International Promotional Brochure. See Sandline International Internet Website for additional image laundry, <www.sandline.com>

22. Ibid; see also *Southern Africa Political and Economic Monthly* (SAPEM).

23. Given the average salary package of each mercenary, it is not surprising that clients are charged so highly by suppliers. A rank and file recruit is paid on average US$3,000–$4,000 per month. A pilot or an experienced armour officer earns up to US$18,000 a month, and each package comes with life insurance. A more fundamental problem, at least in the short term, is the 'safe haven' that this new trend provides for demobilised soldiers in search of material wealth and fulfilment in their lives.

24. Office of the UN High Commissioner for Human Rights, *Report on the Question of the Use of Mercenaries as a means of Violating Human Rights and Impeding the Exercise of the Rights of Peoples to Self-Determination (A/52/495)*, 16 October 1997.

25. See Vol. 11 of Foreign Affairs Committee report, Minutes of Evidence and Appendices, p. 131.

26. See the excellent account of the Bourgainville affair by S. Dorney, *The Sandline Affair: Politics and Mercenaries in the Bourgainville Crisis* (ABC Books, 1998).

27. Plaza 107 is also the name of the suite at 535 Kings Road, Chelsea, London, occupied by Tony Buckingham, coincidentally opposite Tim Spicer's Sandline International.

28. Yves Goulet, 'South Africa: Executive Outcomes Mixing Business with Bullets', *Jane's Intelligence Review*, September 1997, p. 426.
29. 'UN Secretary-General explores potential causes, cures of conflict in Africa', (UN-IRIN *West Africa News*), SG/2045 –AFR/50,SC/6501, 16 April 1998.
30. Ibid.
31. See a recent book co-authored by the UN Co-ordinator of the Peacekeeping Operations in Somalia, Robert Oakley. In it, the authors concede the inability of the UN to cope with the burden of peacekeeping operations due to lack of interest from member nations. See also Robert Oakley, Michael J. Dziedic and Eliot M. Goldberg, *Policing the New World Disorder: Peace Operations and Public Security* (National Defence University Press, 1998).
32. 'Report of the Panel of Governmental Experts on Small Arms' (New York: United Nations, July 1997), p. 25, cited in Abdel-Fatau Musah, *Africa: The Challenge of Light Weapons Destruction During Peacekeeping Operations*, Basic Papers, Number 23, December 1997. Also see Joanna Spear, 'Arms Limitations, Confidence Building Measures, and Internal Conflict' in Michael Brown (ed), *The International Dimensions of Internal Conflict* (MIT Press, 1997) and Stiftung Wissenschaft und Politik, *Consolidating Peace Through Practical Disarmament Measures and Control of Small Arms: From Civil War to Civil Society*, Report of the 3rd International Workshop, Berlin, Germany, 2–4 July 1998.
33. Most of the troops owed their loyalty to the defeated president, Sassou Nguema, who eventually overthrew Lissouba.
34. *Time* magazine, 1 June 1997.
35. Meeting with Michael Grumberg by representatives of the Campaign Against Mercenaries in Africa, London, 30 March 1998.
36. Hans Morgenthau, *Politics among Nations: The Struggle for Power and Peace* (Alfred A. Knopf, 1983), pp. 525–8.
37. SAPEM.
38. See 'Kayode Fayemi and Napoleon Abdulai, 'Somalia: Operation Restore Hope or Operation Restore Colonialism?, *Africa World Review*, November 1993–April 1994, pp. 32–3.
39. Countries like Senegal and others within the French axis, hitherto reluctant of the need for ECOMOG, now actively campaign for its involvement in the recent crisis in Guinea-Bissau. Egypt has officially requested that ECOMOG be adopted as a continental model in Africa.
40. This is the subject of a much larger research work at the Centre for Democracy and Development.
41. *Harmonisation of Conflict Management Mechanisms in West Africa: The Facilitating Role of AFSTRAG*, AFSTRAG RoundTable, Serial No.1, Vol. 2, 1998.
42. See, for example, OAU Doc.ECM/Res.17 (vii), December 1970 (Lagos); OAU Doc.CM/St.6 (xvii) Addis Ababa, 21 to 23 June 1971; OAU Doc.CM/1/33/Ref.1, Rabat, June 1972 among many others.
43. Interestingly, Jon Lee Anderson, Guevara's official biographer, quoted these poignant lines from Che's African diary which detailed his experience in Congo thus: 'The OAU had blacklisted the Congo regime of Prime Minister Moise Tshombe over his unholy alliance with the Belgian and white mercenary forces ... However, it stated that if the mercenaries went, those States aiding the rebels would have to end their support as well. All foreign intervention in the Congo was to cease – and that meant the Cubans too ... .'

Jon Lee Anderson, Che Guevara: A Revolutionary Life (Grove Press, 1997) p. 605.

44. Article 12, Convention on the Elimination of Mercenarism in Africa (CEMA). See Appendix.

45. CEMA, Article 1, (1), (a–f).

2

Private Military Companies and African Security 1990–98

Kevin A. O'Brien

Since the Pretoria-based private military company (PMC) Executive Outcomes Pty Ltd (EO) first emerged publicly in Angola in late 1992, international attention has focused on the role and influence of these companies in stabilising – or, alternately, destabilising – national and regional security throughout Africa. While the focus of attention on PMCs grew initially, and then continually, from an interest in tracking the involvement of EO in African countries, it is unlikely that if EO had not come from the background it did (its personnel were almost exclusively former military and police special forces who served in the South African security forces during the apartheid era), such attention and reproach would not have been focused on the PMC sector as a whole.

It is quite often forgotten that such companies have existed for decades, and have been active in Africa (and other parts of the world) for far longer than EO has been in existence. Indeed, the origins of these companies can be seen in two previous types of international groupings that intervened in the colonial world: the mercenary 'companies' that fought in the Belgian Congo and elsewhere during the 1960s (such as 5 Commando), and the great colonial exploration companies that were the vanguard for colonial occupation and exploitation during the last three centuries (such as the Dutch VOC or the British South Africa Company). Even the secondment of Western military professionals to train post-colonial armies throughout Africa (embodied today, for example, in the British Military Assistance Training Teams) owe some of their origins to these companies. Also, throughout the decades preceding EO's founding in 1989, a number of private and commercial security companies operated across Africa. Although these companies rarely became involved directly in conflicts in the countries in which they operated, they usually acted to train the national armies of these countries or protect the assets and installations of multinational and transnational mining and oil exploration firms on whose behalf they operated.

What is most interesting to note, however, is the response and attitude of the international community to the existence and operations of these companies. While much condemnation is levelled at them, and they are labelled as 'criminals' and 'mercenaries' by developing countries, there has been a more measured response from Western governments, particularly the United States and Great Britain. An example of the lack of interest by both the international community and regional organisations towards these companies is their attitude to the two key resolutions that have been passed against mercenary activity in the past two decades. The UN General Assembly Resolution of 1989 requires 22 signatures by member states to come into effect; at the time of writing, only 16 have endorsed it. Likewise, the OAU's 1977 Charter against mercenary activities has so far been signed by only 22 out of 54 member states; many of the states that have hired PMCs in Africa since 1990 were among these signatories, an indication of the worth of this 'measure' against the hiring of mercenaries.[1]

In reality, it is not the PMCs themselves which should be singled out for criticism – they are, like any good economist, attempting to meet the demand with their supply – but rather the countries who hire them or who allow transnational corporations (particularly in the mining and oil exploration fields) to use them. These PMCs continue to exist and grow in their operations simply because the demand is there. They often supply what the particular state cannot provide: security, whether for the citizens of the state or for international investment. Much of the time they are partnered by domestic, often government/military, investors of the state in which they are operating. Thus, much of the vitriolic criticism directed at the PMCs is misplaced.

Similarly, there is a clear difference in the reactions to the companies originating in South Africa, most of which developed out of apartheid security units, and those originating in Western countries. It has been forgotten (or, for political reasons, simply overlooked) that many of the PMCs operating out of Western countries today were, and in many cases, continue to be, closely aligned with and supported by the intelligence services of the countries in which they originate. This can clearly be seen today in the cases of the United States, Great Britain and France.

PMCs and the 'New World Order'

The world's attention since 1993 has, however, been closely focused on the role of EO and companies like it in regional conflicts throughout Africa and the larger developing world for a number of reasons. Firstly, the origins of EO and other relatively new companies have sparked concern over their role in regional security. Often perceived to be operating outside of the rules of warfare (an incorrect but oft-cited assumption) and strictly for pay, as mercenaries have done for centuries, the apartheid (or, indeed, former Soviet bloc in increasing numbers of cases) background of these professional warriors has brought into question their motives for becoming involved in these conflicts. This question was never raised with pre-1990 'Western' companies.

Secondly, the failure of the international community to successfully intervene in African conflicts, growing more numerous by the year, in the face of an international outcry to 'do something' to halt the suffering, has led to a blasé attitude about many of these conflicts and a paradoxical unwillingness by the international community to pledge the necessary military, humanitarian, or financial resources to solve these crises. Yet, when PMCs become involved, they are criticised for being a 'band-aid solution' to the crisis of the day without a long-term stabilisation and peace plan being implemented – which is really a failure of the international community and not of the PMCs themselves.

Thirdly, the fact that many of these firms operate with at least the acquiescence of major Western governments and their security services, as was demonstrated, for example, in Britain's so-called 'Sandline Affair' in 1998,[2] brings into question just what policy exists in Western government circles vis-à-vis these companies.

Finally, the perception that many of these companies are operating on behalf of Western mining and oil firms with little interest in the national wellbeing of the countries in which they operate has further tarnished the image of these firms, painting them as neo-colonial exploiters of Africa. One exception to this image appears to be the case of various international demining firms, many of whom are PMCs in their own right, engaged in demining and demining training in many parts of Africa; this particular aspect of the PMCs does not form part of the focus of this discussion.

When taken altogether, these perceptions paint an extremely negative image of the PMCs and their activities in Africa over the course of this decade. Is this the reality, however? Why did the PMCs emerge so strongly in Africa when they did, and who has supported their operations? Is there any difference between the activities of the more 'traditional' mercenary-type operations of yesteryear which still raise their head every now and then in Africa, and the commercial-military ventures which these PMCs are engaging in this decade? What came before the emergence of EO in Angola in 1992?

The answers to these questions would show that the international perceptions of PMCs as outlined above may not ring true when examining the continued rise of PMCs in Africa over the 1990s. Indeed, it may be seen that, in some cases but not all, PMCs have been much more effective in resolving conflicts in many African countries than has the international community; that they take a much more direct interest in the wellbeing of the countries in which they intervene to halt conflicts than the international community; and that – most often – it has not been the fault of the PMCs that long-lasting stability and peace has not been achieved in the countries. This has rather been the fault of either the international community, which failed to step in to secure peace (as for example, in the case of Sierra Leone in 1997), or of the participants to the conflict who were unable to agree to a post-conflict peace settlement following the withdrawal of the PMC (as in the case of Angola). By examining the activities, year-by-year, of these firms and the clients (governments, multinational businesses) for whom they work, a much clearer picture may be found of the trends in African conflict resolution over the past decade.

PMCs before 1990

In the period preceding 1990, the overwhelming majority of PMCs involved in foreign military operations around the world originated in Britain, with the focus of their operations being largely, but not exclusively, in the Middle East. Many of these companies did also operate in Africa, Latin America and the Far East. The most infamous of these were the operations by Sir Percy Stilltoe, a high-ranking British counter-espionage operative from World War II, who was hired in the 1950s by Harry Oppenheimer of De Beers. Oppenheimer contracted Sir Percy to establish intelligence networks in Sierra Leone and use mercenaries to eliminate diamond smuggling, forcing all to sell to De Beers.[3] This could be considered as the first modern manifestation of the private military company in Africa.

The best known of these companies in Britain was formed by members of the British Special Air Service (SAS), including its founder, Colonel Sir David Stirling. In the late 1960s, Stirling founded WatchGuard International, a company which employed former SAS personnel to train the militaries of the sultanates of the Persian Gulf (most frequently, their palace guards and special forces units), as well as providing support for their operations against rebel movements and internal dissent. WatchGuard was to become the model for all future PMCs. Out of it developed Kulinda Security Ltd, which in the 1970s engaged in contracts in Kenya, Zambia, Tanzania and Malawi. WatchGuard was also generic to another aspect of PMCs, that of close connections to Western governments. At the time it was founded, it was hoped by many that it would become the 'unofficial arm' of the British Secret Intelligence Service (SIS), but Stirling did not move in this direction, probably because he preferred to remain outside government squabbling and interference.

WatchGuard carried out many operations throughout the Middle East; one of its few operations in Africa involved providing a force for a counter-coup in Zambia between 1967 and 1969, as well as guarding the Zambian borders. A similar operation was conducted in Sierra Leone in the early 1970s.[4] In 1986, nearing the end of his life, Stirling formed KAS Enterprises with Ian Crooke to conduct anti-poaching operations in Southern Africa, often in co-operation with the South African Defence Forces and intelligence, and with the support of the World Wildlife Federation. KAS was based in Pretoria and was closed down in early 1990, just months before Stirling's death.[5]

Other companies formed during this period included KMS (thought to mean 'Keenie Meenie Services', SAS slang for covert operations) and its subsidiary Saladin Security. Ex-SAS officers Jim Johnson and David Walker formed KMS in the early 1980s after leaving Control Risks Group, which they had founded with Arish Turle. While KMS undertook international contracts, Saladin originally only fulfilled domestic contracts. With the demise of KMS in the early 1990s, however, it began to operate internationally.[6]

The Corps of Commissionaires had operated since the end of the last century, providing security support services to British and foreign companies. By the 1980s it had begun to expand its operations into Africa. In 1975, John Banks established Security Advisory Services Ltd to help recruit mercenaries to fight with Holden Roberto's FNLA in the Angolan civil war; however, SAS Ltd never took off as anticipated and slowly withered.[7] Other British commercial firms established security wings to assist in their operations: a good example is the British-based multinational Lonrho operating in the midst of Mozambique's civil war throughout the 1980s. Lonrho hired companies of Gurkhas to protect its assets and operations in Mozambique throughout that decade.[8] Longreach Pty Ltd, a South African company formed by Craig Williamson as a front company for the South African Police Security Branch's foreign operations in 1986 and believed to have been involved in a number of assassinations, provided military intelligence support to the Seychelles from 1986. Finally, a number of hitherto unknown security firms such as Falconstar and Intersec operated out of Britain in the 1980s, providing armed personnel for security work throughout Africa.[9]

By far the biggest of all such companies to emerge out of Britain's special forces community is Defence Systems Ltd (DSL), founded in 1981 by Alastair Morrison. Throughout the 1960s and 1970s, small 'security consultancies' had been established in garrison towns throughout Britain, especially in Hereford where the SAS is based, often enticing serving SAS personnel and their 'wannabe' circle of other British Army personnel away from their national service and into quick and lucrative contracts for foreign potentates, usually in Africa or the Far East. DSL acted as a clearing house for these groups by ensuring that an internationally recognised company would be waiting to hire former SAS personnel into legitimate contracts as security consultants, military trainers, or to support commercial enterprises with assets in regions of conflict. Quickly gaining a reputation for professionalism, integrity and good business, DSL rapidly shed any insinuations of 'mercenarism'. It was hired throughout the 1980s and into the 1990s by international organisations such as the World Bank, the UN and various humanitarian NGOs to protect their personnel and assets in regions of conflict. DSL's operations in Africa throughout the 1980s included Mozambique, Sudan and Kenya, to name a few.[10]

In the meantime, the overwhelming majority of all other private military operations in Africa was characterised by ad hoc groupings of former soldiers from Western countries fighting for or against rebel movements in African countries wracked by conflict, labelled the 'Wild Geese' syndrome. It must be remembered that, throughout the 1970s and 1980s, the vast majority of conflicts in Africa were subsumed within the global bipolarity of the Cold War. One side in the conflict would be supported by the United States and its allies (often including South Africa), and the other side by the Soviet bloc or Communist China.

With the ending of the Cold War in the northern world, the conflicts were left to stumble on without external support and, not surprisingly, in many cases the antagonists turned to private sources to support their efforts. Where American, British, Soviet (including Warsaw Pact) or Chinese intelligence or special forces officers had been present in these countries to 'advise' on the conflicts and oversee the import of military equipment to support the conflict, now private military consultants filled the vacuum, often with strong ties to the international arms trade community. Alongside the winding down of the Cold War in Africa was the withdrawal of Portuguese colonial forces from 1975 onwards and a gradual downgrading of French post-colonial involvement on the Continent. Either PMCs or other national actors have increasingly filled this vacuum.

The final element in this mix was the ending of apartheid in South Africa. This event did not happen in isolation, but followed on the heels of the ending of international involvement in the war in Angola (with the withdrawal of Cuban and South African forces) and the termination of South African colonial rule over Namibia. These three events placed thousands of personnel from the South African Defence Force, as well as the South African Police and intelligence, into the potential pool for private military firms eager to expand their operations in the new, post-Cold War Africa. To this end, Eeben Barlow, a still-serving South African Special Forces officer, founded EO in September 1989, initially, it is believed, as a front company for breaking the arms embargo then in place against South Africa. Since its inception, EO has generated more controversy and international press coverage than any other such entity.

EO in 1990–92

The first evidence of EO emerged in 1990, when former Civil Co-operation Bureau (CCB) personnel, in the course of the Harms Commission investigation into hit squad activities in South Africa, began to leave the SADF for 'private consultancy'. At the time, although the name Executive Outcomes was not known, it was evident that CCB and other SADF Special Forces personnel were leaving the security forces in anticipation of the coming transition. Subsequently, there were secret reports that the SADF Special Forces were being trained by these 'private consultants' in counter-intelligence and other tactics for internal operations in South Africa, following the withdrawal from Angola and Namibia. In later years, EO would state that its initial contract was with the SADF to undertake this training.[11] However, the reality of EO's origins prior to its initial operation in Angola in late 1992 was much different.

Luther Eeben Barlow was born in Northern Rhodesia but his parents moved to South Africa when he was still young. Joining the SADF in 1974, he became established in engineering. In the late 1970s and early 1980s, Barlow became one of the golden boys of 32 Battalion, the SADF unit operating in Angola and composed of former FNLA black Angolan soldiers fighting the MPLA government. Barlow served under Willem Ratte as the second-in-command of 32 Battalion's reconnaissance wing. There is evidence to suggest that Barlow spent two years in the British SAS jungle-training camps in Malaysia between 1981 and 1983 before returning to South Africa and transferring to the Military Intelligence Directorate of Covert Collection in 1983. This followed rumours that Barlow had served as a covert operator in Rhodesia in the months prior to its 'fall'. There is also speculation that he engaged in a number of covert operations for ARMSCOR, the South African arms manufacturer, during this period. He served in the Directorate Special Tasks in the Division of Military Intelligence from 1983 to 1986, assisting RENAMO in their operations in Mozambique. In 1986, he joined the SADF Special Forces Civil Co-operation Bureau, which had evolved out of 3 Reconnaissance Regiment, itself a patchwork of former Rhodesian and Portuguese special forces personnel. Barlow worked for the CCB until September 1989.[12]

It is believed that EO was originally founded by Barlow (who headed the European Section of the CCB) as a front company for the CCB, allegedly to facilitate the covert procurement of weapons systems for the SADF in violation of the arms embargo against South Africa. With the break-up of the CCB and other covert units of the SADF from 1990 onwards, many covert specialists joined forces with Barlow to enhance the scope and capabilities of EO. Over the proceeding three years, EO took in the majority of former members of 32 Battalion, as well as members of other SADF Special Forces regiments, and former members of the Rhodesian Selous Scouts/SAS and the SAP counter-insurgency unit Koevoet.

Thus EO grew in numbers and capabilities, moving from training SADF personnel to training Military Intelligence Division personnel and, allegedly, even Umkhonto weSizwe personnel as part of bridge-training to prepare the ANC guerrilla force for integration with the SADF. During this time, Barlow and Lafras Luitingh were engaged in secret discussions with Tokyo Sexwale of the ANC, offering to 'quietly remove' from South Africa former special operators who could potentially endanger security during the 1992–94 transition. The ANC agreed but subsequently denied any such discussions.[13] At the same time, Barlow was in contact with National Intelligence Service (NIS) personnel in Pretoria, acting (it is believed) as a trouble-shooter in operations for the NIS and other South African agencies.[14] Word of Barlow's capabilities began to filter out to both Britain's SIS and to a number of South African mining houses who were looking for individuals to help protect their assets throughout Southern Africa.

In these two developments lay the origins of the evolved EO. In 1991 and 1992, Barlow was contracted by De Beers and other mining houses to undertake covert reconnaissance missions throughout Southern Africa, particularly in Botswana, Namibia and Angola; these contracts included training the security elements of De Beers. By late 1992, his name had been brought to the attention of Simon Mann, a former British SAS officer, by a South African NIS officer. Mann mentioned Barlow to Anthony Buckingham, a British businessman with worldwide mining interests and himself a former British Special Boat Service officer.[15] Buckingham contacted Barlow in mid-1992 with the request to 'do a job' for some of Buckingham's business interests in Angola.[16] Thus, the contemporary EO was born.

These origins were intentionally kept unclear: a much-mooted British defence intelligence report records that 'Executive Outcomes was registered in the UK in September 1993 by Anthony Buckingham, a British businessman and Simon Mann, a former British officer'; Barlow's and Luitingh's names were kept hidden, at least in Britain.[17]

At the same time, other companies continued their activities in Africa. Secrets, created by Paul Barril, a former French Gendarmerie Captain, signed its first three-month contract in 1992 to train the guard of the Cameroonian President, Paul Biya. Another French company, Eric SA, was established by Jean-Louis Chanas, formerly of the DGSE (Direction

Générale de la Sécurité Extérieure), at around the same time to provide security to Algerian oil installations.[18]

PMCs in Angola 1993–97

The explosion of the number of private military companies engaged in Africa began in 1992. By mid-1997, it was estimated that there were more than 90 private armies operating throughout Africa, the majority of them in Angola. In many senses, Angola has been the testing ground for the development and evolution of PMCs in Africa.

In October 1992, Buckingham and Mann commissioned Barlow (on behalf of Sonangol, the Angolan parastatal, acting as the intermediary to Chevron, Petrangol, Texaco and Elf-Fina Gulf) to recruit a force of South African veterans to secure Soyo, one of the centres of the oil industry, which was in the hands of UNITA. This operation was carried out with the acquiescence of Luanda, although not on their behalf. Its primary aim was to secure a US$80 million computerised pumping-station owned by the oil companies. A small force of approximately 80 personnel succeeded in early spring of 1993 in securing the station, but UNITA recaptured Soyo when the South Africans left.

The Angolan government then, in July 1993, requested a larger force, allegedly guaranteeing Buckingham's companies future considerations in return. Ranger Oil West Africa Ltd, a Canadian firm 'associated' with Buckingham through its joint sharing of oil-blocks in Angola with Heritage Oil & Gas (Buckingham's company), provided a guarantee of US$30 million for the operation. Buckingham and Mann introduced EO's personnel to President Dos Santos and FAA Chief of Staff João de Matos; Dos Santos subsequently authorised the operation.

EO maintain that, although Ranger financially supported the operation, they did not pay EO; the Angolan oil parastatal Sonangol provided the payment. There is evidence that De Beers may have provided US$7 million to the government in Luanda in return for offshore prospecting rights, enabling Luanda to pay EO. Other reports point to IMF debt rescheduling to provide these funds.[19] It would have been very much in De Beers' interest to have EO active in Angola, especially in the north-east of the country, in order to halt diamond smuggling. There has been a long-standing suspicion that the support to EO operations by Heritage and Branch Energy (another Buckingham company) was given partly at the behest of De Beers.[20] Barlow himself admitted that the diamond fields in many of these operations would be the first natural target as it would mean that EO would be guaranteed payment by the host government: securing the treasury first appears to be a wise move in many of these very unstable countries.

In this second operation, with approximately 500 men, EO routed UNITA and secured the whole oil region of Angola; the finalisation of the contract was followed by substantial concessions in oil and diamond mining to companies associated with Barlow's newly-formed (1994) Strategic Resources Corporation (SRC). At the same time, EO retrained the 16th Brigade of Forcas Armadas Angolanas (FAA), which began inflicting heavy casualties on UNITA. During this time, according to some personnel, EO 'assisted' the FAA in retaking the rich diamond fields of Saurimo and Cafunfo in Luanda Norte province, the source of much of UNITA's funding for its war effort. In reality, EO did most of the fighting, launching a number of heliborne assaults on UNITA positions. UNITA was defeated by EO due to the use of tactics (used also by supporting FAA units) that UNITA was unaccustomed to in its confrontation with the FAA. These included night fighting, air-land assaults, and long-range reconnaissance missions by EO personnel. EO's contracts were worth US$40 million in 1993, US$95 million in 1994, and an undisclosed monthly fee (estimated at US$1.8 million) for the remaining months between 1995 and their withdrawal in January 1996. Altogether, more than 1,400 EO employees were cycled through the Angolan contracts.[21]

During their time in Angola, EO also established numerous affiliates and subsidiaries. Among these were Saracen International, a commercial security venture, and Bridge International, which specialised in construction and civil engineering. EO was also responsible for bringing in numerous foreign support companies to assist in their operations. These included Capricorn Systems Ltd, whose name was subsequently changed to Ibis Air International and is run by South African Director Crause Steyl. Ibis has offices in Johannesburg, London, Luanda, Malta and Freetown; its principal office is in Guernsey, while it is registered in the Bahamas.[22] Branch Mining Limited was also established by Barlow as part of SRC; Falconer Systems, set up as a front for Executive Outcomes to provide logistical supplies to 'United Nations-related organisations' and other firms continued to operate in many of the countries from which EO 'withdrew'.

When EO's contract was terminated on 12 December 1995 and EO personnel left in January 1996,[23] Saracen International was used to guarantee EO a continuous presence in Angola, both directly through SRC's control over Saracen and indirectly through former EO personnel continuing to work for Saracen. The company provided demining work around Soyo, as well as establishing a special rapid reaction police unit in Angola. It was later awarded a US$10 million contract by FINA to protect their petroleum installations in Soyo;[24] Mechem, a former South African military intelligence front company often associated with EO, was awarded a US$6.5 million contract to demine the roads and highways in Angola.[25] In August 1996, the same month as Carson Gold

(Robert Friedland's corporation) bought out the shares of Buckingham's Branch Energy, LifeGuard Management was formed in Angola to provide security at the mining installations controlled by the Branch group of companies. In October of that year, Carson Gold was renamed DiamondWorks Ltd.

Around the same time, former EO employees were establishing other private/commercial security firms to compete in the lucrative Angolan market. These included Alpha-5, established by FAA Chief of Staff de Matos and former EO personnel; Stabilco, formed by Mauritz Le Roux and Neil Ellis; Omega Support Ltd, formed by John Smith, the former South African defence attaché in Luanda. Longreach Pty Ltd, Panasec Corporate Dynamics, Bridge Resources, COIN Security and Corporate Trading International all provided security to mining and other commercial facilities in Angola.

EO itself established Shibata Security in Angola (under the SRC umbrella) with Portuguese partners to provide security for mining facilities and demining operations throughout 1996, as well as demining operations, following EO's withdrawal. Shibata later became involved in demining in Mozambique, but dissolved in 1997. Teleservices International, originally formed by Gray Security of South Africa and Executive Outcomes in 1996,[26] but now a partnership between Gray and senior Angolan military personnel, has been providing security to Branch's mining sites in Angola since 1996, specifically in Luo and Yetwene, and was often deployed to secure these individual mines during the fighting in 1995–96.[27] Alpha-5 was contracted by Endiama, the Angolan mining parastatal, in 1996, to provide security to the mines that it controls.[28] Many of EO's former employees, when not establishing PMCs themselves, became managers and directors in the Branch-Heritage group. A good example of this is J. C. Erasmus, another former member of the CCB, who worked with EO in Angola. Branch Energy subsequently hired him as their head of security in Luo; he also served as Branch's Angola Manager for a period.[29]

Other companies, many of which have continued to operate in Angola, include Britain's Defence Systems Ltd, who began providing security to mining and petroleum sites for a number of transnationals, as well as protecting diamond routes for De Beers, in 1996. DSL was specifically contracted to provide protection to the Soyo offshore oil platforms from UNITA (which it did, using former British SAS and Special Boat Service personnel). It also provided security for four embassies, the Angolan parastatals Sonangol (on behalf of Shell, Mobil, Amoco and Chevron) and Endiama (on behalf of De Beers and Anglo-American), a cement factory, and the main hotels in Luanda. According to ex-DSL advisers, the company also had an arrangement to train the Angolan rapid-reaction police.[30] By October 1997, DSL had more than 1,000 personnel working in Angola, the majority of whom were Angolans;

however, Luanda terminated this contract on 24 December 1997, with the order that all non-Angolan DSL personnel had to leave the country by February 1998. The Angolan government maintained that this was because DSL was 'operating illegally' by not paying sufficient taxes for all the earnings the company and its employees were making and siphoning out of Angola.[31]

In 1992, it had become illegal for foreigners to own PMCs in Angola. Most often, to circumvent this restriction, international directors (as happened with many EO spin-off firms) would partner Angolan nationals, usually politicians or military leaders, to form Angola-based PMCs (Defence Systems Angola was an example of this). DSL's operational command was, however, just a little too 'British'. This statute clashes almost directly with the Angolan government's demand that foreign investment companies provide their own security at their installations. Luanda currently spends more than 70 per cent of its annual revenue (approximately US$1.6 billion in 1996) on security; therefore, encouraging others to provide it makes financial sense.[32]

One of the largest PMCs in the world, Military Professional Resources Incorporated (MPRI),[33] has made numerous attempts to become active in Africa, with little success so far. Its most high-profile operations have been in the Balkans. In December 1995, during the visit of President dos Santos to Washington, it was rumoured that President Bill Clinton had 'encouraged' dos Santos to terminate his contract with EO and hire MPRI in its place.[34] However, negotiations between Luanda and MPRI, which began in March 1997, continued for more than 18 months without result. One story has it that MPRI was unable to deal with the 'African way of business', another that the US State Department, which must license all US foreign military activities, had refused MPRI a licence for Angola.[35] The MPRI contract would have involved providing basic training to officers and non-commissioned officers (allegedly for two elite paratrooper brigades of the Angolan military), training army engineers to repair roads and bridges, carrying out mine clearing, and conducting instructions on administration and an English language programme.

For PMC operations in Sierra Leone between 1991 and 1998, see Chapter 3.

PMCs in Central Africa 1996–1998

Following the genocide in Rwanda in the spring of 1994, the new Rwandan government attempted to consolidate its domestic and border security and return Rwanda to a semblance of normalcy for its citizens. To achieve this, the new Rwandan Patriotic Front contacted a number of PMCs and foreign states to help them develop their security capacity. One of the first was the American company Ronco, which supplied both demining expertise and technology, as well as limited training to the Rwandan forces; Ronco provided similar expertise in Mozambique from 1996 to 1997.[36]

During the crisis in Zaire (now the Democratic Republic of Congo) in 1996–97, a number of PMCs operated within the country on behalf of either some of the neighbouring countries involved in the war (i.e. Rwanda, Uganda, and especially Angola), the Kinshasa government, or the multinationals supporting Laurent Kabila's rise to power.

The first of these groups was the so-called 'White Legion' raised by former GSPR Colonel Alain Le Carro, former Gendarme Robert Montoya, and Serbian commando Lieutenant Milorad Palemis,[37] with the assistance of Belgian mercenary leader Christian Tavernier. Numbering approximately 300 personnel, the unit was composed of Serbian, Moroccan, Belgian, Angolan, Mozambican, South African, French and British mercenaries fighting for President Mobuto Sese Seko.[38] The Legion was assisted in its operations by the French DGSE intelligence service,[39] who used the telecommunications company Geolink to provide funds to Tavernier and buy arms from the Former Yugoslavia. Tavernier, in turn, hired Le Carro and Montoya, who put the force together.[40] Geolink had been used by the DGSE before and its directors admitted quite openly to having worked for French intelligence for more than a decade.[41] With no support from the Armed Forces of Zaire (indeed, a number of mercenaries were killed by FAZ soldiers during the battles around Kisangani), the Legion disappeared into Congo-Brazzaville in early 1997, where it fought for the besieged government there.[42] According to reports from *Le Monde*, Montoya had been the EO link-man in Zaire, as well as acting for the company in Togo; EO strongly denied this.[43]

As the crisis in Zaire worsened, Mobutu considered hiring many other PMCs to rescue him.[44] In October 1997, EO admitted to having been approached by both President Mobutu and Laurent Kabila during the conflict to support their operations against each other in Zaire. EO refused on the grounds that it viewed each antagonist as 'politically suspect'; however, former EO personnel undertook contracts for both the Zairean government and transnational mining companies.[45]

What is more likely is that Mobutu was unwilling to pay the fee that EO had listed for their services. According to reports, EO had offered him two options: the first, for US$21 million, would have provided a small strike force; the second, for US$59 million, would have provided a combat-ready battalion. Zairian generals had discussed these options with EO in February 1997 but, ultimately, the deal was never consolidated.[46] There is some possibility that these deals were pre-empted by Le Roux's Stabilco's proposals to Mobutu.[47] Mobutu approached other companies such as MPRI, but the US government turned down MPRI's request for a licence to assist Mobutu. At this time, Kabila had already made his deals with America Mineral Fields, Branch Energy's chief rival, and their associated security firm International Defence and Security (IDAS). It is also thought that Omega Support had provided Kabila with training and equipment from late 1996 through 1997.[48]

There is evidence that the US military and American private security firms had been secretly supplying the new Rwandan government with military equipment, in violation of the arms embargo on Rwanda/Burundi, as well as training elements of the Rwandan Patriotic Army which, together with Ugandan assistance (another country that had been receiving US military support), later supplied and supported Kabila's rebel alliance against Mobutu.[49]

The US has admitted that its Army Special Forces personnel have been active in Rwanda since early 1995 and began conducting ongoing special forces training for the Rwandans in July 1996. PMCs mentioned in this connection include BDM International, which has long provided the US military with specialised training in information technologies, information warfare and intelligence, as well as providing interpreters and translators to US operations overseas, including Africa. Another is Betac Corporation, which is involved with US Special Operations Command and has assisted it with clandestine operations throughout the world. Sources have reported that, prior to the launch of Rwandan and Ugandan assistance to Kabila's rebellion, the US 3rd Special Operations Group had been training the Rwandan and Ugandan armies in that area, supported by a number of US private security firms assisting with the training. The same US special forces personnel appeared again in early August 1998 on the Rwandan-Congolese border, accompanying Rwandan military forces entering the Congo.

Similarly, other firms had been making approaches to Burundi, which was caught up in an internal war and the Central African crisis. South African employees of the GMR Pty Ltd offered to supply arms from Somalia to both the Burundian government and the rebels. People alleged to have been associated with this offer included Craig Williamson (recently associated with Saracen International's operations in Angola)[50] and Willem 'Ters' Ehlers (former secretary to P. W. Botha), now running GMR. Ehlers had previously supplied the Rwandan Patriotic Front with more than 80 tons of armaments in 1994; he had also supplied arms to Hutu extremists in Zaire in 1996, as well as UNITA. Human Rights Watch alleged that EO had provided military training to Burundi's rebel National Council for the Defence of Democracy (NCDD) group around the same time. This was strenuously denied by EO.[51]

As the crisis spread from Zaire to the Cabinda enclave of Angola and into Congo-Brazzaville, other PMCs already active in the area became involved. The Luanda government was employing a number of different private firms to secure their oil interests in Cabinda and northern Angola, as well as the diamond fields of eastern Angola, and to retrain their rapid reaction forces. One of these was International Defence and Security (IDAS), registered in the Dutch Antilles, staffed by Belgians and Israelis, which also engages in its own mining operations. It was particularly instrumental in establishing a blocking force between UNITA forces in Luanda Norte province and their supply route from Mobutu's forces across the border in Zaire.

IDAS is closely connected with Boulle's America Mineral Fields,[52] which had pushed Anglo-American and De Beers out of their mining contracts in Zaire, signing a US$1 billion deal with Kabila to assist him in his rebellion in return for immediate mining rights in the Katanga province.[53] Mining sources have stated that the ties that exist between EO and the Branch-Heritage group are reflected in similar ties between IDAS and America Mineral Fields, with similar contracts being pursued. In fact, when EO left Angola, IDAS simply took up their contracts in Luanda Norte province protecting the diamond fields.

Once UNITA had been temporarily immobilised, the secondary target of Luanda was the Front for the Liberation of the Enclave of Cabinda (FLEC), a secessionist group that has been struggling for independence since the 1960s. In October 1997, this operation was accomplished with the aid of AirScan, a new PMC providing security assistance to the Angolan government in the Cabinda region. Florida-based AirScan was hired by Luanda in the autumn of 1997 on the recommendation of Cabinda Gulf Oil Company, a subsidiary of US Chevron Oil (which owns jointly many of the oil assets in Cabinda), to provide protection against guerrilla attacks from FLEC. The AirScan operation, led by US Brigadier (retd.) Joe Stringham, was to have been granted to MPRI, but this firm pulled out of negotiations with the MPLA government at the last moment. As in suspected links between DSL and IDAS, AirScan is believed to have links with MPRI, and is still employed in Cabinda, guarding Chevron's (via Cabinda Gulf Oil Company) wells.[54]

The war in Central Africa soon engulfed Congo-Brazzaville, resulting in the overthrow of the democratically elected government there. The Brazzaville government had used the Israeli firm Levdan to train the Presidential Guard and the armed forces.[55] Levdan, which is owned by Kardan Investment, an import-export company active in the diamond trade, was sub-contracted by Kardan in 1994 on a US$50 million contract with the government in Brazzaville to train the local army and Presidential bodyguard. For three years, approximately 65 former Israeli military personnel under the command of Colonel Zev Yerine trained the Brazzaville Presidential Guard.[56] At the same time, former EO director Lafras Luitingh was organising arms supplies to the Angolan government in Cabinda and to the Brazzaville government from his base in Pointe-Noire.[57] He was not the first former EO commander to supply Lissouba's government in Brazzaville; Ian Liebenberg had provided twelve South Africans and an unknown number of other mercenaries to serve as a 'special force' to Lissouba prior to his overthrow.[58]

During the year that Kabila consolidated his power in Kinshasa, a number of PMCs secured deals with his government. These included DSL, whose subsidiaries USDS and Sapelli SARL are protecting the installations belonging to the SOCIR oil refinery of Congo-SEP (the Belgian oil company Petrofina's subsidiary) and the US embassy. MPRI had reportedly attempted to negotiate contracts with Kabila, once again without success. Toronto-based International Consulting Services, run by former SAS officer Alan Bell, had provided military training services for some of Kabila's new army, as well as commercial security and related consultancy work for Canadian mining companies developing new interests in the DRC.[59]

Kabila's support in the region did not continue for long. In early August 1998, a new rebellion began in the east of the DRC (exactly where Kabila had begun his) and quickly gained ground against an ill-disciplined Congolese army. Kabila immediately began to look for outside assistance. One of the companies that he approached first was EO, who began negotiations with Kabila in mid-August. EO's contract talks reportedly included supplying personnel and installation protection to Kabila's government, logistics and air support against the rebels, and securing of the hydroelectric damn at Inga, which supplies the capital with electricity. However, as soon as these reports were made public by the South African Broadcasting Corporation on 24 August, the discussions were reportedly cancelled.[60]

Silver Shadow, a security company led by retired Lieutenant-Colonel Amos Golan, at this time made an offer to Kabila to build up a special protection unit; however, the Israeli government ordered the company to terminate negotiations due to the instability of Kabila's government.[61] Angola's initial direct support for Kabila has, however, meant that Angolan-based PMCs (many of which are EO spin-offs or are run by former EO personnel) have begun to support Kabila as part of their operations against UNITA, including through aerial reconnaissance, intelligence-gathering and operations planning. In the final week of August, it was reported that more than 100 South African and French troops arrived in Katanga province to defend mining installations in areas of rebel activity.

The question remains as to which PMC is protecting the DRC's oil fields at Nkossa, estimated to account for more than 60 per cent of Kinshasa's revenues. Given the state of the DRC armed forces and the active rebellion in the country, these installations must be protected by a PMC.

Although little appears to be known about them, the South African *Mail & Guardian* reported that South African-based PMCs are also providing assistance and support to the rebels in the DRC. Johan Niemöller, a former CCB operative, has been recruiting former South African soldiers to fight for UNITA since early 1997; at the same time, he was employed by former Mobutist General Kpama Baramoto to establish arms caches in Central Africa and hire forces to support a move against Kabila. Baramoto had attempted something similar in January 1997 when he hired Ellis's and Le Roux's Stabilco, based in South Africa, to employ 500 South African private soldiers to halt Kabila's final advance; the plot was never concluded due to disagreements between Mobutu's generals and Stabilco's directors.[62] Le Roux later established SafeNet in South Africa to provide Baramoto with personal security both in South Africa and during his secret travels abroad. SafeNet originally provided the basis for Baramoto's attempts to provide military support against Kabila in 1998. Baramoto split with Le Roux in January 1998, however, and began negotiations with a second, unknown PMC.[63]

PMCs in Angola and Sierra Leone 1998

By mid-1997, PMCs from South Africa, the United States, France and Britain had become intricately involved in supporting and supplying both UNITA and the MPLA in their operations. This has been particularly the case with South African firms supporting UNITA, due to the past links established between apartheid Military Intelligence and the rebel movement.

In January 1997, a new South African PMC began recruiting former soldiers for operations in Angola and elsewhere. Called Executive Research Associates, it is believed by South African intelligence to be an EO front-company.[64] Since late 1997, a number of South African PMCs have been running supplies to UNITA, taking advantage of the extremely lax (almost non-existent) border controls in Angola and using UNITA-controlled airfields for many operations. Some of this support involves providing military hardware and training in return for diamonds. Two of the biggest PMCs believed to be involved in these operations are Omega Support Ltd (closely tied in South Africa to Strategic Concepts Pty Ltd, established by former military intelligence operative Sean Cleary at the end of apartheid) and Stabilco.

In June 1998, a hitherto unknown group began to recruit South African and British forces to assist UNITA in its operations; by the middle of the summer, it had recruited more than 300 personnel, the majority of whom had previously worked for EO.[65] Known as IRIS, it is composed of former SADF commanders, former MK and SADF soldiers, and has the backing of British, American and South African businesses in Angola. While at the time of writing little is known about IRIS, it is believed to have been established in late 1997 as a rival organisation to the EO group and associated with other mining and oil firms in Angola.[66]

Sierra Leone presents yet another picture of PMCs developing their links with the mining and oil companies on which the governments of

Africa rely so heavily. Here, an operation was planned by Tim Spicer, Tejan Kabbah and – allegedly – British High Commissioner Peter Penfold, with at least the acquiescence of British intelligence, to support the ECOMOG operations in February to overthrow Johnny Paul Koroma's junta that had, earlier in May 1997, overthrown the Kabbah administration. Sandline subsequently provided intelligence, logistical and air support during the operation, as well as 35 tons of military equipment, purchased in Bulgaria, to the ECOMOG forces.

The subsequent investigation in Britain into the assistance provided by Sandline and the acquiescence of members of the British government failed to reveal how deeply Sandline had been involved in the continuing operations between the coup and its overthrow. The involvement of British and American intelligence in supporting this involvement was also not revealed. Rupert Bowen, Sandline's representative in Conakry during the coup period and Branch Energy's representative at a number of meetings held between Sandline and Foreign Office officials prior to the overthrow of the coup in February 1998, is a former MI6 career intelligence officer who retired in 1993.[67] This double role, similar to that played by Buckingham and Grunberg in EO's operations since 1993, points quite clearly not only to the links between Sandline and the Branch group, but to the intelligence services' links with the EO group.

In Freetown today, PMCs are proliferating at an astounding rate. One of the dangers becoming apparent in Sierra Leone is that of former EO employees going to work for rival firms and rival firms working for mining companies in competition with the Branch-Heritage group. Reports emerging now that former EO personnel have started flying for the rebels are disturbing. The EO/Sandline personnel in Freetown freely admit to knowing the individuals involved, saying that they are former colleagues. Russian helicopters supporting the RUF have been seen flying across the north of the country, their base thought to be in either Liberia or Guinea. Other worrisome signs are that Jean-Raymond Boulle spent a good part of August 1998 in Monrovia, thought to be raising money for the RUF as part of an attempt to destabilise the DiamondWorks-owned mines in Kono.

Other PMC Activity in Africa

As of 1998, Sandline confirmed having engaged in approximately eight operations internationally since 1993; the total number has been reported to be as high as 34. Sandline is believed to have evolved out of Executive Outcomes, an entity registered in September 1993 by Buckingham, Barlow and Mann, but never active. In December of 1996, prior to its operation in Papua New Guinea, Sandline was formally incorporated by Barlow, Buckingham, Michael Grunberg, Mann, Nick van den Berg and Spicer. During the Papua New Guinea inquiry, Heritage Oil & Gas was named as 'owning' Sandline (although Sandline now states that Adson Holdings, based in Guernsey, does). Buckingham was named as chairman, and Grunberg as chief financial officer.

Since then, Sandline has begun to operate as the British successor to EO generally, especially following EO's termination on 1 January 1999. Its management and personnel are largely the same, as well as the personnel base from which it draws its soldiers. It is part of the greater EO-Branch-Heritage network. There is every likelihood that, as EO will no longer operate in its present form, Sandline is set to become the legitimate successor to EO in Africa and elsewhere. Many feel this is a result of the Branch-Heritage/DiamondWorks groups wishing to avoid the continuing negative attitudes towards EO's operations evidenced by Washington and London.

Other countries in which EO have been involved include Kenya, where Barlow established a joint-venture security consulting company with Raymond Moi, son of the President, in 1995; Moi and Barlow are also linked through Ibis Air, which is in partnership with Simba Air in Kenya, partly owned by Moi.[68] In Uganda, Saracen Uganda (an EO subsidiary) was established to provide security protection for Branch Energy's (Uganda) gold mining operations. Saracen Uganda was established in co-operation with Major-General Salim Saleh, the half-brother of President Museveni;[69] Saleh owns 45 per cent of Saracen Uganda and a 25 per cent share in Branch Energy (Uganda)'s operation in Uganda.[70] He was also instrumental in securing import-export licences for military hardware and technology brought into Uganda by Saracen.[71] There have even been allegations that EO was involved in fighting the Lord's Resistance Army in Uganda when one 'EO employee' was found dead on the Sudanese border after a supposed reconnaissance mission, although it is more likely that this was a Saracen employee.[72]

As EO itself and many of its former employees have established PMCs throughout Africa, what must be remembered is that, because of the growth in their reputation, any white 'mercenary' spotted today in any conflict in Africa is automatically assumed to be from EO, especially if they are South African. This becomes even more difficult to follow as many former EO employees (or even those who were never employed by the company) continue to use EO's name as their affiliation, feeling it gives them more of a powerful image in promoting contracts. Examples such as dead 'EO employees' in southern Sudan or northern Uganda point to this: it is most likely that the individual was working for Saracen or even another totally unrelated PMC.

EO has also carried out training projects in Mozambique, Malawi and Zambia (between 1995 and 1997); it has provided military training to the Algerian government, the militaries of Botswana and Madagascar, and acted as a security force (now through EO spin-offs) for the gold fields in Nigeria. Finally, contrary to Barlow's profession to have turned down a request to work in the Sudan, there are suspicions that EO has been providing protection for Canadian oil interests there since 1995, although this has become embroiled in the whole myth that has developed around Arakis Energy and its security teams.

What is known is that the Branch Group (of which Buckingham is principal) has, for a number of years now, looked at expanding their operations into the Sudan. In 1996, Buckingham wanted to deploy EO in the Sudan to secure potential concession areas against the rebels in the south, but no confirmed move towards this was made at that time. In mid-1998, however, it was reported that Branch began acquisition moves in the Sudan to acquire some of the undeveloped mining sites in the Sudan; it would be expected that Sandline would provide security to the sites.

At its height, EO was reported to have links with more than 30 countries, 70 per cent of which were in Africa; as at January 1998, EO stated that they were not involved in any country outside of Sierra Leone and South Africa. By August 1998, this was down to South Africa alone: deterring cattle-rustling in the Eastern Cape. On 2 December, EO announced that it was terminating all operations and shutting down on 1 January 1999. It is thought that threatened prosecution under the new Regulation of Foreign Military Assistance Act, passed in South Africa in May 1998, was the impetus for this move, as well as the aforementioned increase in Sandline's activities.

Throughout the 1980s and 1990s, DSL has also provided security to many of these same clients, as well as to the United Nations, the World Bank, and many NGOs in countries such as Botswana, Rwanda and Malawi (BATT), Kenya (infrastructure support), Mozambique (mine clearance), and Somalia, Sudan and Uganda (infrastructure support). Its French subsidiary, CIAS, is now trying to get a contract to protect the projected 1,100km pipeline between the Southern Chad oil fields and the port of Kribi in Cameroon. DSL was recommended to T & L Sugar Estates in Mozambique by now-UN Secretary-General Kofi Annan in 1994 to provide security assistance on behalf of the World Bank, which had also discussed refugee protection with DSL over Rwanda in 1994.[73] Annan has continued to consult DSL on a variety of peacekeeping operations around the world.

The American firm Vinnell has been training the Egyptian military on information systems for a number of years now.[74] MPRI has also been involved in training the military of Liberia. AirScan has reportedly been involved in arms trafficking from Uganda to the southern Sudan to support the Sudanese People's Liberation Army there.[75] This allegation further solidifies the view that a number of such private security groups are operating in this area of Africa.

The oil company Arakis Energy contracted with AirScan to protect their oil ventures in southern Sudan. At the same time, the SPLA is convinced that EO has been at work in Sudan training ethnic Nuer to protect the pipeline that will have to traverse Nuer country in the Upper Nile. The Khartoum government denies this, saying that 'Canadian security personnel' are responsible for the security at Arakis's sites.[76] It is clear, however, that at least two and possibly more private security groups are operating in the Sudanese region centring on the oil fields of the south.

Other countries where commercial security firms operate include the Ivory Coast, Ghana, Nigeria, Rwanda and Namibia. The ex-gendarme Robert Montoya, mentioned above, set up a company called Service and Security which provides training to the Togolese anti-riot forces, as well as advising President Eyadema on security issues.[77]

PMCs in South Africa since 1991

Private military and security companies abound in South Africa; currently, there are more than 130,000 private security personnel (40,000 of whom are armed) in South Africa, operating on behalf of more than 2,700 companies. Only Russia surpasses South Africa in these numbers.

PMCs which have emerged in South Africa generally have their origins in the apartheid security forces, often (like EO) having been established as front companies by covert units of the security forces. These include Honey Badger Arms and Ammunition, originally established by Eugene De Kock in 1992; Combat Force; Investment Surveys; Shield Security; Longreach Pty Ltd; Omega Support Ltd; Panasec Corporate Dynamics; Bridge Resources; Corporate Trading International; Stabilco; and Strategic Concepts Pty Ltd.

Panasec was originally established in 1995 by Daan Opperman and Kobus Schultz, both former employees of South Africa's National Intelligence Service; they were joined by Maritz Spaarwater, the NIS officer who was responsible for co-ordinating secret negotiations with the ANC in the late 1980s; Spaarwater later left. Its operations centred mainly in South Africa, providing companies with private intelligence, although they have also been involved in Angola and Mozambique.

Executive Outcomes and Branch-Heritage: The EO-Branch Network

Executive Outcomes attracts and deserves the most attention not only because of its background in the apartheid security forces but, more importantly, because EO and its alliance with the Branch-Heritage group is the strongest manifestation of the rise in power of the late twentieth-century version of the great colonial exploration companies of the nineteenth century.

EO's links not only to the Branch-Heritage group, but to British and South African intelligence, also demonstrate the sea change that is being witnessed in the post-Soviet world. The end of the Cold War confrontation has meant that Western powers can no longer overtly or even clandestinely undertake interventions in regions where their interests are not at stake: the public will not accept this. Instead, to support national and business interests, the Western powers are turning to the private security sector to undertake, often covertly on their behalf these interventions to secure these interests. For this reason, it has been right to focus closely on the EO-Branch-Heritage group (including the Branch-Heritage group, Plaza 107 Ltd and Strategic Resources Corporation) as the prime example of this.

The Branch-Heritage Group

In the beginning, there was Heritage Oil & Gas Ltd, established by Anthony Buckingham in 1992. Heritage was involved in Africa in a number of joint-venture oil exploration projects with other Western oil companies (particularly Chevron and Ranger Oil West Africa) co-ordinated through the Angolan parastatal Sonangol. The problem was, however, that the facilities for the pumping and refining of the oil at Soyo in northern Angola were under the control of the UNITA rebel movement. The Angolan army had been unable to remove UNITA from Soyo, so the oil did not flow.

Buckingham was not to stay solely interested in oil, however. When he formed Branch Energy Ltd in 1993, its aim was to exploit strategic resources in regions of conflict. To secure these resources, however, Buckingham needed two things: the agreement of the host government in the country where the resources were, and the ability to secure those resources from the conflict around them. By being able to offer the second as the incentive to the host government, Buckingham was able to achieve the first.

In this manner, what was later to become the Branch-Heritage group – so called because Heritage Oil & Gas forms the biggest of the various independent companies controlled by Branch Holdings Ltd, of which Buckingham is the principal and Michael Grunberg the primary director – began to work as the facilitator for the PMCs (at first, EO and, later, Sandline International, LifeGuard Management and all the other EO spin-offs) to be introduced into the country of conflict. The Branch-Heritage group would put forward the money initially to pay for the solidification of the PMC contract as the purchasing price for concessionary rights over certain mines or oil fields in that country. This happened first in Angola, later in Sierra Leone, and now in other African countries. The money would then be used to pay for PMC start-up operations.

The result is clear: in ensuring the government a 30 per cent ownership (the general rule-of-thumb) in the liberated or restored mining operation, the government would make money out of the operation as would the Branch-Heritage group. At the same time, the Branch-Heritage group has been able to state that they have never received mining concessions in return for any Executive Outcomes-related contract, nor have they funded any Executive Outcomes-related operation. The direct payment – on the record – for the PMC operation would be from revenues gained for the restarting of the mining operation by the mining company (in this case, the Branch-Heritage group) associated with the PMC. This is what Grunberg once described as the second of 'two hands: one hand is involved in mineral exploitation, the other with security; but both work together'.[78]

At the same time, the host government has been able to contract with the PMC directly to secure its own stability through not only the retraining of its armed forces against the rebel organisation in the country but also through the direct use of the PMCs (in these cases, Executive Outcomes and Sandline International, along with their spin-off companies) in military operations to defeat the rebel organisation. For example, Branch Energy received the Koidu lease in Sierra Leone on 22 July 1995; Executive Outcomes' involvement in Sierra Leone to secure these fields began in April 1995.

As David Shearer pointed out, the mining company needs the mines to be secured in order to exploit its reserves; the host government needs stability in the country to attract the international investment that comes with mining concessions; and both need the revenue from the mines and oil fields. The use of the PMC in this case ensures all three of these.

Branch Energy has expanded its operations greatly in the last few years. The Branch-Heritage group has grown to include affiliated companies in Uganda, Kenya, Namibia, Angola, Sierra Leone, Sudan, Algeria, and South Africa, to name a few. Companies trade positions and directorships through the umbrella of Branch International, the holding company for the Branch-Heritage group. Heritage Oil & Gas is the senior partner of the Branch-Heritage group. Other companies included Branch Africa Holdings Ltd, which controls Branch Energy (Namibia) Pty Ltd, which in turn owns Indigo Sky Gems in Namibia; Branch Energy (Uganda), which bought Branch Mining from Barlow in late 1995[79] and now explores the Kidepo National Park gold reserves while Heritage Oil & Gas is exploring Lake Alberta;[80] and Branch Energy (Kenya) and Branch Energy Holdings.

These are a few examples of the Branch-Heritage group's current operations in Africa. There are even suspicions in some circles that, given Barlow's links with De Beers early in his private career, the Branch-Heritage group has undertaken operations on behalf of De Beers and the Central Selling Organisation in Africa in order to prop up investment stability throughout Africa in the interest of all companies.

Strategic Resources Corporation

At the same time, Buckingham suggested to Eeben Barlow that EO should not only diversify but also attempt to expand its interests clandestinely in order to further confuse any links between the Branch-Heritage group and Executive Outcomes. For this reason, Strategic Resources Corporation (SRC) was formed by Barlow, Lafras Luitingh, and Nicolaas Palm (EO's financial manager) in 1993. In its lifetime, almost 50 companies are thought to have been associated with it. At its height in late 1996, SRC, as an umbrella holding company, controlled a number of PMCs generally interested in commercial security work rather than private military work, as well as numerous infrastructure rebuilding, investment consulting and general relief assistance companies.[81]

Branch-Heritage officials have stated that SRC simply devoted the profits made by EO in its operations to what the SRC directors considered to be 'worthwhile causes and investments'. Thus a number of these companies were owned directly by Barlow, through his ownership of EO, which filtered its investments through SRC, or through partnerships, while others he simply had an interest in.

Examples of the SRC holdings include Gemini Video Productions (established in June 1994); Applied Electronic Services (counter-intelligence equipment); OPM Support Systems (investments); TransAfrica Logistics (import-export advisory); Bridge International (logistics and infrastructure, 50 per cent owned by SRC); Branch Mining (60 per cent SRC-owned, later transferred to Branch Energy Uganda); Afro Mineiro (40 per cent owned by SRC, mining diamonds in Angola); Wrangal Medical, later a subsidiary of Stuart Mills International, another SRC firm based in South Africa and involved in demining; Ibis Air Ltd; LifeGuard Management; Saracen International; Saracen Uganda; AquaNova (drills for water in Zambia, 66 per cent SRC ownership); Double-A-Design (engineering, 60 per cent SRC ownership); Advance Systems Communications Ltd (British-owned); Mechem (demining joint venture between EO and Denel of South Africa); and Shibata Security (jointly owned by Barlow and Portuguese investors).[82]

Plaza 107 Ltd

In the autumn of 1993, Buckingham and Simon Mann, who had originally, through his British and South African government security contacts, introduced Barlow to Buckingham, registered Executive Outcomes (UK) Pty Ltd in Britain. They secured the services of Michael Grunberg, who had formed Plaza 107 Ltd, based in London, to act as both a co-ordinator for combined operations between the Branch-Heritage group and Executive Outcomes, and a management and financial services provider for all of the companies affiliated through the two groups. Essentially, Plaza 107 Ltd became the linchpin of the whole operation.

Tim Spicer, the Director of Sandline International, has stated that Sandline, although registered in December 1996, had functioned 'in other forms since at least 1993' out of the Plaza 107 Ltd offices. However, Spicer left the British Army only in 1995: what, then, had 'Sandline' in its previous entities done out of Plaza 107 Ltd? Sandline International was, as stated above, the evolution of the EO (UK) Ltd concept. When EO's operations in Africa were becoming too public by the middle of 1997, and the sensitivity of the 'apartheid issue' attached to EO's background had elicited ongoing criticism, the decision was made to remove EO's direct involvement in Africa and elsewhere by pushing Sandline as the new operational unit. With EO's 'demise', it appears that Sandline will continue to be the 'favoured son' in the Branch-Executive Outcomes network; somewhat surprisingly, Sandline has managed to emerge virtually unscathed from two scandals (Papua New Guinea and Sierra Leone) in which it has become embroiled.

Links between the Branch-Heritage group and Executive Outcomes

Throughout all the speculation surrounding the existence of corporate links between the Branch-Heritage group and Executive Outcomes, Buckingham and Barlow have been at pains to dispute any conclusion that corporate links exist between the two. Buckingham has stated categorically that 'there is no corporate link between Executive Outcomes and the Branch Heritage [sic] group'. It is not, however, directly through corporate links that the relationship can be traced, but first, through personal affiliations and, second, corporate spin-offs. Regardless of embellishment or sensationalist stories in most media reports on the two since 1994, it is true that links exist between them.

SRC was, according to Executive Outcomes statements, terminated in July 1997, the same month that Barlow and Luitingh left EO. It is believed that at this time a number of companies were transferred from the EO umbrella to that of Branch-Heritage, which included Sandline International. For example, LifeGuard Management was incorporated separately in Pretoria but continued to use EO's management, personnel and financial services, as well as drawing from EO's force pool; however, that did not divest LifeGuard of any direct corporate connection to the Branch-Heritage group. Barlow proceeded to disappear from the scene, living some of the time in Southampton and other times in Africa, continuing to give attention to his non-African ventures (in the Middle East and Far East) as well as to new African ones.

The first solid indication of this is the existence of Plaza 107 Ltd and the incorporation of Sandline International in late 1996. If 'Sandline has existed in a different form since 1993' and EO (UK) Ltd was established in the autumn of 1993 by Buckingham, acting on Barlow's behalf in Britain, Sandline draws the first direct link between the Branch-Heritage group, which is also housed under Plaza 107 Ltd's umbrella for management and financial services, and Executive Outcomes.

The second link is the development of a number of affiliated companies such as LifeGuard Management, Ibis Air, Alba Marine, and Stuart Mills International. These companies were all owned or controlled by Barlow either as direct subsidiaries of Executive Outcomes (in the case of LifeGuard Management) or as associated companies under the SRC umbrella. In early 1997, for example, Barlow stated that he 'owned LifeGuard Management'; in mid-1998, Spicer stated that LifeGuard was 'our commercial security entity'.[83] This connection is further enhanced when reading the planning documents for Operations 'Oyster' and 'Oracle', the Sandline operations in Papua New Guinea. In these documents, signed by Spicer, Grunberg and Nick van den Berg (EO's last managing director), it is stated quite clearly that Alba Marine, Stuart Mills (and its subsidiary Wrangal Medical) are subsidiaries of Sandline. Yet all of these companies had been either established by Barlow under EO/SRC or were controlled by SRC.

The third, and most revealing to this image of non-affiliation, lies in the press statement of 1 July 1998 by DiamondWorks Ltd (formerly Carson Gold, which bought Branch Energy Ltd in August 1996) to the effect that 'in the past, there has been a relationship between Branch Energy, Ibis Air, Alba Marine and LifeGuard Management'.[84] Ibis Air began its life as Capricorn Air Systems, an SRC company, and is currently housed in Plaza 107 Ltd; it has been EO's and Sandline's air logistics and air support company since 1993.[85] LifeGuard itself received its first security contract in Angola in August 1996, the same month in which Carson Gold bought Branch Energy for US$50 million.[86]

Since all of these companies were at one time or another owned and controlled by Barlow and Executive Outcomes (either directly or through SRC), the Branch-Heritage group of companies, including Diamond-Works, very clearly has corporate links with Executive Outcomes. Barlow himself even served as a director of Branch Energy prior to its being bought by Carson Gold. Therefore the statement that there is no corporate link between the Branch-Heritage group and Executive Outcomes is untrue: a further examination of the shareholdings and directorships of all SRC and Branch-Heritage companies over the past four years would dispute this claim even further.

What still remains unclear is the involvement of Robert Friedland, the former owner of Carson Gold.[87] Friedland had been partnered with Jean-Raymond Boulle in December 1993 in the company Diamond Field Resources. Following a dispute between the two, Boulle severed his relations with Friedland and established American Mineral Fields as a follow-on to his pre-1993 company, Sunshine Mining. Friedland took Carson Gold and, through an introduction by Grunberg to Buckingham, bought out Branch Energy Ltd in August 1996, changing its name to DiamondWorks on 11 October 1996.[88] While Buckingham continues to control 29 per cent of DiamondWorks' shares (either directly or through Hansard Management Services and the Hansard Trust Company, based in Guernsey), Friedland has made every attempt to distance himself from any pointed affiliation with Executive Outcomes or even any control over DiamondWorks (he maintains that he currently controls 8.5 per cent of DiamondWorks' shares),[89] lest it should affiliate him with Executive Outcomes through the Branch-Heritage group.

This intense rivalry between Boulle and Friedland has continued to grow since their split, resulting in extremely competitive quests for control over the mining concessions of Angola, Sierra Leone, the Democratic Republic of the Congo (former Zaire), Uganda, and elsewhere (including Papua New Guinea's Panguna Mine). In the summer of 1997, Sandline had approached Boulle to offer him a military contract aimed at overthrowing the coup plotters in Sierra Leone and re-establishing control over his mining interests in that country. Boulle refused the offer, it is thought, because of his intense rivalry with Sandline's patron Buckingham.

At the same time, however, America Mineral Fields had also begun to develop very close contacts with International Defence and Security (IDAS), a company believed to be a Belgian-Dutch spin-off of EO, which had secured numerous mining concessions in north-east Angola (Luanda Norte and Luanda Sud) in co-operation with AMF.[90] This belief is probably misplaced, given the intense rivalry between Buckingham's group of companies and Boulle's group (AMF-IDAS), especially in view of AMF's reported close ties with London-based Defence Systems Ltd (DSL), the major British PMC, which is also locked in bitter competition with Sandline.

What this growing scenario does present, however, is the nightmare scenario of rival mining companies hiring rival PMCs to fight for control of strategic resources in countries where there is little, if any, governmental security control. This is especially the case in countries such as Sierra Leone where an army does not exist, various PMCs control the various militias, and the mining companies are falling over one another to offer the government security services in return for mining concessions. Determining what each mining-security network is involved in or looking to become involved in is central to being able to analyse and assess the impact of these companies on both the host country and regional security. The change and mutation of the EO-Branch-Heritage network is the key to all difficulties in tracking this; personal holdings, board directorships, and personal relationships are central to understanding and tracking this ongoing evolution of the PMCs.

Conclusion

What is the future of these companies? There are clear differences in the relations between PMCs based in Britain, the United States and France, and those based in South Africa and Israel. The US government often pushes for its national companies to obtain contracts; Britain does the same, albeit much more quietly. France uses its intelligence services (especially the DGSE) to facilitate such operations by French firms in Africa. In Israel and South Africa, parliamentary and popular pressure and condemnation (especially in the South African case) have forced the governments to place controls on the provision of such services by national companies.

In reality, though, what is the difference between these countries? Is it solely the apartheid affiliation that South African PMCs retain in the popular perception (similar to perceptions of Mossad interference by the Israeli PMCs throughout Africa and Latin America) that marks South African and Israeli companies as somehow being 'dirtier' than British or American ones? Or is it the fact that South African and Israeli companies state outright that they will involve themselves in services that include military operations, while American and British firms talk much more subtly of training and support? Is it the ANC attitude towards South African PMCs as having been 'the worst of the worst of apartheid's security forces', a view that has generally been accepted internationally? Or is it simply a failure by the public to appreciate that American firms have been directly involved covertly in almost all of America's dirty wars of the 1960s, 1970s and 1980s? For example, how many US companies were tied to the CIA and DIA in the 1960s and 1970s – companies such as Vinnell which in 1970 the CIA called 'our own private army in Vietnam'?

These companies will continue to be with us for a long time to come. Even though the resource pool of many of them (particularly the South African ones) is ageing rapidly, they will continue to find recruits from national and sometimes international force pools. This is due to the fact that, firstly, they pay considerably more than a national soldier is paid and, secondly, because they offer the kind of life that many professional soldiers desire, not the dreariness of routine duties and constant training for an operation that may never come.

UN Secretary-General Kofi Annan warned that the overthrow of the constitutional government in Congo-Brazzaville and the involvement of foreign forces in the conflict could become a pattern in Africa. In many of these cases, the private warriors may have done what the UN would have been unable to do: take sides and achieve stability, however transient, through the forced termination of the conflict in the region. While the future of peace in Africa does not look good, the future of PMCs in Africa does. While many have noted that EO drew many of its personnel from the old SADF, and that these personnel are ageing rapidly, there is nothing to stop South African and other PMCs from drawing on personnel who are rationalised or have resigned from the SANDF, just as DSL and similar British companies will continue to draw from the British military forces. Therefore, there is no fixed lifetime for these companies; as long as conflict persists, so will the PMCs.

The international community has demonstrated time and again its unwillingness to become involved in regional conflicts where Western foreign policy concerns are not threatened directly; this gap will continue to be filled by the private military company.

Note on Sources

As this is a compilation of both documented and unattributable information derived from both open and proprietary sources, I have endnoted this chapter as much as possible. Where further information is required, see the bibliography that follows the endnotes.

Notes Chapter 2

1. United Nations Economic and Social Council, *The Right of Peoples to Self-Determination and its Application to Peoples under Colonial or Alien Domination or Foreign Occupation* (Special Rapporteur: Enrique Ballesteros), 1998, p. 62. See Appendix II.
2. For example, see Richard Norton-Taylor, 'FO officials in spotlight over Sierra Leone arms', *Guardian*, 18 May 1998; Raymond Bonner, 'US reportedly backed British mercenary group in Africa', *New York Times*, 13 May 1998.
3. Peter Klerks, 'South African Executive Outcomes or Diamonds Are A Grunt's Best Friend', *Intelligence Newsletter*, No.55 (10 March 1997); Pratap Chatterjee, 'Mercenary Armies and Mineral Wealth', *Covert Action Quarterly* (Fall 1997).

4. Alan Hoe and David Stirling: *The Authorised Biography of the Creator of the SAS* (Little, Brown & Company, 1992), pp. 371–5; 389; 405–13.
5. Ibid., pp. 478–9; 486–7.
6. Interview, former KMS officer, August 1998; Michael Smith, 'Big business for officers of fortune', *Daily Telegraph*, 27 March 1997.
7. Kim Richard Nossal, 'Roland Goes Corporate: Mercenaries and Transnational Security Corporations in the Post-Cold War Era', *Civil Wars*, 1:1 (Spring 1998) p. 25.
8. Ibid., p. 26; Alex Vines, 'Gurkha Security Guards', in J. Cilliers and R. Cornwell (eds), *Peace, Profit and Plunder: The Privatisation of Security* (forthcoming).
9. Nossal, 'Roland Goes Corporate', p. 27.
10. Interview, DSL officer, London, June 1998.
11. Interview, EO officer, Pretoria, April 1998.
12. Biographical information on Luther Eeben Barlow compiled from personal interviews (London, Durban and Pretoria, Spring 1998); Jan Breytenbach, *They Live By The Sword: 32 'Buffalo Battalion'* (Lemur Books, 1990); Peter Klerks, 'South African'; Laurence Mazure, 'Lucrative reconversion des mercenaires sud-africains', *Le Monde Diplomatique*, October 1996, p. 22.
13. Lindsay Murdoch and Phillip van Niekerk, 'Mercenaries who show no mercy', *Mail & Guardian*, 25 February 1997; David Greybe, 'Mercenary contact explained', *Business Day*, 27 January 1997; 'ANC denies having endorsed mercenary activities', South African Press Association (SAPA), 26 January 1995.
14. Interview, former Executive Outcomes officer, London, April 1998.
15. Tony Buckingham's own background is of great fascination and speculation in this regard, including his time in the Special Boat Service (as compared to Special Air Service) and his links with the British establishment and various financial interests. For a profile, see Paul Lashmar and Andrew Mullins, 'Will Tony Buckingham be the next Tiny Rowland?', *Independent*, 3 August 1998.
16. Interview, former Executive Outcomes officer, London, April 1998; Klerks, 'South African'.
17. K. Pech and D. Beresford, 'The Corporate Dogs of War who grow fat and make the anarchy of Africa', *Observer*, 19 January 1997.
18. Francois Misser and Anver Versi, 'Soldier of Fortune – the mercenary as corporate executive', *African Business Magazine*, December 1997.
19. Edward O'Loughlin, 'SA mercenaries conquer Africa', *Mail & Guardian*, 4 August 1995.
20. Kate Dunn, 'The Loaded Continent – Africa's Big Share of World Minerals', *Christian Science Monitor*, 25 March 1998.
21. 'Angola: The militarists on top', *Africa Confidential*, 18 February 1994, p. 6; Misser and Versi, 'Soldier of Fortune'; Klerks, 'South African'; 'Hired Help: Reining in the Dogs of War', *African Express*, electronic download, 10 May 1997.
22. Stephen Laufer, 'Mercenary group recolonialising Africa', *Business Day*, 17 January 1997.
23. 'Angolan Ambassador commends quartering of police and repatriation of military advisors as key step in peace process', SAPA, 11 January 1996.
24. 'Datafile Portugal – Africa Special', 13 November 1995 (electronic download).

25. Al J. Venter, 'Mercenaries fuel next round in Angolan civil war', *Jane's Intelligence Review*, 1996.
26. Chris Gordon, 'Mercenaries grab gems', *Mail & Guardian*, 9 May 1997.
27. DiamondWorks, 'Africa Confidential', Press Release, 3 April 1998; <www.diamondworks.com>
28. Mercedes Sayagues, 'We know war is not our only fate', *Mail & Guardian*, 19 September 1997.
29. Chris Gordon, 'Mercenaries'.
30. Misser and Versi, 'Soldier of Fortune'.
31. Anna Richardson, 'British firm in talks over expulsion from Angola', Reuters, 13 January 1998.
32. Sayagues, 'We know war'.
33. For a good description of MPRI, see 'Generals For Hire', *Time* (147:3) 15 January 1996.
34. 'Executive Outcomes starts withdrawing from Angola', SAPA, 2 January 1996; 'Private Armies: Soldiers for sale', *Africa Confidential*, 19 July 1996, p. 4.
35. Interview, Lt.-Gen (retd.) Ed Soyster (MPRI), Washington, 10 October 1998; also see 'Private US companies train armies around the world', *US News and World Report*, 8 February 1997.
36. Colum Lynch, 'For US firms war becomes a business', *Boston Globe*, 1 February 1997.
37. For an examination of Serb mercenaries in Zaire, see James C. McKinley, 'Mercenary who came to Zaire and left a bloody trail', *New York Times*, 19 March 1997.
38. 'La France aurait sountenu Mobutu a Kisangani', *Le Soir* (Brussels), 3 May 1997.
39. Some believe that the 'Legion' was actually composed – in part – by French troops from the élite unit Commando de Recherche et Action trained in Kota Koli (Zaire) and based in Kisangani, with Tavernier simply providing the 'cover' to this operation by the French government: see *La Gauche*, 6 December 1996.
40. 'Mercenary chief detained in Zaire', *ReliefWeb*, 24 March 1997; <www.reliefweb.net>
41. Raymond Bonner, 'French "covert actions" in Zaire on behalf of Mobutu', *New York Times*, 2 May 1997.
42. 'Congo Minister Speaks of Ukrainian, Serb Mercenaries', ITAR-TASS, 10 November 1997.
43. Stephen Laufer, 'Report links SA mercenary group to Zaire conflict', *Business Day*, 1 August 1997; Klerks, 'South African'.
44. Peta Thornycroft, 'Mobutu couldn't afford SA mercenaries', *Mail & Guardian*, 18 July 1997.
45. 'Verbal clashes over South Africa anti-mercenary law', Associated Press, 13 October 1997, <www.nando.net>; Khareen Pech and James Tomlins, 'Soldiers of fortune mint money in Mobutu's war', the *Sunday Times* (South Africa), 15 January 1997.
46. Pech and Tomlins, 'Soldiers of fortune mint money'.
47. Thornycroft, 'Mobutu couldn't afford'.
48. Misser and Versi, 'Soldier of Fortune'.
49. 'Dogs of War', *Africa Research Bulletin*, 1–31 January 1997, p. 12553; Mark Malan, 'US Response to African Crises: An Overview and Preliminary

Assessment of the ACRI', *ISS Papers* (Institute for Security Studies – Midrand, South Africa), No.24 (August 1997), p. 5.

50. Pratap, 'Mercenary Armies'.
51. Stefaans Brümmer, 'SA arms "stoke" Burundi fire', *Mail & Guardian*, 5 December 1997.
52. 'Diamond Mercenaries of War', *Background Briefing Transcript – ABC*, 4 August 1996.
53. Chris Gordon, 'Mercenaries grab gems', *Mail & Guardian*, 9 May 1997.
54. Al J. Venter, 'US Forces Guard Angolan Oilfields', *Mail & Guardian*, 10 October 1997.
55. 'Congo: A lifeline for Lissouba', *Africa Confidential*, 18 February 1994, p. 5.
56. Misser and Versi, 'Soldier of Fortune'; Adam Zagorin, 'Business Soldiers For Sale', *Time* 149:21, 26 May 1997.
57. Khareen Pech, 'SA sold arms to war-torn Congo', *Mail & Guardian*, 15 August 1997.
58. Thornycroft, 'Mobutu couldn't afford'; Khareen Pech, 'SA "soldiers" sent home', *Mail & Guardian*, 10 October 1997.
59. Ian Mulgrew, 'Gunning for business', *Vancouver Sun*, 31 January 1998, p. C1.
60. 'Comment by Executive Outcomes on a report in the *Mail & Guardian*', 28 August 1998: <www.eo.com>
61. Misser and Versi, 'Soldier of Fortune'.
62. Thornycroft, 'Mobutu couldn't afford'; Klerks, 'South African'; Stefaans Brümmer & Ann Eveleth, 'Showdown over Mobutu's generals', *Mail & Guardian*, 13 February 1998.
63. Thornycroft, 'Mobutu couldn't afford'.
64. 'South Africa: EO to be out-manoeuvred?', *Africa Research Bulletin*, 1–31 January 1997, p. 12527.
65. 'SA Mercenaries in Angola', *ZA*NOW – Electronic Mail & Guardian*, 3 August 1998.
66. Peter Alexander, 'South Africa's veterans recruit army of outlaws', *Daily Telegraph*, 6 April 1997.
67. Nicholas Rufford, 'Ex-spy's link on arms deal turns heat on Cook', *The Times*, 5 July 1998.
68. 'South Africa: EO to be out-manoeuvred?', *Africa Research Bulletin*, 1–31 January 1997, p. 12526.
69. 'Uganda II: War in the North', *Africa Confidential*, 24 May 1996, p. 3.
70. 'Mercenaries: A Kenyan Connection?', *Economic Review*, 24 February 1997: <www.africaonline.co.ke/AfricaOnline/ereview/079224/mercenaries.ht ml>; Klerks, 'South African'.
71. Stephen Laufer, 'Mercenary group recolonising Africa', *Business Day* (South Africa), 17 January 1997; Klerks, 'South African'.
72. Klerks, 'South African'.
73. Pratap, 'Mercenary Armies'
74. Misser and Versi, 'Soldier of Fortune'.
75. Al J. Venter, 'US Forces Guard Angolan Oilfields', *Mail & Guardian*, 10 October 1997.
76. Pratap, 'Mercenary Armies'; Pratap Chatterjee, 'Activists Condemn Oil Company's Operations in Sudan', *IPS*, 26 August 1997.
77. Stephen Laufer, 'Report links SA mercenary group to Zaire conflict', *Business Day*, 1 August 1997; Klerks, 'South African'.

78. Edward Mortimer, 'International charities and mercenaries are stepping into the spotlight in shaping foreign policy', *Financial Times* (London), 24 September 1997, p. 18.
79. Klerks, 'South African'.
80. Laufer, 'Mercenary group'.
81. Klerks, 'South African'.
82. Laufer, 'Mercenary group'; Klerks, 'South African'.
83. '60 Minutes', 1 June 1997; Khareen Pech, 'Too late for the mercenaries', *Mail & Guardian*, 30 May 1997; Interview, Tim Spicer, London, 18 February 1998.
84. DiamondWorks, *South Africa Monthly Regional Bulletin*, Press Release, 1 July 1998: <www.diamondworks.com>
85. Klerks, 'South African'.
86. Laufer, 'Mercenary group'; Klerks, 'South African'.
87. For a discussion of Robert Friedland and his background, see 'Robert Friedland: King of the Canadian Juniors', *Background Briefing – ABC Radio National Transcripts*, 6 April 1997.
88. 'Man with the Midas touch linked to mercenary bosses', *Sydney Morning Herald*, 7 April 1997.
89. DiamondWorks, *Setting the Record Straight – Statement and Media Advisory*, 21 September 1998: <www.diamondworks.com>
90. Misser and Versi, 'Soldier of Fortune'.

3

A Country Under Siege: State Decay and Corporate Military Intervention in Sierra Leone

Abdel-Fatau Musah

The civil war in Sierra Leone has already claimed close to 20,000 lives and displaced over 1.5 million people – mainly innocent civilians. Sparked by the complete breakdown in the internal negotiation process, the war nonetheless traces its roots to the state's peripheral status in the global economy, a status that has been exacerbated by illegitimate methods of resource appropriation. The fundamental link between imperial order at the centre and perpetual disorder at the periphery has thus become central to our discussion of the war, not least because of the role that resource appropriation played in igniting the war and the motives behind foreign and private military involvement in it.

The war is a classic post-Cold War conflict, as it owes little to traditional superpower rivalry and much to the post-Cold War reconfiguration of regional power balances. Nonetheless, it is deep-rooted and does not fit the 'mindless violence' paradigm into which neoconservative experts would want to pigeonhole all of today's low-intensity conflicts.[1] The extensive deployment of mercenaries and other foreign troops in the conflict has raised many disturbing questions, not least about the appropriateness of the chosen tools for conflict management in an internal war that is essentially about the control of political terrain and, consequently, of dwindling national resources. It has also reopened the debate in Africa about the relationships between regime security and state security, governance and conflict, weak microstates, the questions of privatisation, sovereignty and the threat of corporate recolonisation.

This chapter examines the role of mercenary outfits, particularly Executive Outcomes (EO) and Sandline International (SI), and other foreign forces in the Sierra Leone conflict within the context of an integrated and comprehensive security paradigm. To achieve this goal, it has been necessary to analyse the crisis and the concomitant foreign intervention by querying two interrelated factors: First, the role of the internal political economy, and also the social and generational contradictions, in the explosion of violence. Secondly, the influence of the market-driven global political economy on these internal processes. Finally, the question is asked whether the drama unfolding in Sierra Leone can be classified as an attempt to assert sovereignty or as capitulation to corporate recolonisation.

Early Private Security in the Political Economy of Sierra Leone

Since 1930, the economy of Sierra Leone has relied heavily on the extraction industry. The encroachment on mineral resources and revenue by corrupt state officials, in collaboration with informal mining groups such as illegal diamond hunters and buyers mainly from the Middle East, constituted a legitimate security threat to the state. Up to 1968, diamonds alone had generated US$200 million in profit to the economy of Sierra Leone and had provided 70 per cent of foreign exchange reserves.[2] By 1987, however, government revenue from legal diamond exports had dwindled to a paltry US$100,000.

Private security involvement in Sierra Leone has always been linked to mining operations. Expatriate mining concerns, and at times the government on the insistence of the former, hired private security firms to guard mines against illegal encroachment or pilfering by employees. The use of private security in the extraction industry in Africa traces its roots to the diamond marketing giant De Beers which, in the 1950s, hired mercenaries under British World War II intelligence veteran, Sir Percy Stilltoe, to fight diamond smugglers and middlemen who undermined its operations in Sierra Leone. Extensive deployment of 'mining' mercenaries, however, did not become an issue until 1991, when the rebel Revolutionary United Front (RUF) ignited a civil war in the country. Thus began the influx of mercenary forces, mainly from the UK, into Sierra Leone.

Among the precursor mercenary outfits were the German firm, Specialist Services International, the British firms Marine Protection Services, Frontline Security Forces and Special Projects Services Ltd, as well as the American outfit Sunshine Boulle.[3] 1995 marked a watershed in the influx of organised mercenary armies into the country, beginning with the British outfit the Gurkha Security Guards (GSG).[4] Mercenary and foreign military intervention in the Sierra Leone war will be discussed in greater detail later. For now, however, it is pertinent to attempt to understand the systemic factors that midwifed the Sierra Leone crisis, leading to the proliferation of mercenaries.

Sierra Leone – A Portrait

Freetown, the capital of Sierra Leone, was founded in 1787 as a settlement for black ex-servicemen who had fought under the British Crown in the American War of Independence, as well as freed slaves who expressed the wish to go back to their roots. The formal British colonisation of the territory occurred in 1808. In 1896 the hinterland was declared a British protectorate. Thereafter, Sierra Leone became a resettlement territory for slaves freed from transit camps and slave ships on the high seas[5] by abolitionist missions. Sierra Leone, along with Liberia,[6] thus symbolises the complex, usually tortuous relationships between Africa and the Trans-Atlantic, beginning with the holocaust of the slave trade and followed by colonisation; these exploitative relations continue today through post-independent neo colonial economic arrangements.

Sierra Leone is made up of some 13 ethnic groups, the two dominant ones being the Mende in the south and the Temne in the north. With a population of about 4.4 million, the country is endowed with rich forest, sea and mineral resources; it boasts rutile, titanium, bauxite and some of the finest diamond deposits in the world. Indeed, more than 50 million carats of diamond gems, the quality of which is second only to Namibian diamonds, have so far been produced in the country.[7] These alluvial diamond deposits and prized kimberlite dyke concessions are mined in three main fields. The oldest is the Koidu-Yengema deposits in the Kono region. The others are the Tongo concessions, spanning eight Mende chiefdoms, and the Zimni fields along the Mano River down to the border with Liberia.

The resources have, however, not translated into national wellbeing due to a plethora of factors.

The Colonial Imprint

The British colonial order in West Africa was premised on the system of indirect rule, whereby colonies were allowed a certain level of local self-government under powerful traditional rulers.[8] These local chiefs were subservient to the colonial governor and had no control over any major political, economic, taxation or security policy in the colonies. They could, however, be a thorn in the side of liberation movements.[9]

Sir Milton Margai's Sierra Leone People's Party (SLPP), which emerged as the most powerful movement in the transition to self-rule in Sierra Leone, was conservative and elitist, drawing much of its support from powerful paramount chiefs. It was under the SLPP that Sierra Leone attained formal independence on April 27 1961. From then on, Sierra Leone became a parody of bipartisan democracy, with the socialist All People's Congress (APC) led by Siaka Stevens providing the main parliamentary opposition to the SLPP. Sir Milton Margai succeeded to a large extent in welding together the diverse ethnic and regional groupings in Sierra Leone in a delicate balancing act that involved wooing the various paramount chiefs, and instituting an informal national pecking order based on loyalty to political patrons across the ethnic divides.

After Margai's death in April 1964, two former protégés, the Deputy Leader of SLPP and heir apparent John Karefa-Smart from the north, and Albert Margai, brother of Milton, from the south, emerged as the key candidates for the throne. In the event, the popular Karefa-Smart lost out to Albert Margai, prompting accusations of tribalism and rupturing the delicate ethnic harmony in the country. The SLPP split along ethnic lines, and the opposition APC gained in popularity as a consequence. To cling on to power, Margai consolidated the positions of his fellow southern Mendes and cronies in government, the economy, the bureaucracy and the army, while playing the powerful chiefs against the opposition. He resorted to draconian measures to clamp down on the press, restrict freedom of association and disrupt opposition activity. Finally, he toyed with the idea of a one-party state under a republican consti-tution, but could not see this project through. Siaka Stevens's APC defeated the SLPP in the 1967 general elections, an event that resulted in Sierra Leone's first coup d'état.

External exploitation in the colonial and post-colonial era has thus combined with socio-economic mismanagement and graft under ethnic and rent-seeking politics since independence to impoverish and destabilise the country. The result is the seemingly intractable civil war that has ravaged the country since 1991, and which has effectively put the state under foreign occupation today.

Military/Defence Structures in Sierra Leone

In terms of the organisation of pre-colonial defence structures, and also the setting up of the colonial army, Sierra Leone was no different from most countries of Africa.[10] The main ethnic entities that made up Sierra Leone incorporated groups whose functions encompassed military, police and social duties, and which invariably bore the local name for 'Hunters'.[11] In the south among the Mende, they were called the Kamajoisia (Kamajors). In the north-east, populated by the Temne and among the south-eastern Konos and Kissis, analogous formations went by names such as Tamaboro or Kapra. These traditional defence structures are as old as the organised societies. Myth surrounded the exploits of these forces in local folklore. On the Kamajors, Abdullah and Muana wrote: 'The revered and ancient esoteric Mende cult of invincible and heroic hunters was revived [during the civil war] as a communal militia.'[12] These militias did not therefore emerge out of the blue during the recent conflict, but were reorganised to meet the contingencies of the war. The use to which the modern ruling elite put them, however, raises issues pertinent to our discussion.

In Sierra Leone, like in most of Africa, the colonial authorities had idiosyncratic and strategic reasons for sidestepping such traditional structures and supplanting them with new standing armies. One such reason was to meet the demands of the metropole for troops to execute the two intra-European 'World Wars'. The other, more insidious, was to put down local rebellions and to lay the cornerstone for the neocolonialist state apparatus in the emerging nations. According to Miners, colonial armies 'had been created not to defend the inhabitants against foreign attack but to assist foreigners to conquer the country; [and] ... as far as the rank and file were concerned, they were mercenary armies'.[13] This was evident in both the recruitment and indoctrination strategies within the army. As a rule, the economically more backward regions and less influential ethnic groups within the colonies provided the recruitment pool for the colonial army. The colonial authorities looked for special qualities in their recruits: 'The blacker their faces, the huskier their voices, the thicker their neck and darker their skin, and the more remote parts of Africa they came from the better soldiers they made.'[14]

The attributes usually associated with the neocolonial army are apolitical loyalty to the state, professionalism and neutrality. In essence, however, the army has been brought up to be a loyal and professional defender of the conservative political values and ideals of the 'metropole' and the neocolonial disorder in the periphery. Its neutrality has been known to end whenever a radical political agenda has threatened the status quo. Officers moulded from the colonial paradigm were to become a thorn in the flesh of radical leaders across Africa after independence.[15]

The conservative nature of the Sierra Leone political elite within the SLPP was in conformity with the wishes of the British colonial authorities; also the Officer Corps within the Republic of Sierra Leone Military Forces (RSLMF) reflected internal power relations both ethnically and ideologically. The military elite of Sierra Leone, like its counterparts in Senegal or Cote D'Ivoire, shared the same international alliances, cultural and political values as their civilian counterparts. The RSLMF made its first independent incursion into politics in 1967, after Siaka Stevens' APC party had defeated the SLPP in general elections. Chief Hinga-Norman, the current Commander of the Kamajors and a major actor in the ongoing civil war, was a participant in that coup. Its aim was to restore the status quo ante in the ethnic and ideological balance of power.[16] A counter-coup by junior officers in April 1968 reinstated Siaka Stevens's government.

The Anatomy of State Collapse

From his restoration until his retirement in 1985, Siaka Stevens demonstrated a remarkable staying power, combining co-optation and coercion with political astuteness and bravado in his attempt to heal the political and ethnic rifts, tame the army and fight economic decline. On assuming office he struck a coalition deal with his opponents, the SLPP, paving the way for one-party rule in 1978 through a stage-managed referendum.

Some analysts claim that the ethnic differences that were so openly preached in the elections of 1967, 1973 and 1977 were considerably subdued (under Stevens), and that political confrontation was no longer inter-party, but among individuals within the one-party framework.[17] While there may be some truth in this assertion it must, however, also be noted that the one-party system actually midwifed the present confrontations, firstly by denying space for healthy competition between ideas and among personalities, and secondly, by destroying the nascent structures of checks and balances within the body politic. Having out-manoeuvred his civilian rivals, Siaka Stevens turned on the army. He coopted the army into his party, courting the loyalty of the military top brass through selective promotion and corruption. He consciously started dismantling the army through downsizing and demobilisation, while creating and strengthening his own private army, the State Security Department.[18] By the beginning of the civil war in 1991, the national army had been trimmed to a mere 3,000 personnel.

Economic Decline and Mounting Tensions

As was stated earlier, the extraction industry imposed its stamp on the Sierra Leone economy in the 1930s. It attracted interest and capital from transnational mining giants such as De Beers. The mining boom offered alternative sources of income to otherwise semi-marginalised rural youth, the so-called *njiahungbia ngorgesia*, many of whom could now shuttle between subsistence farming and the mines, either as temporaries of foreign companies or as individual diamond hunters. The oppressive rural life offered few opportunities to youth due to depressed prices for raw materials and locally produced food. Alongside ineffective state formation based on the overriding concerns for regime security and self-enrichment came the neglect of support for farming, traditional occupations and, with it, any advancement of a crumbling educational system. Instead, from elite to local, the thirst for quick money (diamonds) and Western merchandise further impoverished and tore apart traditional communities and regard for traditional means of subsistence; even rice, the staple food, was imported. As a consequence, many of the youth had drifted to the cities in search of non-existent jobs and ended up on the fringes of urban society as *rarray man dem* – petty criminals, pimps or 'mercenaries' for political patrons and criminal bosses. For sections of this youth, the mines offered a reverse migration route to the hinterland as labourers in the mines or illicit miners, *san-san boys*. This subclass was to constitute the fertile pool from which all the parties to the Sierra Leone conflict – the RSLMF, the RUF and the Kamajors – were later to recruit fighters.

The transition to one-party politics in 1978 coincided with a downturn in fortunes on the minerals front. The prices of minerals on the world market crashed as a consequence of global recession. The transnational companies which exploited mineral resources in Sierra Leone felt that mounting extraction costs were eating too deep into profits. Marampa mines closed in 1975; by 1982 De Beers had stopped its operations in the country.[19] Industrial mining had virtually come to a standstill, to be effectively replaced by small-scale, usually illegal mining. Meanwhile, competitive producer prices for diamonds and other raw materials across the border in Liberia and Guinea had made smuggling rampant and haemorrhaging of scarce foreign exchange became terminal.

Siaka Stevens retired in 1985. By then, Sierra Leone's GNP had dwindled to a mere US$965 million, one of the lowest in Africa. (This figure has continued to drop, reaching US$646 million in 1994.[20]) In the 1974/75 financial year 15.6 per cent of government expenditure went into education, 6.6 per cent into health and 4.8 per cent into housing. By 1988/89, these figures had dropped to 8.5, 2.9 and 0.3 per cent respectively.[21] The levels of impoverishment, social and generational exclusion had fuelled social unrest, and state repression had reached alarming proportions.

Stevens was succeeded by his creation, Major-General Joseph Momoh. In spite of the scale of economic mess engulfing the country, Momoh allowed the army to swell from 3,000 to 8,000, without the means to equip and sustain it. This was in a vain attempt to contain mass rebellion against the inept administration. At the same time, he had declared in a speech at Kailahun that education was not a right, but a privilege.[22] The state found itself effectively on a life-support machine. While the regime solicited financial handouts from the international community, complex networks of politicians and rogue Lebanese businessmen siphoned state resources into private savings through bogus financial transactions and illegal diamond deals.[23]

According to Bangura, 'Momoh and his government at this time waged a successful [sic] campaign in the UN system to redefine the status of the country from that of a low income country to an LLDC [least of the less developed countries] in order to qualify for more concessionary loans and grants.'[24] The usual IMF prescriptions aimed at downloading the state cut little ice in Sierra Leone because the bloated civil service and the political establishment constituted the feeding ground for the kleptocracy.

Challenges to the Neocolonial State

Civil society challenges to the neocolonial edifice took their first tentative steps in Sierra Leone in the mid-1970s with labour and student agitation. In the wings, elements within the RSLMF were monitoring the situation with the view to creating a political space for the army. Student protests, co-ordinated by the radical group, Gardeners' Club, in 1977, snowballed into mass positive action for fundamental change in the 1980s, spawning several radical movements. Among these were the Pan-African Union (PANAFU), which was later to forge links with the leftist Movement for Justice in Africa (MOJA) in neighbouring Liberia, the Gaddafi-inspired Green Book Study Group and the Socialist Club.[25] At the cultural level, radical writings that graphically exposed the peripheral status of Africa in the global political economy and the parasitic nature of the ruling elite had become the Bible for radical, and often rebellious, students.[26] For these students, and also the semi-literate underclass that found itself at the receiving end of socio-generational exclusion, lyrics by artists such as Bob Marley and Fela Anikulapu Kuti provided handy slogans for the struggle against the hated comprador ruling elite and their external paymasters.

Simmering discontent within society had begun putting excessive strains on the fragile neocolonial state structure that the ruling elite had helped erect. The one-party legacy bequeathed to Momoh was on its last legs and the artificial unity of the nation, encouraged by Stevens through co-optation, was giving way at the seams. Something had to give. The Revolutionary United Front (RUF), the guerrilla movement at the centre of the Sierra Leone civil war, was a product of these popular challenges to the state.

Emergence of the Revolutionary United Front[27]

An attempted coup d'état against Siaka Stevens in 1971 led to the arrest and incarceration of army officers. One of those detained was Corporal Alfred Foday Saybana Sankoh, who was later to become the leader of the RUF. On regaining his freedom, Foday Sankoh took to itinerant photography in the southeast.[28] His long treks through the forest and familiarity with the terrain were later to inform the hit-and-run guerrilla tactics of the RUF during the protracted civil war. The various radical student movements born out of the struggles of the 1970s, and later the student disturbances at Fourah Bay College (FBC) in 1984, coalesced into a loose populist movement called, the Mass Awareness and Partic-ipation (MAP), which henceforth became the vanguard of student–government confrontations. The MAP leader, Allie Kabba, and four of his colleagues were arrested in April 1985, detained and later expelled from FBC. Student activism was banned on campuses, but this only led to a more qualitative stage in mass awareness as the centres of resistance shifted to the mines, factory floors and the recreational centres (potes) for fringe dwellers in Freetown, Bo, Kenema and Koidu.[29]

Allie Kabba eventually sought sanctuary in Ghana, allegedly with Libya's financial support, and enrolled in the University of Ghana, Legon. It was from there that the initial theoretical and logistical preparations for an armed insurrection were co-ordinated, with the recruitment of volunteers for training in Libya. Foday Sankoh and two future leaders of the RUF, Abu Kanu and Rashid Mansaray, were recruited for training in Benghazi in July-August 1987. The Libyan connection opened the way for the creation of the RUF as well as the alliance between the RUF and Charles Taylor's National Patriotic Front of Liberia, another Libya-supported guerrilla movement that was instrumental in launching the initial RUF incursion into Sierra Leone.

The RUF insurgency is very much an explored territory. The expla-nations proffered for the extreme violence that has characterised its campaign, however, remain sketchy. Much of the discourse about the RUF and violence, particularly among Sierra Leone's intellectuals, has centred on the low level of education, and hence the intellectual bankruptcy, of RUF leaders; their lack of vision for Sierra Leone and the preponderance of crime-prone lumpen elements among RUF recruits.[30]

Not much attention, however, seems to have been paid to such factors as the collapse of the national negotiation process, the neglect of regions, the mismanagement of the political economy and the block on avenues to peaceful change by the ruling elite. These were factors that pushed thousands of young people, peasants and workers to the fringes of society, hungry and angry. More importantly, however, no serious analysis has been made of why the radical intellectuals surrendered the insurrectionist initiative to the 'less worthy'. Challenges to the post-independence West African states have almost always followed a well-known pattern: they have been engineered from the campuses, extended to the nascent labour movement and surrendered to military adventurists.[31] That the Ghana-based Allie Kabba group and PANAFU activists dismissed the RUF as adventurists does not come as a surprise.[32] The programme document of the RUF, *The Basic Document of the United Revolutionary Front of Sierra Leone (RUF/SL): The Second Liberation of Africa*[33] was originally drafted by PANAFU as an agenda for social democracy in Sierra Leone based on an indictment of the rent-seeking political system. It was reworked by the RUF leadership to incorporate the armed phase of the RUF revolutionary vision for Sierra Leone, and adopted on the eve of the invasion. Tinged with radicalism and populism, the document saw the RUF as the Sierra Leone wing of a Pan-African liberation umbrella army fighting to overthrow the neocolonial system and replacing it with peoples' governments.

Sierra Leone watchers have tried to understand the contradictions between the eclectic theoretical basis of the RUF insurgency as contained in *Footpaths to Democracy*, and the wanton destruction and violence that underpin the movement's campaign. For example, the RUF programme claims that it is in harmony with the countryside, a position that is captured in the borrowed Maoist phrases: 'We moved deeper into the comforting bosom of our mother earth – the forest' and 'The forest welcomed us and gave us succour and sustenance.'[34] How does one reconcile such proclamations with the fact that the rural population bore the brunt of RUF atrocities?

One may need to go further than the lack of academic credentials within the RUF leadership in explaining the violence which, it must be noted, was indulged in by all the factions in the war. Solomon Berewa, the Sierra Leone Attorney-General and an avowed opponent of the RUF, for example, warned Tejan Kabbah about the 'threat of the RUF's persuasive ability to captivate the imagination of the young and non-commissioned officers' rank and file'.[35]

A better understanding might be provided by the inability of the RUF leadership to always control its followers due to the contingencies of the war and the sheer frustration with the population for apparently supporting foreign invaders, particularly the South Africa-based Executive Outcomes (EO). More convincingly, however, the explanation lies in the influence of resource appropriation methods that applied in the Sierra Leone war. The RUF had access to diamonds, which it could easily exchange for arms and logistics across the Sierra Leone borders with Guinea and Liberia.[36] This trade provided the RUF with an extended economic space, thus greatly reducing its dependence on the rural population for supplies. Had RUF manoeuvrability been confined only to Sierra Leone, to what Rufin terms 'a closed economy',[37] where it was restricted to only weapons and logistics it could acquire from its interrelationships with the rural dwellers, it would have been compelled to harmonise relations with the peasants. As it were, the forest literally 'welcomed the RUF and gave it succour and sustenance', without the intermediary of the rural population.

Some Sierra Leoneans had been recruited by Charles Taylor's NPFL for the civil war in Liberia in 1989. Here, they received their baptism in guerrilla warfare, later to be applied at home in 1991. The RUF invaded the Kailahun District of eastern Sierra Leone from inside Liberia in March 1991, with the aim of toppling the Momoh government. Among the invading guerrillas were NPFL veterans from Liberia and self-styled freedom fighters from Burkina Faso. Charles Taylor supplied weapons and logistics. In the heat of the guerrilla insurgence a section of the national army, led by Captain Valentine Strasser, overthrew the Momoh government in April 1992, accusing it of paying lip service to the war efforts and starving the front troops of weapons and logistics.

Strasser set up a military junta, the National Provisional Ruling Council (NPRC) and promised a quick victory over the rebels. By 1994, in an attempt to push back the RUF advance, the NPRC had increased the army strength to 14,000, largely by conscripting the lumpen elements among the urban youth. Children as young as eleven were conscripted by both factions. The army soldiered on without proper training and adequate pay. The AK-47 became a form of currency, used by both the rebels and the official soldiers as a blank cheque to harass civilians, pillage villages and engage in illicit mining of diamonds. Many government troops only pretended to perform their normal duties, but in practice they were indistinguishable from the so-called rebels in terms of their violence. They applied a scorched earth policy to towns and villages, took liberties with defenceless women and private property and helped themselves to illegal diamonds; this earned them the neologism *sobels* – 'soldiers by day, rebels by night'.[38]

Thus, by 1995, Sierra Leone was exhibiting all the symptoms of a failed state, just like its neighbour, Liberia, from where the war had spilled over. The country had slipped down the UNDP table of living standards, occupying the fourth from last place. At 37 years its life expectancy was the lowest in the world.[39] What used to be Sierra Leone had disintegrated into fiefdoms controlled by warlords and bandits as central authority had virtually collapsed. The RSLMF was underpaid, undisciplined and incapable of enforcing security. Foreign companies and the besieged government turned to the market for mercenaries.

The Gurkha Security Guards

In February 1995, the then ruling junta, the National Provisional Ruling Council (NPRC), through the British weapons manufacturer J&S Franklin, contracted the services of the Gurkha Security Guards. GSG was tasked with providing security for the possessions of the US-Australian mining concern, Sierra Rutile, and offering training to special forces and officer cadets of the Republic of Sierra Leone Military Forces (RSLMF) in the prosecution of the war against the United Revolutionary Front (RUF).[40] GSG considered the offer as a major breakthrough, and accordingly boasted in its sales brochure: 'It is extremely unusual for any nation-state to entrust the training and welfare of its future military officers to a commercial company composed solely of foreigners.'[41] Led by two European officers – the former British Gurkha officer James Maynard, ex-British sergeant of the Coldstream Guards Andrew Myres – and American Colonel Robert Mackenzie,[42] GSG immediately despatched 58 Gurkhas to the country.

At the time, it had become obvious that the unruly and incompetent RSLMF troops, a reflection of the political order characterised by inept governance and fiscal indiscipline, were in no position or mood either to take on the RUF insurgents or to protect civilians against excesses. Indeed, they were a part of the problem of violence against non-combatants. The acute lack of security had prompted the mobilisation of a *levée en masse* by the civilian population to defend settlements against both the rebels and the army. By the end of April 1995, two major mining centres at Gbangbatok and Makanji had fallen to the RUF, who already controlled vast areas of southeastern Sierra Leone and were bearing on the capital, Freetown. The Gurkha Security Guards was thus called in to help reverse the tide against the rebels.

The GSG adventure was short-lived. On 24 February 1995, while seven members of the group and a platoon of RSLMF infantry were undertaking a scouting mission in the Malal Hills, they stumbled into an ambush near an RUF training camp. It has been suggested that elements within the RSLMF tipped off the guerrillas about the GSG expedition. In the ensuing confrontation, as many as 20 troops of the scouting party lost their lives, amongst them Mackenzie, Myres and Major Abou Tarawali, the aide-de-camp to the NPRC leader Captain Valentine Strasser. Their bodies were never recovered, as they were allegedly mutilated and put on display in rebel-controlled areas.[43]

Typical of romantic mercenary literature, Tim Ripley gave a glamorised but misleading account of the annihilation of the group in his glossy book *Mercenaries: Soldiers of Fortune*. He wrote: 'Bob Mackenzie, who was on a freelance contract in the country [Sierra Leone] in 1995, was shot by the government soldiers he was supposed to be leading and left to be captured by rebels, who reportedly ate him.'[44] After that incident, GSG continued with their operations but refused to engage in combat. It eventually bowed to the dangers, withdrawing from Sierra Leone in April 1995.

Besides GSG, other mercenary outfits explored opportunities for their lethal trade in Sierra Leone around this time. They included Defence Systems Limited, Control Risks Group, J & P Security Ltd, Rapport Research & Analysis Ltd and Group 4.[45] A Group 4 officer of South African origin was among the RUF hostages released and handed over to the then International Alert intermediary, Akyaaba Addai Sebo,[46] as part of the process leading to the first attempts at international mediation in the conflict. Following the influx of British and other European mercenaries, the RUF began to target expatriate nationals in Sierra Leone for hostage-taking in the course of their insurgency.[47] This was aimed at advertising their cause and piling pressure on the international community which was largely supportive of the NPRC junta. The response of the UK government was, on the one hand, to distance itself from the 'mining' mercenaries and, on the other, give tacit approval to their activities. In March 1995, the Foreign and Commonwealth Office issued a statement saying:

> We have not been giving any military assistance to the Government of Sierra Leone. We have made this perfectly clear to the RUF, but if people who are no longer in the British Army decide to sell their services elsewhere, we cannot stop them. They can do what they like with their specialist knowledge as long as they don't break British laws [sic].[48]

The mercenary outfits which took a cue from the above statement in a big way were Executive Outcomes and Sandline International.

Executive Outcomes

Following the fate of GSG, virtually all the companies that explored the freelance military market avoided direct combat, that is, with the exception of Executive Outcomes (EO). It was against this background that, in April 1995, Captain Strasser contracted their services. Strasser hoped for strategic advantage in the war; EO, by controlling the security balance in Sierra Leone, hoped to be able to influence internal politics as a leverage to pry financial and mineral concessions on the cheap from the enfeebled government. Thus, in May 1995 EO deployed 100 mercenaries in the conflict. At the height of their operation they had at least 300 troops in Sierra Leone, backed by one Mi-17 and another M-24 helicopter leased by the Belorussian firm Soruss Air. In addition to engaging in direct combat, EO also trained and deployed the local Kamajor militia and RSLMF units in anti-RUF operations.

Brigadier Julius Maada Bio, Strasser's deputy, doubled as the liaison between EO and the NPRC and as the commander of the EO-trained Special Forces within the army. The links between Maada Bio and EO went even deeper. His Soviet-trained graduate brother, Steven Bio, was a partner in Soruss, the Belorussian company that leased aircraft to EO. By the end of 1995, the connections between Brigadier Maada Bio and EO had given him enough muscle to become an autonomous political player. In January the following year, he toppled Captain Valentine Strasser as leader of the NPRC in a palace coup, with the help of the EO-trained units under his command and making full use of his links with the mercenary outfit.[49] The outwardly respectful corporate military company, EO, was re-enacting in Sierra Leone the game of musical chairs, played so effectively by vagabond mercenaries like Bob Denard in the Comoros Islands and Seychelles and the sophisticated ousting of 'uncooperative regimes' by Western intelligence services during the Cold War.

On the battleground, EO's intelligence capacity and air power, combined with the superior terrain knowledge of the Kamajors and the increased efficiency of reorganised units of the army, had tilted the balance in the civil war in favour of the NPRC government, dealing telling blows to the RUF on several fronts. Within nine months of their arrival, EO had led the anti-RUF coalition to recapture all major mining centres including the Kono diamond mines and Sierra Rutile possessions. The coalition had also formed a *cordon sanitaire* around the capital, Freetown, checking the advance of RUF guerrillas and permitting a semblance of normalcy to return. The RUF guerrillas were forced to retrace their steps along bush paths into the forest to regroup, their war machine having been greatly weakened.

The Peace Agreement

The successes of the Kamajors and EO played a part in convincing the rebels to talk peace even if for the purpose of buying time. The London-based non-governmental organisation, International Alert, played the part of facilitator by acting as the conduit between the RUF and other actors, namely the United Nations, OAU and the Economic Community of West African States (ECOWAS). The series of negotiations eventually yielded the Abidjan Peace Accord. Included in the demands of the 28-article agreement were the following:

- the demobilisation and resettlement of combatants
- the transformation of RUF into a political party
- the setting up of a UN-controlled Neutral Monitoring Group and a joint Monitoring Group made up of the warring parties to oversee compliance
- the withdrawal of foreign forces (EO in particular)
- the holding of a Citizens' Consultative Conference to chart the return course to representative politics
- the holding of elections
- a package to undertake socio-economic reforms.[50]

Preceding, and then parallel to, the peace process was the raging debate within civil society about democratic elections. This 'peace before elections' or 'elections before peace' debate developed a life of its own quite independent of the peace negotiations. Consequently, the Abidjan Accord had no influence on the timing of the elections that took place following two national consultative conferences held in Bintumani in August 1995 and January 1996 by representatives of Sierra Leone's civil society groups. The new military leader, Brigadier Maada Bio, was obliged to pass on the hot seat that the leadership of Sierra Leone had become. He succumbed to the enormous pressures that would have been brought to bear on him had he chosen to hang on. Civil society was demanding a constitutional government, and the Americans and the British had warned him not to contemplate aborting the planned elections.[51]

In a classic case of putting the horse before the cart, elections were hastily arranged in March 1996 in the midst of a raging war, without prior attempts at any meaningful cease-fire or demobilisation of combatants. Meanwhile, at the negotiating table, mistrust, partisan games and a lack of goodwill fostered intransigence among the warring parties. As if this was not enough handicap, the international community, whose resolve should have pushed the belligerents to an agreement, further slowed down the talks while wrangling and turf battles raged between the OAU on one hand, and the UN and Commonwealth on the other. By holding elections first and thinking later about key elements of conflict settlement such as demobilisation, reintegration and rudimentary reconstruction, further progress in the peace process became contingent upon the goodwill and selflessness of the incoming administration.[52] Lacking both, the newly installed civilian administration soon became a party to the conflict.

Post-Election Traumas

The Sierra Leone People's Party (SLPP) won the ensuing disputed elections and headed straight into a security crisis. The leader of the party, Ahmad Tejan Kabbah, had been disqualified from holding public office by a national commission of enquiry under a previous administration, a disqualification that had compelled him to leave the country and take up an appointment with the United Nations Development Programme. Though leader of SLPP, Kabbah lacked an independent constituency in the Sierra Leone body politic. He was effectively an outsider to the SLPP Mende power base and vulnerable to the whims of party stalwarts.

Immediately, the imminent expiry of EO's contract posed the biggest security conundrum for Kabbah. EO's exit would lead to a deterioration of the precarious security situation, providing an opportunity for the military to have another go at power. There was also the dilemma of pushing the peace process forward. As indicated above, the Abidjan Peace Accord had demanded the withdrawal of foreign troops from Sierra Leone and the dismantling of all sub-state armed formations, understood to include the RUF and the Kamajors. However, the parties to the conflict had other ideas. The political structures of the RUF, described as 'an armed movement with a political objective',[53] lagged far behind its military organisation, and the movement was not in a hurry to commit suicide by disarming, as it lacked political cadres on the ground.

Kabbah's government, on the other hand, was most distrustful of the RSLMF and had neither the intention of disbanding the Kamajors – the façade of national pride – nor sending home the real backbone of his security – EO, Nigerian and Guinean troops. In the event, one of Kabbah's first moves as president was to extend EO's presence in the country and strengthen the Kamajors. The SLPP also solicited aid from international financial institutions to meet the immediate security and reconstruction needs of the country. The International Monetary Fund, alarmed by the staggering pay-offs in cash and mineral concessions to EO and its partners, made the termination of the outfit's contract a precondition for bailing out the collapsed economy.

The Price of EO intervention

The cash contract for EO from May to December 1995 was US$13.5 million.[54] Ahmad Tejan Kabbah, convinced that EO's exit would definitely spell his demise, hurriedly renewed the latter's contract from April 1996 for 20 more months at a fee of US$35.2 million. By July 1996, diamond mines in the Kono District, where the EO offensive against the RUF was mainly concentrated, kimberlite pipes and other mineral assets along the Sewa River in Koidu, had been granted to

Branch Energy,[55] one of the several corporate mining entities fronting for EO and Sandline International. In 1996, the majority shares in Branch Energy were transferred to DiamondWorks, another front organisation, 30 per cent of which is controlled by Tony Buckingham, the main financier behind the EO-Branch Energy conglomerate.[56] When EO was forced to leave Sierra Leone in January 1997, they left behind their mercenary field brigade, LifeGuard Systems, to protect Branch Energy's possessions in Koidu.

The Report of the Commission of Inquiry into the Engagement of Sandline International in Papua New Guinea had cited Tony Buckingham's company, Heritage Oil & Gas, as having 'extensive interests in Sierra Leone after EO were invited there by the Government to put down the rebel movement in that country'.[57] Some of these concessions were earmarked for transfer to Jupiter Mining Corporation, owned by Samir Patel and Rakesh Saxena, and linked to the mercenary organisations, as part of the deal to finance Sandline's intervention in the civil war (to be discussed later). In a fax communication with Patel, Saxena stated:

> [DiamondWorks] are quite willing to part with virtually all the Sierra Leone properties, with the possible exception of Kono, on which some kind of deal could be discussed ... The issue here is that the real people in control of the deal are the Branch Energy people in London ... I will be in touch with Branch (especially Tony Buckingham) later this week and will inform you.[58]

These are some of the real costs of private conflict intervention that some private security watchers deliberately ignore when they refer to the 'strategic advantage and cost effectiveness' that mercenary groups bestow on those who hire them.[59] Indeed, the hidden cost of corporate intervention in current civil wars in Africa, besides the mortgaging of natural resources, is sovereignty. In Angola and Sierra Leone private military intervention has at best only secured partial victory for the hiring party in conflict and consequently only transient and fragile peace in the countries.

Corporate Mercenaries in the Kabbah Administration's Strategic Thinking

Ahmad Tejan Kabbah was elected to office at a time when Sierra Leone had undergone almost full reconfiguration. Violence had been institutionalised in the country. As pointed out earlier, Tejan Kabbah was considered an outsider within the traditional Mende power base of the ruling party,[60] and had to come under immense pressure from contending interests. The presence of foreign troops responsible for regime security was to change the balance of forces within the SLPP government. EO's (and Nigeria's) control over internal security eased Kabbah's dependence on traditional power brokers, but increased his reliance on foreign forces to remain in power.

The foreign security shield and the successes chalked up by the EO-led coalition against the RUF were, however, to play a role in Tejan Kabbah's choice of strong-arm tactics over reconciliation in his post-election dealings with the RUF. That Sierra Leone was still effectively at war at the time of the hurriedly arranged elections should have convinced Kabbah to initiate a healing process, but he chose otherwise, continuing in the disastrous tradition of winner-takes-all politics of division and brinkmanship. Regime security therefore took precedence over reconciliation and inclusive security soon after the 1996 elections.

Two contending security strategies, one based on Nigerian troops and the other on mercenary outfits, informed regime security thinking at this time. Both implied ceding the protection of the regime and natural resources to foreign forces.

The arrangement based on Nigeria crystallised in the renegotiations of the military pact between the two states. The new agreement, Status of Forces Agreement (SOFA),[61] was signed in March 1997 on behalf of the two governments in Lagos, Nigeria by the Sierra Leonean Deputy Defence Minister, Captain (Retd.) S. H. Norman and the former Chief of Defence Staff of Nigeria, General A. A. Abubakar. The open-ended agreement was essentially an 'arrangement for the provision of military and security assistance for the sustenance of the sovereignty and territorial integrity of the Republic of Sierra Leone'.[62] The confusion over which interventionist forces are currently in Sierra Leone – ECOMOG or Nigerian – and the reluctance of other sub-regional states to send in troops to assist the Kabbah regime, partly flow from this agreement.

Of greater interest to the subject matter of this chapter, however, is the other strategy based on corporate mercenaries, which is similar in many respects to SOFA. Soon after the elections of 1996, Solomon Berewa,[63] an SLPP old guard and confidante of Tejan Kabbah, placed before the latter a policy document intended as the blueprint for the future security of Sierra Leone. The document was entitled *Strategies to Adopt – Restructuring the Army, Role of Executive Outcomes and Control of the RUF Forces.*[64]

On the security situation in the country, Berewa describes the total control over the armed forces by the Chief of Defence Staff, 'whose loyalty is questionable', as 'a time bomb' and calls for a complete restructuring of the armed forces.[65] He then goes on to praise the magnanimity and efficiency of EO:

It should be recalled that the current contract with EO ended on 31 March [1997] and funds to pay the arrears owing [approx. US$18 million] were not available. The Directors of EO, cognisant of Sierra Leone's vulnerability if the RUF should learn of their withdrawal, and as a demonstration of their commitment to help the newly installed Government settle in a calm atmosphere, dipped into their own private resources, and paid for and kept their ground troops, aircraft and pilots ... for the month of April [sic] ... The importance of EO's continued presence in Sierra Leone ... to serve as a deterrent to all undemocratic aspirators, should not be underestimated.[66]

Rebutting the opinion that the Kamajors were the main force that defeated the RUF prior to the elections, and building the arguments for his eventual recommendations, Berewa states:

The success in the war against the RUF is due primarily to EO's disciplined ground forces, communication and surveillanc[ing] equipment installed in their Cessna aircraft and general air supremacy. The aircraft locates the rebel camp, the helicopters pulverise the RUF from the air, before the ground forces move in and conduct a mop-up operation.[67]

Berewa's solution to the security conundrum in Sierra Leone involved 'planning, recruitment, training and streamlining the police, paramilitary forces and the SSD [the Secret Police] by EO, based on complete loyalty to the president'.[68] Regarding the armed forces and RUF, Berewa recommends as follows:

- On the Army and Kamajors
  The experience and supervision of EO [will ensure that] disarming the army and beginning [re]training in earnest will go on with minimal resistance.
  1. EO to restructure and train the army.
  2. Absorb the Kamajors into the army.
  3. The head of the army should be the Army Chief of Staff who, together with the Commander-in-Chief of EO, shall report directly to the president.
- On the RUF
  Berewa warns Tejan Kabbah never to underestimate the threat posed by 'the persuasive ability of the RUF to captivate the imagination of the young ... officers and non-commissioned officers'[69] within the armed forces. In disregard of the Abidjan Accord, Berewa strongly advises Kabbah to make sure that before any further progress in the peace talks could be made:
  1. All RUF forces must be camped in a location ready for disarmament.

2. Access to Freetown should be denied to the RUF.
3. All RUF combatants must be disarmed under the supervision of the Sierra Leone forces including EO personnel [sic].

- On the Air Force and Navy
  1. The air force and navy will be under the control and supervision of the air wing and naval patrol of EO.
  2. All pilots and officers must reapply and be screened by EO and the Defence Secretariat before recruitment into the air force.
  3. All newly conscripted pilots and officers will only use aircraft in the presence of and under the supervision of EO.
  4. The contract [with EO] may even call for EO fast patrol boats to be attached to the navy ... with training and supervision services by the EO.

'The possibilities are endless', Berewa concludes his policy memo. As it came to pass, this act of voluntary surrender of control over national security was temporarily shelved by the military coup of 25 May 1997, the coup this very memo sought to forestall. Consequently, the project did not get off the ground. It was, however, very much alive in the thinking of the Kabbah administration in exile as it courted the services of EO's partner – Sandline International.

On coming to power, therefore, Kabbah began transforming the Kamajors, the predominantly ethnic Mende militia, into a private army, paying them allowances, replacing their traditional hunting rifles with modern equipment and enlisting the support of the Nigerian army and EO to train them. By May 1997, the Kamajors had developed into a 20,000-member standing army. Their traditional commander, Chief Sam Hinga-Norman, was appointed Deputy Defence Minister. The other militia groups representing other ethnic groups, such as the Tamaboro and Kapra, disbanded. Parallel to the modernisation of the Kamajors, Kabbah launched a downsizing operation on the Sierra Leone army, retiring several officers and retrenching others without adequate compensation or a back-up reintegration programme. His aim was to cut down the army from around 18,000 to 3,000. Next, Kabbah enacted the draconian Media Practitioners Law aimed at muzzling the feeble independent and opposition press.[70]

Eventually, Kabbah apparently developed a plan of gamesmanship aimed at undermining the RUF. In March 1997, he allegedly advised the Nigerian dictator, Sani Abacha, to detain Foday Sankoh, who had travelled to Abuja to discuss the stalemate in the peace process. Meanwhile, at home, his regime was encouraging dissent within the RUF, rushing to support an abortive palace coup by a faction within the movement led by Captain Philip Palmer. Thus, by the end of May 1997, bad faith had squeezed the life out of the peace process. Conditions in the country were degenerating to levels before the elections as renewed orgies of violence spread across the country from the east.

The army saw its second opportunity. Led by Major Johnny Paul Koroma, it pushed aside Kabbah's administration and set up the Armed Forces Revolutionary Council junta in partnership with the RUF. This time, the junta was even more bestial. Joined by Freetown's underclass, it set off indiscriminate looting, violence and cruelty against private property, women and children. Tejan Kabbah and his cabinet fled to Conakry in neighbouring Guinea.

Reactions to the 25 May Coup

Thus seen, the situation prevailing in Sierra Leone in the period between the elections and the coup d'état of 25 May 1997 was anything but democratic and peaceful. Distrust among the parties to the conflict, unfulfilled undertakings under the peace agreement, the hurriedly organised and flawed elections as well as the attempts by Tejan Kabbah to supplant the army with mercenaries and the Kamajors, meant that the country was effectively still at war. Nonetheless, the 25 May coup attracted immediate and angry condemnation not only from the international community, but also from ordinary Sierra Leoneans, partly because of the scale of destruction and cruelty that ushered in the AFRC junta.

The immediate chorus of international condemnation of the coup cut a sharp contrast to the usually tepid international reaction to violent subversions of constitutional order in Africa. After all, an even more cynical putsch occurred in Congo-Brazzaville barely four months after the AFRC coup, but attracted only muted concern and international acceptance as a fait accompli.[71]

Two reasons may explain the contrasting responses. First, the coup in the Congo favoured the new Anglo-Saxon alliances with the emerging leaders of the Great Lakes region, following the fall of Mobutu. That most of the Great Lakes leaders supported the Nguesso coup believed to have been financially backed by French oil giant Elf Aquitaine, was cause enough for restraint in the West. The second reason was the African response to the Freetown coup. In contrast to the events in the Congo, the coup in Sierra Leone, though a country of marginal interest to the West, nonetheless had the misfortune of accidentally coinciding with the OAU summit in Harare. The forum provided the needed courage and platform for a collective voice of condemnation, qualities not usually associated with African leaders in the aftermath of coups d'état.[72] If Sierra Leone and Burundi (after the Buyoya coup) seemed to mark a positive shift in OAU politics, Congo-Brazzaville shattered any lingering illusions to that effect.

The OAU on 4 June 1997 unanimously condemned the coup in Sierra Leone, appealed to the international community not to recognise the junta and urged ECOWAS to assist the people of Sierra Leone to restore the status quo ante. Noteworthy was the call by the OAU for the implementation of the Abidjan Peace Accord, describing it as the 'viable framework for peace, stability and reconciliation in Sierra Leone'.[73]

The UN, in the spirit of the new-found entente with the OAU, followed with Resolution 1132 which, among other things, expressed support for the efforts of the ECOWAS Committee and encouraged it to 'continue to work for the peaceful restoration of the constitutional order, including through the resumption of discussions with all parties to the crisis'.[74] It also reaffirmed the view that the Abidjan Accord would continue to serve as a viable framework for peace, stability and reconciliation in Sierra Leone. Of relevance to our discussion, the resolution went on to prevent:

the sale or supply by all states to Sierra Leone, by their nationals or from their territories, or using their flag vessels or aircraft, of petroleum and petroleum products and arms and related materiel of all types, including weapons and ammunition, military vehicles and equipment, paramilitary equipment and spare parts for the aforementioned, whether or not originating in their territory.

The Plot behind an African Coup

Nigerian troops laid siege to the Freetown port to enforce the OAU and UN embargoes as diplomatic efforts were initiated to find a peaceful settlement to the crisis. While Tejan Kabbah was pretending to be negotiating with the junta under ECOWAS auspices, his government in exile and corporate mercenary organisations were putting together a plan for the invasion of Sierra Leone to oust the junta. In an interview after coming to office and terminating EO's contract, Kabbah said of the mercenary outfit:

We had to pay these people over a million dollars a month, and by the time I became president we owed them a lot of money, millions. And so what we had to pay every month and what had accumulated, was becoming a serious burden on our economy.[75]

Despite this appreciation of PMC impact on state resources, Kabbah was impatient to return. It was obvious that Kabbah, the ex-UN official had decided a priori that either the junta would not relinquish power voluntarily or that the peaceful road to the restoration was too long and laborious. Meanwhile, ECOMOG was not making any headway against the AFRC junta by its campaign of counter-terrorism underpinned by low morale, sheer incompetence and unclear goals. Disarmed of any airworthy warplanes, they laid siege to the port of Freetown, indiscriminately firing mortars into the suffering city, adding to civilian casualties and destroying property. The rag-tag rebels had even managed to capture and disarm scores of Nigerian troops.

Sandline International

The services of Sandline International were proposed to Tejan Kabbah by Peter Penfold, the UK High Commissioner to Sierra Leone. The man behind the initiative was Penfold's friend Rupert Bowen, a former British diplomat and intelligence operative who now acted as Sandline representative in the West African region. In July 1997 a plot was set in motion for the invasion of Sierra Leone to dislodge the AFRC. A three-way communication network was set up between the exiled government represented by the ex-Presidential Affairs Minister Momodu Koroma, Jupiter Mining Company represented by Rakesh Saxena and Samir Patel and the mercenary outfit Sandline International represented by Tim Spicer.[76] The role of Sandline was:

- To train and equip some 40,000 Kamajor militia,[77] some of whom were based in camps inside Guinea and others involved in skirmishes with the AFRC-RUF alliance
- Plan the strategy and co-ordination of assault on Freetown
- Provide arms, ammunition, transportation and food for the assault coalition
- Co-ordinate with 20,000 ECOMOG (mostly Nigerian) troops based in Lungi Airport near Freetown for the assault on Freetown.

In mid-July 1997, Tim Spicer flew to Conakry to file a situational report on the project, and to assess the needs of the Kamajor militia.[78] Jupiter Mining Company, on behalf of Tejan Kabbah's government, financed this trip at the cost of US$70,000.[79] In addition, Rakesh Saxena was to underwrite the acquisition of arms, ammunition and operational costs to the initial tune of US$1.5 million.

Jupiter Mining Company (Rakesh Saxena) and Sandline International had one deal between them and a separate arrangement with the beleaguered government of Tejan Kabbah. While details of the cash part of the deal with the government are yet to emerge, it is obvious that the counter-coup would lead to the expansion of mineral concessions controlled by the two groups in Sierra Leone. The negotiations between Saxena and Sandline on financing the coup also involved the transfer of some of the mines controlled by DiamondWorks/Branch Energy in Sierra Leone to Jupiter Mining Company.[80]

The British government was in the know about Sandline's project in Sierra Leone, including its intention to broker the shipment of arms in contravention of UN resolutions. Tim Spicer informed the Foreign and Commonwealth Office by telephone on 5 January 1998 that 'Sandline had signed an agreement to give support worth US$10 million to President Kabbah and the Kamajors'.[81] Also, Mr Penfold had handed a Sandline document codenamed 'Project Python Military Operations' to the FCO on 29 January 1998. In the document, Sandline clearly stated its objectives in Sierra Leone as 'to return the democratically elected government to Sierra Leone by means of direct action [combat], procurement and delivery [of arms and logistics]'.[82]

The British government clearly wanted to fulfil its pledge to return Kabbah into office, but at minimal cost in terms of taxpayers' money and overt involvement in a coup. In addition, it clearly did not want Nigeria to take credit for the coup. Sandline effectively served in Sierra Leone as the government's forces and intelligence network. It also doubled as the eyes and ears of the US. This privileged position further emboldened the mercenary outfit.

In view of the novel ethical foreign policy and the EU Code of Conduct on arms transfers, which the Labour government was championing, Sandline was discouraged from ferrying arms from the UK. Bulgaria, arguably the most notorious arms merchant in Europe, filled in. Recent recipients of Bulgarian arms include Iraq, Rwanda's Hutu militia at the height of the genocide, both sides of the conflict in Angola and Croatia.[83] For the coup, Sandline brokered the shipment of 35 tons of AK-47 assault rifles, ammunition and mortars into a country already awash with weapons. The weapons were shipped from Bulgaria to Kano in Northern Nigeria on a relay to Sierra Leone via Ibis Airline, a company partly owned by the mercenary network. The moment the plane landed in Kano, Nigeria took the baton and ferried them to Sierra Leone to arm the assault partnership.

On 18 February 1998, combined forces of Nigerian troops, the Kamajors and about 200 mercenaries of Sandline International, launched an air and ground assault on Freetown. The AFRC-RUF alliance was no match for the combined forces and, within a day, most of its guerrillas had tactically surrendered the capital, Freetown, to the invading forces and fled into the bush and neighbouring countries. The key role played by Sandline-EO in the counter-coup was underlined by the fact that they were the only forces to fly a warplane in the operation.[84] Tejan Kabbah's government returned to Freetown from exile.

Motives behind the Assault Partnership

Sandline International

As outlined above, Sandline International and Executive Outcomes are an organic part of an intricate mercenary and mining network that revolves around Tony Buckingham, an astute modern-day Cecil Rhodes who proposes crude and transient military security shield to beleaguered African leaders in exchange for their people's resources. The other key personalities, all with SAS pasts and strong connections within the intelligence services, are Tim Spicer, Eeben Barlow, Simon Mann, Rupert Bowen and Michael Grunberg. LifeGuard Systems, usually deployed to guard conquered mines, is only one of the several above-ground 'security companies' that are essentially field brigades of the two mercenary armies. EO and Sandline are the northern and southern axes, respectively, of the same mercenary family, and share the same database of mercenary soldiers.

The business aspect of the network is carried out under companies such as Heritage Oil, Branch Energy and DiamondWorks. The mutations, mergers and absorption of companies, (Branch Energy and Diamond-Works for example), as well as the occasional reshuffle of personnel, such as the widely publicised 'resignation' of Eeben Barlow as head of EO in 1997, are all part of strenuous ploys and decoys aimed at covering their trails and leading inquisitive eyes into blind alleys.

Plaza 107, headed by Michael Grunberg, is the clearing house and operational command centre for the entire network. (See the figure at the front of the book. Note that this is just *one* network of several made up of military and mining wings.) It monitors public perception of the network's activities, churns out the slick propaganda and public relations literature, hires academics and politicians for damage limitation and image-building exercises. It also has a keen eye for business opportunities, oversees contracts and maps out strategy to fight off rivals. The unity of this network was demonstrated in the abortive contract in Papua New Guinea in 1996. It was likewise demonstrated in Sierra Leone.

There, the network did not take kindly to the abrupt termination of the EO contract in 1996 by President Tejan Kabbah, who was under enormous pressure from international donors to send the mercenaries packing. According to Michael Grunberg, EO warned Kabbah that his government would be overthrown within 100 days if he carried out the decision to expel them.[85] Kabbah was overthrown on the 95th day after EO's 'exit'! Indeed there is evidence to suggest that 'LifeGuard, the security wing of both EO and Sandline into which many EO personnel were absorbed after EO officially left Sierra Leone, was responsible for supplying a shipment of arms (comprising RPG-7 rockets, AK-47 ammunition, mines, mortar bombs and a range of other small arms) to the AFRC in Sierra Leone.'[86] EO officially quit Sierra Leone in January 1996, leaving behind unfinished business. The Freetown government owed the organisation US$19.5 million at the time of the 25 May coup. There was no way the AFRC junta, against whom EO had fought, was going to honour that debt. Tony Buckingham had sunk US$20 million[87] into Branch Energy mining concessions that were awarded by the Sierra Leone government as part payment for EO's activities. The 25 May coup threatened to swallow the investment, as the junta had captured the mines. These factors constituted enough grounds for Sandline to be interested in any anti-AFRC crusade.

There was, however, another compelling factor. Strategically, Sierra Leone constituted a suitable bridgehead for operations in West Africa. The government of Guinea, another country rich in gold and bauxite, is a potential client. Jupiter Mining Company which, together with Rakesh Saxena, funded the initial Sandline reconnaissance work in Sierra Leone before the intervention, has concessions in this country[88] as well as in Sierra Leone. For years now, insurgents led by Ahmed Touré, son of the former Guinean president, Sekou Touré, have been threatening Lansana Conte's government from bases inside rebel-controlled areas of Sierra Leone and the north-east of Liberia. Conflict is simmering in the Casamance Province of Senegal. Gold and diamond-rich Ghana could be a future client. Ultimately, unstable oil-rich Nigeria could implode, providing tantalising prospects. Finally, Sandline was aware that the Labour government in UK had made the restoration of the Kabbah government to power a crusade. This was an opportunity for the outfit to indirectly serve country, make money and salvage an image severely battered by the Papua New Guinea fiasco, even as it ultimately had little to do with the actual overthrow of the AFRC junta.

The UK Government

When the Labour Party came to power in May 1997, it proclaimed 'ethics' as the flagship of its foreign policy. The UK, then president of the EU, had put together the draft of a restrictive Code of Conduct on EU Arms Transfers. The 25 May coup in Sierra Leone and General Sani Abacha's pariah regime in Nigeria offered immediate challenges to this ethical foreign policy in Africa, not least because both states were former British colonies. Sierra Leone offered the terrain to stretch Labour's new-look foreign policy to its limits. The Labour Party had made no secret of its determination to restore Tejan Kabbah to power. In exile in Conakry, the British High Commissioner to Sierra Leone, Peter Penfold, took Tejan Kabbah under his wing, counselling him and weighing various tactical options with him. Tejan Kabbah was a special guest at the Commonwealth Heads of Government summit held in Edinburgh not long after the coup.

The UK was wary of Nigeria's role in the conflict, but could not completely sideline Abacha because of the pivotal role into which the dictator had steered Nigeria. To hide their embarrassment, British officials kept referring to ECOWAS and ECOMOG any time they wished to cite the forces confronting the AFRC junta in Sierra Leone. However, in Sierra Leone, ECOMOG was synonymous with Nigerian troops. It was the attempt to avoid indirectly bestowing international recognition to the Nigerian role that pushed the UK to dice with another devil, Sandline International. In so doing, the UK was setting double standards within its ethical foreign policy by effectively privatising foreign policy towards a country considered of little strategic relevance.

In the aftermath of the arms-to-Africa scandal, UK Foreign Secretary Robin Cook cut a figure of someone who knew next to nothing about the goings-on in his ministry, and who appeared to have been set up by the mandarins in the Department of Trade and Industry, Foreign and Defence Ministries. Behind this benign façade of apparent discomfiture, however, was the determination to protect British intelligence involvement in the Sierra Leone coup. When it appeared that the Labour government was letting Sandline carry the can, the latter came out with a letter revealing the extent of their dealings with the British (and US) intelligence services.[89] In coming out with the letter, Sandline was motivated more by the desire to protect their corporate future in Africa (which could be threatened by possible criminal proceedings) than their desire to confront the government.

Thus, the British government's motive in Sierra Leone was to demonstrate its new foreign policy in practice by restoring a democratic government to power and, in the process, undermining another brutal regime with regional hegemonic designs (Nigeria). The UK could have achieved this goal more honourably if it had used the UN and OAU structures to pile more pressure on the junta that had, in principle, agreed to hand over power in April 1998 but was prevaricating, testing international resolve. This option was all the more plausible given the fact that, the UK, UN and US had applied it only a few months before in Iraq. Commenting on the eventual climb-down by President Saddam Hussein over the inspection of presidential palaces by the UN team for biological weapons Kofi Annan, the UN Secretary-General said that the best way to use force was to demonstrate it and not actually use it.

Sierra Leone obviously is not considered of enough strategic importance to warrant such an approach. What seems to have tilted the balance in favour of Sandline was perhaps the fact that this group was already on the ground in Sierra Leone and providing intelligence information to the government. By tacitly endorsing the use of Sandline in this project, however, the UK achieved the exact opposite of its objectives: it was obliged to collaborate with General Abacha's forces through Sandline; indirectly, it helped prolong instability in Sierra Leone and condoned the exploitative motive of Sandline's involvement.

General Sani Abacha's Strategy

A clear distinction must be drawn between the Nigerian forces and ECOMOG troops in Sierra Leone. Nigerian military presence in Sierra Leone predates the present crisis. Upon the invitation of former president Joseph Momoh, a battalion of Nigerian troops was stationed in Freetown and Bo in 1991 as part of the war efforts against the RUF. Under the NPRC junta, the deployment was enlarged to include military trainers in 1994 to form the Nigerian Technical Assistance Group (NATAG). This was a purely bilateral agreement in which ECOWAS had no input. NATAG transformed into a 10,000 Nigerian interventionist force that effectively assumed the mantle of ECOMOG. For the same purposes and under a separate agreement, a battalion of troops from Guinea-Conakry has also been deployed in the country since 1991.

If Lenin's adage that foreign policy is an extension of domestic policy were to be set aside, it would be right to say that Nigeria's foreign policy since independence has been consistently progressive. Despite the appalling human rights records of successive military dictators, Nigeria contributed immensely towards the anti-apartheid struggle and has consistently opposed paternalistic outside interference in African affairs. Nigeria's attitude to West Africa is captured in its 'concentric foreign polity' doctrine as propounded by former leaders, Generals Murtala Muhammad and Olusegun Obasanjo. It consists of 'a pattern of concentric circles, at the epicentre of which are the national politic and economic interest of Nigeria, which are inextricably tied up with the security, stability and economic and social well being of [our] immediate neighbours'.[90]

Under General Ibrahim Babangida, and more so General Abacha, this doctrine underwent a vulgar reconfiguration. For example, in 1989, a study commissioned by Babangida into ways of turning the tide against the spate of junior officer coups in West Africa recommended keeping junior officers busy in peacekeeping operations outside Nigeria.[91] Under Abacha, Nigeria's West African policy projected the leader's bloated ego and insatiable quest for riches.

Abacha's personalised agenda in Sierra Leone had three main objectives. First, he wished to conclusively establish Nigeria's undisputed hegemony in West Africa and as such the natural African representative at a yet to be enlarged UN Security Council. Besides, he had a score to settle with the AFRC, which had humiliated the Nigerian troops by easily toppling a Kabbah administration that the Nigerian forces were supposed to be protecting. The second reason was aimed at laundering the junta's image within the international community. By acting as the promoter of peace and 'democracy' in Liberia and Sierra Leone, he wanted the international community not to judge his regime by its internal policy of repression and banditry, but by its external role as stabiliser-in-chief in the West African sub-region.

The third reason was economic. With the continuing slump in world oil prices and the pervasive corruption that characterised Nigeria, Sierra Leone's diamonds presented another appealing proposition to diversify sources of income to the Nigerian military bureaucracy and their civilian collaborators, at the head of which was Abacha himself. A Nigerian newspaper has stated with authority that Abacha's energetic policy in Sierra Leone was underpinned by the offer of a staggering diamonds concession.[92] As a fall-out, these diamonds also offered an opportunity for the poorly paid and restless Nigerian junior officers to line their pockets. There were reports of looting, arms sale and diamond hunting by Nigerian officers both in Liberia and Sierra Leone.[93] Finally, as Olonisakin observed, 'battle-tested troops that have spent long periods in an area of operation are a potential source of instability from the [Nigerian] regime's point of view'.[94] Nigeria's military dictators were wary of bringing home troops with combat experience from Liberia, particularly as their counterparts from other countries had returned home to topple their leaders.[95] Sierra Leone provided an opportunity to keep the soldiers occupied elsewhere.

The Vultures Swoop: A Country under Mercenary Occupation

The script of the Berewa project remained essentially the same after the restoration of the Kabbah regime, except that the principal characters had swapped places. The RSLMF was disbanded and a new army, with a key role for the Kamajors, was being mooted. Officially, the Kamajors and Nigerian troops under the umbrella of ECOMOG were in charge of security in Sierra Leone. Brigadier-General Maxwell Khobe, the Commander of the Nigerian (ECOMOG) forces in Sierra Leone, was appointed Minister of Defence and Chief of Defence Staff, but he and other ECOMOG commanders were ferried across Sierra Leone jungles in an EO-manned helicopter. Key political figures were soon at each other's throats over who should control the ethnic militia and how soon it should be absorbed into the army.[96]

In spite of the scandal that surrounded the involvement of Sandline International in the counter-coup, mercenary outfits still dominated the security landscape of Sierra Leone. The hub of frenetic mercenary activity began to shift from Luanda to Freetown, where private military companies are falling over themselves to market their deadly expertise to the state and international entities alike. The Sandline-EO alliance is at the head of the pack. At the time of writing, it continues to support Nigerian operations with its helicopters and supplies pilots – including veterans like Joup Joubert, Fred Morafono, Neil Ellis, Ukrainians and Russians – for the Nigerian AlphaJets and EO's Mi-17 gunships.[97] In conjunction with Nigerian forces, the alliance has stepped up its training programme for the Kamajors. It also provides escort facilities for NGOs, such as World Vision, United Nations agencies and foreign businesses, including Lebanese diamond dealers. Through the entrenchment of LifeGuard in Sierra Leone, the Sandline/EO transnational mercenary conglomerate with its mining wing, the Branch-Heritage group, has made itself indispensable to the efficacy of Nigerian troop presence and the very survival of the Kabbah administration.

With this influence, the network dominates and dictates terms in the security market. It has first refusal on lucrative mining and security deals, such as the Sierra Rutile and Koidu concessions. Through its links with Jupiter Mining that owns rich goldmines in neighbouring Guinea, the Sandline-EO group is expanding its presence in the sub-region. The fear is that Sierra Leone, which has relatively limited resources compared with some other states of the sub-region, seems to serve mainly as a launching pad for future mercenary incursions into West Africa. Nigeria and Ghana, among others, represent far richer pickings. This is one of the key security challenges facing ECOWAS.

Other individual mercenaries and private security companies are either fighting for crumbs or, more dangerously, offering their services either to the AFRC/RUF resistance or rival mining companies. Perhaps, the most dangerous mercenary today remains Carl Alberts, a former EO top gun who has switched allegiance to the rebels. Russian helicopters and Eastern and Western European mercenaries have been spotted making sorties for the rebel alliance from bases inside Liberia.[98] Jean-Raymond Boulle, a rival mining financier to Tony Buckingham's empire, is alleged to be raising funds for the rebels from his base in Liberia. He has an eye on wresting control of the Branch-Heritage concessions, such as the DiamondWorks mines in Kono, from his rival Tony Buckingham. Such a move could provoke an inter-mercenary war for the control of Sierra Leone's resources (see Chapter 5).

Besides the Sandline-EO group, other security companies are active in the country. These include Defence Systems Ltd., which provides security to UN and NGO humanitarian convoys and the American groups International Chartered Incorporated (ICI) and Military Professional Resources Incorporated (MPRI).

The January 1999 rebel occupation of Freetown and the counter-attack by ECOMOG forces threw further light on the extent of mercenary involvement in the conflict.[99] In the course of the counter-offensive, ECOMOG fighter jets and naval gunboats attacked a ship ferrying mercenaries from Liberia into rebel-controlled territory in Sierra Leone.

Several mercenaries – alleged to be mainly Ukrainians, Liberians and Burkinabe – perished in the attack.[100] An Israeli mercenary, Reserve Lieutenant-Colonel Yair Klein, contracted by Tejan Kabbah as military advisor in August 1998, turned out to be a double agent who previously trained the Medellin cartel's death squads in Columbia and who also shipped weapons from Libya and Ukraine to the RUF.[101] In a related development, Colonel Fred Rundels (until recently an operative of EO) and Nico Shafer (who worked for the former Colombian cocaine baron Pablo Escobar), are reported to have set up an international consortium with Liberian President Charles Taylor and the RUF that involves diamonds, narcotics, arms and mercenary procurement.[102]

The convoluted networks sketched above show the links between mercenary activity and arms/narcotics proliferation. They also expose the insurmountable difficulties to be encountered in any attempts to harness or streamline mercenary activity for peace.

Is Sierra Leone any Closer to Peace and Democracy?

It had been hoped that the restoration of the elected government in March 1998 would give President Tejan Kabbah the much needed strategic space to draw the line under the civil war by initiating a programme of reconciliation and reconstruction. However, continued violence put paid to this hope, at least for the population outside Freetown. On the other hand, the rhetoric of reconciliation contained in Kabbah's post-restoration speech was soon replaced with a triumphalist agenda. Bolstered by ECOMOG troops, mercenaries and the Kamajor militia, and diplomatically buoyed by near total international support, the SLPP government decided to prosecute the war to a victorious end. It soon became an all or nothing campaign as revenge killings and witch-hunting replaced dialogue and reconciliation.

In this context, the government caved in to demands from SLPP hawks and supporters for the blood of members and sympathisers of the RUF and the deposed Armed Forces Revolutionary Council (AFRC) regime, and hurriedly tried in partisan courts and executed dozens. Twenty-four soldiers were executed by firing squad after being found guilty before a partisan court that did not allow for appeal. Death sentences were imposed on 26 civilians, including a 75-year-old woman. On 23 October 1998, the death sentence was passed on the RUF leader, Foday Sankoh. For fear of mob justice, no Sierra Leonean lawyer dared come forward to represent the rebel leader both at his trial and appeal. Ironically, a team of British lawyers made up of former Agriculture Minister Douglas Hogg, Charles Buckley and David Hood took up his defence appeal.[103]

At least 43 people were waiting on death row and 1,200 continued to be detained at the Pademba Road Prison by December 1998. In the inner cities and the countryside, irate SLPP supporters led by the Kamajors were hounding suspected rebel sympathisers. In a take-no-prisoners campaign, they shot suspected rebels and collaborators on sight, dousing their bodies with petrol and setting them ablaze in an orgy of revenge killing.[104] In conjunction with Nigerian and mercenary forces, they harassed journalists and human rights activists around the country. The voices of moderation and reason had been subdued and silenced as these were interpreted as sympathies with the rebel cause. Worryingly, a disproportionate number of people being persecuted seemed to come from northern ethnic groups. This has further polarised the already divided community, exacerbating the conflict beyond control. The media emerged as a major casualty. Objective journalism was effectively outlawed in the country and journalists who dared publish uncensored material were dragged to detention camps and kangaroo courts. In the end, it became impossible to distinguish between the violence of the AFRC-RUF alliance and the constitutional tyranny of the elected SLPP government.

President Kabbah was held hostage to countervailing power centres in the motley interventionist alliance and within the SLPP. The SLPP, complacent in the knowledge that regime security was being provided by foreign forces, all but abandoned any efforts at nation rebuilding. Instead, political infighting for control of resources and power within the SLPP hierarchy had dominated the agenda of leading SLPP figures since the restoration. Kabbah was forced to send out a letter to SLPP followers warning against the demands for resource windfalls after the February 1998 'victory'.[105] The state coffers were empty and natural resources had been mortgaged against regime security. In August 1998, President Kabbah expressed doubts about standing for a second term in office in 2001, sparking a power struggle among the hawks inside the party. The struggle between the Deputy Defence Minister, Chief Hinga-Norman and Interior Minister, Charles Margai, for control of the Kamajors should be seen within this context.

Having boxed themselves into Tejan Kabbah's corner, the British and American authorities have been obliged to support the SLPP's unwinnable war. There is a high-profile presence of American and British military advisers in the military restructuring programme of President Kabbah's administration.[106] Characteristic of the new post-Cold War international alliances being forged, the Kabbah administration has also asked China to join in the military restructuring process.

Meanwhile, skirmishes were still going on all over Sierra Leone. The remnants of AFRC and RUF regrouped in the forests and across the border in Liberia for a long drawn out battle. Indiscriminate looting, murder and sabotage were carried out by the rebels under operations code-named 'Operation No Living Thing' and 'Operation Pay Yourself'. These atrocities were concentrated mainly in the rebel stronghold of the

Koidu diamond-mining area in the east. Early in December 1998, in what was seen as an escalation in the fight back, rebel forces launched an attack, the second in two months in a hit-and-run operation, on the strategic town of Mange, 140 km north of the capital, Freetown. The attack, with the help of rocket-propelled grenades and automatic rifles, killed 35 people and forced most of the 25,000 inhabitants of Mange and nearby Port Loko to flee.[107]

At the same time, the RUF-AFRC alliance had infiltrated reconnaissance units into Freetown at a time the government and its backers saw rebel violence in the east as the death-throes of the defeated and splintered movement. A well-planned rebel assault on Freetown on 6 January 1999, therefore, once again caught the Kabbah regime and ECOMOG-mercenary protection forces by surprise. The rebels, led by their field commander Sam Bockarie, stormed the Pademba Road Prison to free all inmates before going on to capture State House, the seat of government. For nearly two weeks, they held on to half of the capital. Kabbah and his humiliated government were evacuated to Lungi airport under the protection of ECOMOG.

Before withdrawing back to the hills following a regrouped ECOMOG counter-attack, the rebels had succeeded in their strategic goals. First, they had transformed their leader, Foday Sankoh, from a prisoner awaiting execution into a key political player. Secondly, they had demonstrated that they could strike at will at any time. Finally, they had succeeded in advertising their cause by demonstrating that they were a credible and resilient force.

Weapons Continue to Dominate Sierra Leone

The weapons that have fuelled the Sierra Leone conflict have poured in from diverse sources. The initial weaponry in the hands of the RUF came from sympathetic neighbouring states, mainly Libya, Liberia and Burkina Faso. As the war intensified, the RUF bartered diamonds mined from areas under its control for arms. The other main source of weaponry has been the mercenary organisations, the clearest example being the 35 tons of Sandline-brokered arms to Nigerian and Kamajor forces in February 1998. EO was reported to have introduced anti-personnel landmines into the conflict. The RUF accused EO of laying landmines around the Kono mines after the latter had recaptured them from the former in 1996.[108]

One of the Kabbah administration's first post-restoration diplomatic moves was to lobby the UN, with the active support of the UK government, to lift the arms embargo imposed on Sierra Leone under Security Council Resolution 1132. Immediately, the arms superhighway linking supplier and transit countries with Sierra Leone became fully operational. This happened in a country already awash in weapons and plagued by close to 4,000 child soldiers. A secret airstrip in Eastern Kenema manned by the RUF has become the main entry point for illegal weapons – AK-47 assault rifles, 60 mm portable mortars and even surface to air missiles, mainly from Eastern Europe – to the anti-government forces.[109] Boats are also used to ferry weapons from Liberia. Among the air companies ferrying weapons to the conflict zone were Ibis Air, Soruss Air and two British companies, Sky Air and Occidental. The last two reportedly shipped nearly 400 tons of arms and ammunition from the Slovak Republic to the RUF forces in defiance of the official UK stand in the war.[110]

The RUF-AFRC alliance augments its arms stockpiles by capturing weapons from state armoury and pro-government forces. In a surprise attack on the ECOMOG base in Kono in December 1998, the RUF routed the ECOMOG forces and captured all their weapons, including three armoured tanks.[111] At the time, 70 per cent of ECOMOG weaponry was stationed at the Kono base and Nigeria had to fly in urgent replacements. With no peace in sight, arms proliferation, together with the illegal diamond trade and narcobusiness, has developed a life of its own in the region, fuelling violence and criminality.[112] A major concern of regimes in the sub-region, such as Jerry Rawlings's Ghana, is that these weapons do not find their way to their shores to destabilise their illiberal democracies.[113]

Conclusion: Winners and Losers

The role of mercenaries in Sierra Leone (and also in Angola), has pushed the debate about the political recolonisation of Africa from the realm of fantasy to that of real possibility. Missionaries and merchants served as the advance parties of classical European colonising missions to Africa. Today, mercenaries seem to constitute the advance shock troops in the new scramble for Africa by transnational extraction companies and finance institutions, with the tacit support of powerful governments in the West. The conflict has once more exposed the dialectical contradictions between an artificially imposed neocolonial super-state and traditional core values of an essentially communo-patrimonial society in transition.[114]

It is also one more conflict that has exposed the yawning gulf between the smooth rhetoric of a grandiose post-Cold War global security agenda and concrete and effective deeds of conflict prevention and resolution by the international community.[115] In Sierra Leone, the international community simply abdicated its responsibility in favour of private concerns. Within Sierra Leone, it is obvious from the above analysis that key elite competitors vying for power are far more concerned about resource appropriation and regime security than the wellbeing of the population. One warning Solomon Berewa fired in the direction of Tejan Kabbah in his memo referred to above was that 'once power goes, it has the habit of staying gone for good'.[116] The coup of 25 May 1996 failed to corroborate this prediction as a result of the counter-coup. With the Kabbah government back in power, they are determined not to relinquish it.

However, the January 1999 rebel incursion into Freetown exposes the self-delusion of those who think that victory has been achieved. The rebel movement, which now constitutes a new reality in the Sierra Leonean body politic, cannot be wished away. The tacit admission of the Kabbah administration that a peaceful exit strategy remains the viable option in the conflict, attests to this reality. Weapons diffusion into society, mounting economic problems and intransigence will only fuel violence. For young people, deprived of education and any hope for the future, the AK-47 or M16 offers the only means of expression, empowerment and livelihood. Revenge cannot be a substitute for reconciliation, and the SLPP and RUF leaderships are reluctantly coming to the realisation that a military approach is the longest road to a lasting peaceful settlement to the conflict.[117]

It has become obvious that the huge cost of the Nigeria-style regional policing that involves deploying a quarter of the entire national army in foreign countries can only be borne by an unaccountable government such as the Abacha junta. With the transition to democracy in Nigeria, the incoming administration will be hard-pressed to justify ongoing large-scale engagement in regional peace enforcement in the light of the myriad social and economic problems facing Nigeria herself.

The only foreseeable alternative to a negotiated settlement in Sierra Leone remains the increased dependence on mercenaries.[118] This option looks plausible if considered against the jostling for power within the SLPP. However, as the Kabbah regime has discovered and as this chapter demonstrates, the government has no monopoly over the hiring of mercenaries. Besides, the ability of mercenaries to engage in long guerrilla warfare is dubious. Finally, mercenaries can only be contracted at the expense of national wealth, the unfair distribution of which lies at the bottom of the conflict.

Ultimately, the continuing war is an indictment of the assertion that mercenary outfits are more effective than the international community in ending wars in out-of-area regions. Just like in Angola, mercenary outfits have only pacified areas where their mining companies can exploit resources. Instability, the key demand factor for mercenary engagement, continues. Today's mercenary companies, such as EO and Sandline, are military corporations with business wings and not business corporations with military wings.

Until African sub-regional bodies come up with genuine structures to contain conflicts and render mercenary organisations redundant, and until, in the medium and long-term, mechanisms are set up to ensure inclusive politics and accountable governance, states like Sierra Leone and Angola face the danger of becoming arenas of inter-mercenary corporate wars for control of internal politics and resources. Sierra Leone today, far from asserting its independence and sovereignty, is actually undergoing corporate recolonisation. The winners are the fledgeling mercenary outfits, gun runners and their internal and external collaborators. There is only one loser: the people of Sierra Leone.

Notes Chapter 3

1. See for example, R. D. Kaplan, 'The Coming Anarchy', *Atlantic Monthly*, February 1994, pp. 44–76.
2. W. Reno, *Warlord Politics and African States* (Lynne Rienner Publishers, 1998), p. 116.
3. For more on the early private military outfits, see Chapter 2 in this book.
4. Gurkhas are special troops from Nepal formed into brigades within the British Army along the lines of the French Foreign Legion. After the Cold War, the Nepalese regiments were downsized by about 70 per cent. The mercenary group GSG emerged a result of these cuts.
5. P. Richards, *Fighting for the Rainforest: War, Youth and Resources in Sierra Leone* (Villier Publications, 1996), p. 37.
6. Liberia was also originally founded by the United States as a settlement for freed slaves.
7. The Conciliation Resources, *Resources, Primary Industries and Conflict in Sierra Leone*, Special Report, No. 3, September 1997.
8. K. Nkrumah, *Africa Must Unite* (Panaf Books, 1985), p. 18.
9. Wherever a radical movement emerged as the dominant anti-colonial political force in a dependency, as in the Gold Coast (Ghana) in the late 1940s, the colonial authority set about playing such local chiefs against the movement.
10. For further discussion of this issue; see E. Hutchful and A. Bathily (eds), *The Military and Militarism in Africa*, CODESRIA (1998); also T. Negash, 'Contradictions Between Western Democracy and African Cultures: Lessons from the Past', in A. Ehnmark and Ka Mana (eds), *Democracy in Africa* (1998).
11. In traditional African societies, hunting was one of the key criteria for determining bravery and manhood.
12. I. Abdullah and P. Nuama, 'The Revolutionary United Front of Sierra Leone', in C. Lapham and J. Curry (eds), *African Guerrillas* (1998), p. 185.
13. N. J. Miners, *The Nigerian Army, 1956–1966* (Methuen, 1971), pp. 2–12.
14. Quoted in Nestor Luanda, 'The Tanganyika Rifles and the Mutiny of 1964', in *The Military and Militarism in Africa*, CODESRIA (1998), p. 176.
15. Examples include General Eyadema against the Ewe (Togo) and General Idi Amin against the Baganda (Uganda). The fact that the Hausa from the north have virtually monopolised the officer corps of the Nigerian army

and, consequently, political power since independence, is more the product of conscious design than a historical coincidence.

16. After the SLPP had been defeated in general elections, Brigadier David Lansana, the Army Commander and close political and ethnic ally of the defeated Albert Margai, interrupted the swearing-in ceremony of Siaka Stevens's cabinet, arrested the new government and temporarily assumed power.

17. K. Koroma, *Sierra Leone: The Agony of a Nation* (Andromedia Publications, 1996), p. 31.

18. 'Sierra Leone's Intractable War', *Africa World Review*, November 1997–March 1998, p. 9.

19. D. F. Luke and S. Riley, 'The Politics of Economic Decline in Sierra Leone', *Journal of Modern African Studies* (1989).

20. UNDP figures, 1996.

21. National accounts: 1970–71 to 1974–75 & 1983–84 to 1988–89 cited in Ibrahim Abdullah, 'Bush Path to Destruction: The origin and character of the Revolutionary United Front/Sierra Leone', *Africa Development*, Vol. XXII, Nos 3/4, 1997, p. 211.

22. *Focus on Sierra Leone*, 1995.

23. For more on the networks of fraudulent deals see Reno, *Warlord Politics and African States*, Chapter 4.

24. Yusuf Bangura, 'Understanding the Political and Cultural Dynamics of the Sierra Leone War', *Africa Development*, Vol. XXII, Nos 3 / 4, 1997, p. 134.

25. For more information on this, see I. Abdullah and P. Muana, 'The Revolutionary United Front of Sierra Leone', in C. Clapham and J. Curry (eds), *African Guerrillas* (1998).

26. These writings included *The Wretched of the Earth* (Frantz Fanon), *How Europe Underdeveloped Africa* (Walter Rodney), *Neo-Colonialism – The Last Stage of Imperialism* and *Africa Must Unite* (Kwame Nkrumah), Ngugi Wa'Thiongo's works, as well as populist ideological tracts such as *The Green Book, The Juche Idea*, and Enver Hoja's writings. In 'Francophone' Africa, works by Sembene Ousmane, Cheikh Anta Diop and others added rebelliousness to youth consciousness.

27. For more details about the RUF, see I. Abdullah and P. Nuama, 'The Revolutionary United Front of Sierra Leone', in C. Lapham and J. Curry (eds), *African Guerrillas* (1998); Paul Richards, *Fighting for the Rainforest*.

28. I. Abdullah and P. Muana, 'The United Revolutionary Front of Sierra Leone', *African Guerrillas*, p. 176.

29. Ibrahim Abdullah, 'Bush Path to Destruction: The Origin and Character of the United Revolutionary Front/ Sierra Leone', *Journal of Modern African Studies*, p. 216.

30. See for example, 'Lumpen Youth Culture and Political Violence: Sierra Leoneans Debate the RUF and the Civil War', *Africa Development*, Vol. XXII, Nos 3/4, 1997, and Y. Bangura's critique of Paul Richards' *Fighting for the Rainforest* (in the same volume).

31. In Ghana, for example, anti-government agitation which culminated in the coups of 1966, 1972, 1979 and 1981 was initiated from the Legon, Kumasi and Cape-Coast campuses.

32. If Allie Kabba and his group had studied the 31 December 1981 coup that catapulted Flt. Lt. Rawlings to power, they would have realised that all

the spade work was done by the June 4th Movement, created by radical student leaders such as Kwasi Adu, Agambillah and Zaya Yeebo. Rawlings was actually co-opted into the movement. The other campus-based radical movements such as the New Democratic Movement refused to participate in the 'revolution', citing the absence of ripe conditions. This, however, did not prevent leading members of NDM (Tsatsu Tsikata, Kwesi Botcwhay) from playing a leading role in the excesses and bloodshed of the resultant PNDC junta, after Rawlings had hounded the JFM cadres into prisons, graves or exile. Allie Kabba and co. arrived in Ghana shortly after these events.

33. The document is contained in the booklet, *Footpaths to Democracy: Towards a New Sierra Leone*, vol. I., undated. See also Paul Richards, *Fighting for the Rainforest*.

34. These phrases contained in 'Footpaths to Democracy' are reproduced from I. Abdullah in 'Footpath to Destruction'.

35. Solomon Berewa to Tejan Kabbah, *Restructuring the Army – Role of Executive Outcomes and Control of the RUF Forces*, Secret Memo to President Tejan Kabbah, p. 7., undated, in possession of author. (See also the section Corporate Mercenaries in Kabbah Administration's Strategic Thinking.)

36. It is an irony that the RUF carried out its trade via Guinea, a state that supported the Sierra Leone government. This trade route was forced on the RUF, as the Sierra Leonean eastern border was controlled by anti-NPFL movements, such as ULIMO.

37. F. Jean and J.-C. Rufin (eds), *Économies des guerres civiles* (Hachette, 1996), p. 21.

38. Ibid.

39. 'Latest Country-by-country Data', *Africa Analysis*, 30 October 1998.

40. The United Revolutionary Front insurgency is covered later in the chapter.

41. See 'G.S.G. Limited: Company Profile' (London, 1996).

42. A. Vines, 'Gurkhas and the Private Security Business in Africa', in J. Cilliers and P. Mason (eds), *Peace, Profit or Plunder: The Privatisation of Security in War-Torn African Societies* (ISS, 1999).

43. See US Department of State, 'Sierra Leone', Country Reports on Human Rights Practices for 1995 (US Government Printer, 1996), p. 232.

44. T. Ripley, *Mercenaries: Soldiers of Fortune* (Parragon, 1997), p. 82.

45. See *Jane's Intelligence Review*, January 1996.

46. Communication with Akyaaba Addai Sebo, a peace facilitator during negotiations towards the Abidjan Peace Accord, 6 January 1999.

47. Akyaaba Addai Sebo, who maintained close links with the RUF, says that the RUF actually rescued these hostages from the line of fire, seeking international benefit from such deed. He also claims the RUF was deliberately blamed for the murder of nuns in 1994 when, in fact, forces of the NPRC junta committed the crime.

48. See *New African*, London, June 1995.

49. See Reno, *Warlord Politics and African States*, p. 129.

50. The Abidjan Peace Accord – Rearranged with Headings, Abidjan, 30 November 1996; also see Document S/1996/1034, UN.

51. Conversation between Brigadier Maada Bio and an international peace-broker (identity confidential).

52. See 'Funmi Olonisakin, *Post-Settlement Reconstruction of Civil Society*, Paper presented at the Regional Security in Global Context Seminar Series, Kings College, London, 27 October 1998.
53. Olu Awoonor-Gordon, 'A Nation Held Hostage', in *For Di People* (Freetown), 12 June 1997.
54. Discussion between the author and Michael Grunberg, Consultant to EO and Director of Sandline International, Kensington, London, 30 March 1998.
55. Branch Energy was partially absorbed by DiamondWorks, a Canadian-based front company with very close links to mercenary organisations. Its directors, like Branch Energy, include people with the closest possible links with EO and Sandline, such as Michael Grunberg and Tony Buckingham.
56. *Africa Confidential*, 29 May 1998.
57. See Report of Commission of Inquiry into the Sandline International Affair in Papua New Guinea, 1997.
58. Fax correspondence from Rakesh Saxena to Samir Patel, 14 July 1997, in author's possession.
59. See David Shearer, *Private Armies and Military Intervention*, Adelphi Paper 316 (IISS, 1998), p. 59.
60. His situation was quite similar to that of Hilla Limann in Ghana (1979). A virtual outsider to Nkrumah's CPP, he was made leader of the party and when elected President, became hostage to party stalwarts, thus considerably weakening his government.
61. *The Status of Forces Agreement between the Government of the Federal Republic of Nigeria and the Republic of Sierra Leone concerning the Provision of Military and Security Assistance to the Republic of Sierra Leone*, Lagos, 7 March 1997.
62. Ibid., preamble.
63. As the current Attorney-General and Minister of Justice, Solomon Berewa oversaw the execution of scores of Sierra Leoneans condemned to death before kangaroo courts for alleged complicity in the RUF/AFRC coup of 25 May 1997.
64. This seven-page secret document, prepared between April and May 1997, was kindly made available to the author by a colleague, Emmanuel Aning, to whom the author is most grateful.
65. Solomon Berewa to Tejan Kabbah, *Restructuring the Army – Role of Executive Outcomes and Control of the RUF Forces*, Secret Memo to President Tejan Kabbah, undated, p. 2.
66. Ibid., p. 4.
67. Ibid.
68. Ibid., p. 5.
69. Ibid., p. 7.
70. See the Media Practitioners Act of 1997.
71. Pascal Lissouba had roundly defeated former president Denis Sassou Nguesso in free and fair elections. However, by November 1997 the latter had violently overthrown the constitutional order with the help of his private Cobra Militia, mercenaries and Angolan forces, after four months of carnage that left over 10,000 people dead.
72. The non-interference clauses in the OAU Charter and the fact that most of the now 'democratically elected' leaders were yesterday's coup plotters

have accustomed them to hailing new leaders and ignoring how they come to power.

73. OAU CM/Dec. 356 (LXVI) Sierra Leone – Doc. CM/2004 (LXVI) – c.
74. S/RES/1132 (1997), United Nations, 8 October 1997.
75. Journeyman Pictures, *Executive Outcomes – The War Business*, Channel 4 Dispatches, April 1998.
76. From fax correspondence between the parties, in author's possession.
77. See fax message from Momodu Koroma to Rakesh Saxena, 8 July 1997, in author's possession.
78. Discussion with Calvin Heard, head of *Defcon*, a small mercenary unit linked to EO, August 1997.
79. Fax message from Rakesh Saxena to Tim Spicer, 14 July 1997, in author's possession.
80. Fax message from Rakesh Saxena to Samir Patel, 14 July 1997.
81. 'Report and Proceedings of the Committee', *House of Commons Foreign Affairs Committee, Second Report, Sierra Leone*, Vol. I, London: The Stationery Office, February 1999, p. xxv.
82. Ibid., p. xxvii.
83. Abdel-Fatau Musah and Robert Castle, *Eastern Europe's Arsenal on the Loose: Managing Light Weapons Flows to Conflict Zones*, BASIC Paper on International Security Issues, No. 26, May 1998, p. 6.
84. Discussion between author and Michael Grunberg, London, 30 March 1998.
85. Conversation between the author and Michael Grunberg, Kensington, London, 30 March 1998.
86. Journeyman Pictures, op. cit. Quoted in P. Trewhitt, *The Business of Killing*, The Parliamentary Human Rights Group, UK, July 1999, p. 21.
87. Conversation between the author and Michael Grunberg, London, 30 March 1998.
88. Fax message from Rakesh Saxena to Samir Patel, 14 July 1997.
89. See letter dated 24 April 1998 from Richard Slowe of S. J. Berwin Solrs, acting for Sandline, to the Foreign Secretary, Robin Cook, extracts of which were published in the *Independent*, 9 May 1998.
90. G. J. Yoroms, 'ECOMOG and West African Regional Security: A Nigerian Perspective', *Issue*, Vol. 21, No. 1–2, 1993, p. 85.
91. See *Tempo* Magazine, Nigeria, 28 January 1999.
92. Ibid.
93. 'Funmi Olonisakin, 'Mercenaries Fill the Vacuum', *The World Today*, June 1998.
94. Ibid.
95. Yahya Jammeh staged a coup against President Jawara soon after he had returned from ECOMOG operations in Liberia.
96. Chief Hinga-Norman, the Commander of the Kamajors and Deputy Minister of Defence, is at war with Charles Margai, the Interior Minister, over the control of Kamajors. See *Newswave* Magazine, October 1998, p. 25.
97. *The Times* (London), Monday, 25 January 1999.
98. 'Ukranians fly Charles Taylor's chopper', *Herald Guardian* and quoted in: <http://sierra-leone.gov.sl/newspapers230399.html>, 27 March 1999.
99. The December 1998 battle for Freetown is covered below.
100. Inter Press Service, 10 February 1999.

101. Reuters, 19 February 1999.
102. <http://www.sierra-leone.gov.sl/newspapers230399.html>, 29 March 1999.
103. AFP Report, IRIN-West Africa Update 346, 25 November 1998.
104. *Sierra Leone: Sowing Terror*, Human Rights Watch Report, Vol. 10, No. 3 (A), July 1998, pp. 23–4.
105. *Newswave* Magazine, October 1998.
106. For further information on military restructuring and civil-military relations, see J.'K. Fayemi, *The Future of Demilitarisation and Civil-Military Relations in West Africa: Challenges and Prospects for Democratic Consolidation*, CDD Occasional Series on Conflict Management (London, CDD, 1998); also in *African Journal of Political Science* (1998), Vol.3, No.1, pp. 82–103.
107. AFP Report, IRIN-West Africa Update, 4 December 1998.
108. 'Diamond Mercenaries of Africa', Background Briefing of *Radio National Transcripts, Australian Broadcasting Corporation*, 4 August 1996. EO chief executive Nick van der Berg denied the allegation but conceded that EO used anti-personnel landmines for training.
109. 'Massacre', *Tempo* Magazine, Vol.12, No. 03, 28 January 1999.
110. *Sunday Times*, 10 January 1999. See also *Tempo*, 28 January 1999.
111. *Tempo*, 28 January 1999.
112. See Abdel-Fatau Musah et al., *Africa: The Challenge of Light Weapons Destruction During Peacekeeping Operations*, BASIC Paper No. 23, Washington, DC: British–American Security Information Council, December 1997.
113. Conversation with a senior official close to the Rawlings administration.
114. Paul Richards in *Fighting for the Rainforest* (1996) defines patrimonialism as the 'systematic scaling up, at the national level, of local ideas about patron-client linkages ... about the duty of the rich and successful to protect, support and promote their followers and friends'.
115. Somalia and Rwanda are two further examples of this gulf.
116. Solomon Berewa to Tejan Kabbah, *Restructuring the Army – Role of Executive Outcomes and Control of the RUF Forces*, undated secret memo to President Tejan Kabbah.
117. See 'Report of the CDD-sponsored RoundTable Workshop on Reconciliation & State Rebuilding in Sierra Leone', Lome, Togo, 21–24 June 1999.
118. Since writing this chapter, a peace deal has been struck between the SLPP and RUF following weeks of peace talks in Lome, Togo, in June–July 1999. Significantly, the Lome Accord calls for the withdrawal of mercenaries. If past experience is anything to go by, PMCs will maintain their presence in Sierra Leone simply by redeploying their mercenary forces into 'security companies'. In other words, simply change their identity from PMCs to PSCs.

4

The Hand of War: Mercenaries in the Former Zaire 1996–97

Khareen Pech

The origins of modern mercenarism in Africa lie in the heart of the continent, in the vast, mineral-rich wilderness of what was formerly the Belgian Congo. This giant territory – discovered by European explorers only at the end of the nineteenth century – has always played a crucial role in determining new epochs in African history. In the early 1880s, a race between two explorers to discover the interior and navigate the Congo river, and thereby secure foreign control of the region, began a frenzied race among European powers for the whole continent.[1]

At the end of this great scramble, the Congo lay in the hands of the richest man in Europe, the Belgian King, Leopold II, who had gambled his entire inheritance on securing a treacherous and unknown territory he believed would hold 'unspeakable' riches. The rest of Africa had been carved up like a cake, 'the pieces swallowed by five rival nations – Germany, Italy, Portugal, France and Britain'.[2] The race had not only brought a divided and colonised Africa that would forever strain at its artificial borders and rage at its imperial masters, but had exacerbated tensions in an increasingly divided Europe that was at the brink of World War I.

Exploration and colonisation were carried out in the name of trade, religion and civilization, but were imposed at the point of a gun. It was a gun often held by hired, armed men – Africa's first foreign mercenaries. These hired guns defended the explorers and the early trade outposts of the colonisers. Many were Africans – recruited in Zanzibar and in West Africa – under white leadership. Fifty years later, when the continent claimed its independence, this lesson would be remembered. This time the gun would be held by Africans, and often they would be assisted by foreign mercenary forces. Once again, the Congo would play a leading role in this epochal shift. In the late 1950s, the Belgian Congo had proven to be richer than King Leopold had ever expected. With the new industrial age, the world's factories hungered for copper, diamonds, cobalt, gold, chrome, uranium and various other strategic minerals – all of which were in abundance in the Congo. To keep this immense wealth in Belgian hands, the administrative authorities refused most of the 18 million Congolese people the right to politics, education and skilled labour. But Belgium's grasp on power was doomed to slip when a forceful wind of change blew a gale through Africa. As Thomas Pakenham wrote in *The Scramble for Africa*, the wind came from the west, carrying:

> the germs of African Nationalism, spread by Nkrumah and the pan-Africanists – Western ideals of the rights of man and the liberty of nations. There was no way Belgium could seal off the Congo from this wind. In 1959, after riots in Leopoldville (Kinshasa) the Belgians suddenly lost their nerve ... They announced a date for a general election, to be followed by independence within eighteen months. Unprepared for party politics, the country split along ethnic and regional fault lines. (The Belgians were surprised by the momentum of the elections and by the swift success of the popular local leader, Patrice Lumumba – whose victory at the polls ended any hopes of having a Belgian-backed candidate in power.) When the Belgians scuttled out of the Congo in July 1960, they had left the country well prepared for civil war and anarchy.[3]

The first paroxysm of anarchy struck only a few days after the Congo gained independence under Lumumba's newly elected government. Twelve days after he was sworn in as Prime Minister, the 24,000-strong Congolese army staged one of the worst mutinies in Africa. The 'Force Publique' had been set up under King Leopold and was commanded by 1,000 Belgian officers, whom the Belgians had expected would remain in control. But the Congolese troops demanded both the dismissal of their Belgian Commander-in-Chief, General Janssens, and the complete Africanisation of the Force Publique. The mutiny spread beyond Leopoldville (Kinshasa) and started a general panic that eventually caused Belgian military intervention and, indirectly, the secession of the Katanga province under Moise-Kapenda Tshombe.[4] Although Lumumba and his President, Joseph Kasavubu, immediately settled the mutiny by agreeing to the Force's demands, the resistance had sparked widespread unrest and a furious struggle between several Congolese political leaders. Each of these rivals was reliant on military support from foreign powers, most notably, Belgium, France, the United States and several Soviet-allied countries. Each country, in turn, had commercial and political interests in the Congo – in it's natural resources as well as its strategic location in Africa. While the Congolese leaders vied for power, the foreign powers jostled anew for political influence and trade benefits in the Congo: a 'jostling' that would precipitate a second scramble for Africa during the Cold War period. To secure the success of their respective protégés, foreign military forces and hundreds of mercenaries were made available by the foreign powers. The mercenary armies were mainly constituted by French, Belgian, German and South African hired fighters. The post-independence, mercenary wars of 1960–65 were played out between four main Congolese leaders.

The popular, Prime Minister, *Patrice Lumumba* whose military campaign with the Force Publique to reclaim Katanga – the keystone of the Congo economy – was backed by a Swedish-led UN peacekeeping force called Operations des Nations Unies au Congo (ONUC), deployed in July 1960. After Lumumba obtained military support from the former Soviet Union in September, he was deposed by the army's Chief of Staff, Joseph Mobutu. Lumumba sought protection from the UN, but he was later captured and flown to Katanga where he was murdered between January and February 1961 – allegedly by a Belgian mercenary in the presence of Tshombe and some of his ministers.[5] After his death, pro-Lumumba rebellions plunged the Congo into renewed cycles of chaos on several occasions.

Joseph Kasavubu, the first Congolese President, was elected into office in June 1960. Two months later, along with Lumumba, he was deposed in a bloodless coup by Mobutu. He was restored to power when Mobutu stepped down in February 1961, after Lumumba was killed. He ruled for four years until Mobutu seized power for the second time in November 1965.

Moise Tshombe was a colourful Katangese politician who led the secession of the mineral-rich Katanga province in 1960. He received backing from the Belgian government, their armed forces and the Belgian mineral firm, Union Minière du Haut Katanga.[6] Tshombe also won support from Rhodesia, Portugal and (Verwoerdian) South Africa – each of which had its own motives for supporting the common cause against the pro-communist Lumumba. Tshombe and his Katangan secessionist backers hired the first large-scale, multinational, foreign mercenary forces in the Congo. He and several mercenary armies waged war for three years before the Katanga rebellion was quelled in 1963 by UN forces. Tshombe exiled himself to Europe where he plotted his return. In June 1964 he was recalled by President Kasavubu to help suppress a new revolt that had started in southern Kivu and northern Katanga. This bloody revolt was led by a pro-Lumumba group, under the command of Pierre Mulele and was backed by pro-Soviet and Chinese forces. These Simba rebels had succeeded in capturing the eastern capital, Kisangani, and again the Congo faced political chaos and violence. Tshombe was appointed Prime Minister, a post he kept for 15 months before his own political ambitions to seize power led to his dismissal in October 1965.

During this time he once again recruited several armies of foreign mercenaries, who helped to defeat the Simba rebellion in January 1965. After his dismissal, he exiled himself to Europe once more and was later kidnapped and jailed in Algeria where he finally died in 1969.

Joseph-Desiré Mobutu came into a position of military control in the early days of independence when he replaced the Belgian General Janssen as Chief of Staff of the Force Publique, which had been renamed Armée Nationale Congolaise (ANC), in 1960. After he deposed Lumumba and Kasavubu, Mobutu was then supported by the UN forces, who themselves hired mercenaries, the British-trained Nepalese Gurkha soldiers, to assist in the UN's offences against the Katanga mercenary forces. Mobutu would later be assisted by French and Belgian-led mercenary armies as well as the US military. After returning power to the elected Kasavubu in February 1961, he continued his military role as Commander-in-Chief of the ANC until 1965. In November, he seized power again and declared military rule. Although he would face several plots against him and would fight a fresh round of mercenary wars between 1966 and 1968, Mobutu retained his grasp on power. After banning all political activity in 1966, he reduced the number of provinces in the Congo, renamed the major cities and established a one-party state with himself as President. In 1971, he renamed the country Zaire and announced a policy of 'Zaireanisation'. Unfortunately this policy proved to be less a system of Africanisation than one of patrimonial rule, which increasingly gave Mobutu and his immediate family personal control of all the means of accumulation available to the giant state and its diverse people. During his 32-year rule, Mobutu used foreign military support and mercenary forces on numerous occasions to quell resistance and maintain the borders of his Zairean state. He was ousted only in 1997 when an old enemy, Laurent Desiré Kabila, the former Lumumbist rebel leader of the 1960s and 1970s, emerged from relative obscurity to head a full-scale rebellion with the backing of Ugandan, Rwandan, American and foreign mercenary forces.

The Congo as Cradle of Modern-day Mercenarism in Africa

Out of the mercenary wars of the 1960s emerged the new state of Zaire and its Western-backed victor, Marshall Mobutu Sese Seko, who would become one of the richest and most powerful leaders in Africa. Apart from its great mineral wealth, Zaire would prove to be one of the most strategically important countries in Africa during the Cold War. As such, it would continue to play a crucial role in determining events in Africa. The political events in the Congo in the 1960s sparked a wave of change and unrest that spread first into Angola, delivering to Africa one of its most protracted and vicious civil wars. These influences would spread further into sub-Saharan Africa under the shadow of the emerging Cold War. In this respect, the words of Frantz Fanon capture the significance of the Congo: he said that the shape of Africa resembles a revolver, and the Congo is the trigger.

Events in the Congo in the 1960s triggered not only a new epoch, but also reinstated an old master – the mercenary commander. In this respect, the Congo gave rise to the first cycle of modern, twentieth-century mercenarism in Africa. These mercenary forces would go on to fight post-independence, secessionist and guerrilla wars elsewhere in Africa. Many would continue their activities into the early nineties – and would resurface once again, in their original fighting posts in the former Congo, 37 years later. Many of these 'dogs of war' would develop well-publicised personalities and, despite official condemnation, they would maintain careers that today, at the start of the twenty-first century, still play a role in determining political outcomes in Africa.

In this sense, a political history of the Congo is also a history of mercenarism in Africa. This history is repeated in revolutionary cycles that are new only in terms of the patterns that evolving winds of change etch into the underlying substructure – which remains the same. It is this permanent substructure, the Congo's geology and geography, its immense natural resources and strategic positioning – which repeatedly attracts the same history, played out by romantic, opportunistic and often greedy protagonists who seek wealth, power and conquest. The cycles of progress, turmoil and chaos in the Congo seem forever to be bound up with the pursuit of empire and with the influence of foreigners who seek both trade and war. Words spoken by the young French explorer Pierre de Brazza, in 1881, to chiefs who owned the land where Brazzaville would later be founded, are still meaningful. He said, 'White men have two hands. The stronger hand is the hand of war. The other hand is the hand of trade. Which hand do the Abanhos want?' The Abanhos answered all together, 'Trade'.[7]

In Africa, war and trade have forever been intricately linked, but seldom has the foreign hand of trade held out general prosperity to the Congolese people. The Congo's 'unspeakable' riches first empowered and enriched a foreign monarch and his distant European nation; then, under Mobutu, they empowered and enriched an African dictator and his family clan. Winds of change that promised reform and liberty in 1960, only ushered in a new oppression.

The Revolutionary Wind of Change, 1996

Almost four decades later, the revolutionary winds of change would come again. In 1996 this 'new wind' confronted Zaire with a series of political and military events that would topple Mobutu Sese Seko and transform the balance of power in central Africa. Once again an epochal shift was delivered at the point of a gun; once again a new paroxysm of violence was brought by allied foreign military and mercenary armies; and once again war was heralded (and enabled) by talk of trade and development. But this time the wind of change blew not from the west, but from within Africa – primarily from Uganda and Rwanda but also from South Africa, where the new, black regime had allied itself with Kampala and Kigali. Although these central African political, military, economic and ideological forces were backed by Western powers – most notably the US – this was a continental shift that seemed for the first time to be driven by Africans for Africans. As such, the promise of reform and economic development, declared by the champions of the new era, gave an otherwise sceptical Zairean people renewed hope and raised widespread international support.

To modernise Pakenham's description of the first wind of change in 1959: the new wind in 1996 carried the germs of a pan-African market system, spread by Museveni, and the new, black leadership of South Africa, Mandela and Mbeki (fostered) African ideals of human rights and a common market that would dissolve economic borders and open the continent to renewed growth, that in turn would produce an 'African Renaissance'. There was no way Mobutu could seal off Zaire from this wind. When he decreed that all ethnic Tutsi would be expelled from Zaire, a full-scale Tutsi-led rebellion was launched. Like its Belgian predecessors, the Mobutu government suddenly lost its nerve and fled six months later. In May 1997, Kinshasa fell to the rebel alliance and its leader, Laurent Kabila, assumed power. Unprepared for a Tutsi-led government, the country split along pro- and anti-Tutsi ethnic lines as well as regional lines. When the Mobutuists scuttled out of Zaire in mid-1997 they had left the country well prepared for civil war and anarchy.[8]

Within months the tensions were evident and less than a year later, renewed war broke out, bringing with it, not an African renaissance, but a pan-African war. As before, when Mobutu had come to power, the new leader, Laurent Kabila, failed to introduce the reforms and freedoms that people had dared to hope for. He had returned the giant state to its original name with a modern twist, the Democratic Republic of the Congo (DRC). Yet, at the time of writing, it is neither democratic nor a republic. It is a quasi-state, split by ethnic and regional rivalries, ruled by warlords, and held ransom by marauding military and mercenary forces who are each, for a price, trying to force a new political outcome out of the present chaos.

Prior to the 1998 war, Kabila's ertswhile sponsor, Ugandan President, Yoweri Museveni, described his new vision of an economic solution and a transnational political federation at Kabila's inauguration in Kinshasa on 27 May 1997. He said that the eradication of Zaire had filled a hole in Africa that would now allow them to unite and construct an African common market. To run open roads through the continent and to prove that the ADFL's victory had 'liberated not only the Congo but also all of Africa'.[9] Both Uganda and Rwanda wanted to form a regional political and economic federation of central African states that would see a geographical reconstruction of borders in the region, greater stability, increased development and renewed trade. For this dream to come true, the involvement of the Congo would be imperative. Its significance – from the Rwandan point of view – was explained by an adviser to the Rwandan military leader, Paul Kagame, in an interview with the *New Yorker*. 'America can live without the Congo. Rwanda cannot ... By itself, ours is a non-viable country.'[10] This simple truth remains at the heart of the matter in the strife-torn Congo. And the country's future is still one of the most important questions for Africa since the end of apartheid in South Africa.

This chapter analyses mercenary events in Zaire between 1996 and 1997. It deals only with the mercenaries hired by the Mobutu government. The Zaire mercenary contracts of 1996/97 introduced new trends to the booming private military and security industry in Africa. These include:

- a greater involvement of foreign financial, mineral and other companies;
- the involvement of a greater number of foreign African mercenaries;
- the first large-scale use of Serb mercenaries
- the unbundling of the giant South African mercenary firm, Executive Outcomes
- the emergence of the supply of military services by the same mercenary group to both sides of the conflict
- co-operation between French and South African mercenaries and mercenary companies
- strategic alliances between rival military companies
- the emergence of a greater number of South African military companies who began operating more widely in Africa
- a more aggressively competitive approach to the war business in Africa
- greater political destabilisation after the 1996–97 war brought greater business to a greater number of military companies.

The mercenary contracts in Zaire were also influenced by a third scramble for Africa initiated in the early 1990s by the end of the Cold War and the start of a new boom in mineral and energy exploration and trade in the continent. This boom was led by a number of foreign junior mining and energy companies. The corporate powers who part-financed the 1996 war constituted the new foreign influence that would jostle, once again, for access to rival leaders who would reward them with concessions to primary resources and new markets. Trade in Africa – and especially the Congo – has always carried great risks for foreigners. In the early 1990s these risks were mitigated by the private military companies who, in the case of Executive Outcomes, constituted a corporate army. The latest cycle of mercenarism in Africa has seen the two foreign hands of trade and war merge into one. Today, this is the only hand that is offered to the Congolese people.

War in Eastern Zaire and the First Tutsi-led Rebellion

War came to eastern Zaire in early October 1996, when a Tutsi-led rebel group attacked a small village, Lemera, in south Kivu, near the border of Burundi on the shores of Lake Tanganyika. The rebel attack left the Zairean Armed Forces (FAZ) stunned at the speed and efficiency of its strike. A small number of rebels first infiltrated the village in civilian clothing and sized up the right location and moment to begin the attack, which they signalled by dropping a mortar shell into the FAZ garrison area. The main rebel army waiting beyond the village then simultaneously advanced on the garrison from several directions, taking care to leave an escape route for the defenders. Their movements were coordinated by the use of hand-held radios. This pattern of attack was to be repeated as the rebel troops grew in strength and slowly advanced over the next six months into Zaire. About twelve days after the first attack, a large column of Tutsi-led, Rwandan troops crossed into south Kivu from Rwanda via Burundi. Once on the Zairean side, they moved north to attack first Uvira and then Bukavu, the capital of south Kivu. After this success, the next major town to fall to the rebels was the north-eastern capital of Goma, on the northern border of Rwanda, on 1 November.

The small town of Goma had been at the heart of the problem of the volatile Great Lakes region in central Africa for two years, ever since over 1.5 million refugees fled Rwanda and entered into Zaire through Goma in 1994. Most of these people clung to a miserable existence in pitiful, crowded refugee camps spread out along the border of their home state on the verge of the Virunga rainforest, until the rebel uprising in late 1996. The refugee camps gave sanctuary to the exiled former Forces Armées Rwandaise (FAR) who had participated in the brutal, genocidal massacre of an estimated one million Rwandans during the 1994 war, that was won by the rebel Tutsi-led Rwandan Patriotic Front (RPF). The exiled, Hutu-majority troops are estimated to have made up 7 per cent of the refugee populations.[11] They continued to train and rearm in the forest and, together with their former allies, the FAZ troops in the region, they presented a great security risk to the new Tutsi-led government. The refugee camps were also used as launching pads for FAR cross-border raids into Rwanda and Burundi. In retaliation, special units of the newly named Armée Patriotique Rwandaise (APR) would launch search and destroy operations on the Hutu refugee camps in Zaire. This ongoing, low-level, cross-border conflict had escalated tensions in the region. While the US military was training the APR on the Rwandan side, Mobutu's foreign military supporters were involved in training and rearming operations on the Zairean side. For these reasons, the Rwandan Vice-President and military chief, Paul Kagame, visited US State Department officials and people in the Clinton administration in New York and Washington in August 1996. He told them that Rwanda would invade Zaire and dismantle the camps if no one else acted soon. The US itself was not prepared to use force to break up the camps and Department of Defense officials had decided, as early as August 1995, not to oppose such action by Rwanda, provided it was a 'clean' operation, meaning one with limited civilian losses.[12] The Tutsi-led APR therefore finalised its preparations to invade Zaire. Between August and September, they infiltrated 1,200 armed Rwandan soldiers into the forests along the north and south-west borders, 200 Rwandan-trained Banyamasisi Tutsi fighters in north Kivu and 1,000 Banyamulenge Tutsi fighters in south Kivu.[13] A significant number were veterans of the RPF war in 1994.

The initial strength of the rebel forces was about 2,500 and they did not constitute a cohesive whole.[14] They included troops drawn from Rwanda, Uganda, Angola, Sudan and Zaire, military-related technicians from Ethiopia and Eritrea, as well as foreign mercenaries. The allied forces were held together by the APR and its command structures. The Zairean 'rebel' backbone constituted fighters from the Banyamulenge and Banyamasisi people, who are ethnic Tutsi and native to the south and north Kivu Provinces. As the rebel army swelled, Katangans, Kasaians, Mai-Mai and other tribal militias joined them.

A few days after their first victory at Lemera, on October 18, the Rwandan-led rebel troops officially named themselves the Alliance of Democratic Forces for the Liberation of Congo-Zaire (ADFL). Their ostensible leader, Laurent Desiré Kabila, was a Zairean citizen who had long opposed Mobutu. His revolutionary credentials went back to the mercenary wars of the early 1960s when he was a follower of Patrice Lumumba. In the 1960s Kabila had established a revolutionary enclave in the hills above Fizi, along the border between South Kivu and Katanga and it was here that he met Che Guevara, the Cuban revolutionary hero who spent six months at Kabila's base. Kabila therefore seemed the most ideal choice to lead the anti-Mobutu struggle. The ADFL was an ethnic and transnational coalition of anti-Mobutu groups, led by Rwandan and Ugandan soldiers and backed by US military assistance.

The capture of Goma on 2 November precipitated, firstly, the destruction of the camps and secondly, a new cycle of chaos and bloodshed in eastern Zaire, as thousands of refugees fled deeper into Zaire.

They were pursued by the Tutsi-led rebels who methodically carried out a series of revenge massacres, killing thousands of people in the next six months. The revenge attacks carried out by the allied rebel forces caused massive social dislocations and untold casualties among civilians – estimates of the numbers vary from a few thousand to tens of thousands.[15] In a report, Human Rights Watch said that these atrocities resulted in the deaths of not only Rwandan refugees but also Zairean Hutus.[16] Once the objective of wrecking the camps and dispersing the exiled FAR army had been achieved, the Tutsi-led rebels continued their military campaign to fulfil an 'underlying' plan created by Vice-President Kagame of Rwanda and President Museveni of Uganda.[17] This involved:

- the continuation of the ADFL's military operations in the south in order to cripple Burundi extremist Hutu opposition to the new Tutsi regime
- the continuation of military operations beyond Goma in a northerly direction along the Ugandan border to cut off access to safe havens in eastern Zaire used by Uganda's armed opposition movements, Tabliq and the West Nile Liberation Front (WNLF)
- the continuation of a campaign to oust Mobutu who had supported the former Hutu government of Juvenal Habyarimana and had given sanctuary in Zaire to the FAR and genocidal death squads, the Interahamwe. This campaign would replace Mobutu with the ADFL leader, Kabila who would facilitate a new, Tutsi-led order in central Africa.

This underlying plan, it was hoped, would bring greater security to Burundi, Rwanda, eastern Zaire and Uganda as well as Angola, in the long run, and this would usher in the promised new era of prosperity. On each count, the Kagame-Museveni strategy of 1996/97 succeeded very well. Kabila's ADFL emerged as the spearhead of a local rebellion that quickly snowballed into a mass movement, against which Kinshasa would be powerless.

After Goma had fallen, Kinshasa sent FAZ reinforcements to the Kivu provinces, but these troops had little effect on the outcome and were more interested in looting than fighting. According to a US defence attaché in Kinshasa, this reinforcement was one of the few major redeployments of FAZ forces to occur throughout the war.[18]

Mercenaries for Mobutu

While the Tutsi-led civil war flared in eastern Zaire, Mobutu convalesced in Switzerland and later resumed residence at his seaside villa in southern France. Here he lived more like an eccentric billionaire than the president of an African country on the brink of violent collapse. Back in Kinshasa, kleptocracy and political confusion ruled. Prime Minister Leon Kengo wa Dondo was in charge, but his de facto authority was limited by Mobutu's personal rule system and the powerful Ngbandi generals. By late November, the rebels had captured several airports in eastern Zaire and had cut off Kinshasa's prospect of reinforcing troops and regaining lost territory in Kivu. Since the roads were badly deteriorated, it would take months for the Zaire army to redeploy and mount a counter-offensive in the east.[19] Kinshasa would have to act swiftly if it was going to stop rebel forces from seizing the strategic Bunia airport and Mobutu's gold mines in the north-east near the Ugandan border.

Bunia was strategic to a military victory at this early stage of the war, because it was a conduit town through which arms and diamonds were being smuggled and was used by allied Zairean and Sudanese troops to back Ugandan insurgents against President Yoweri Museveni. Partly for this reason, the Ugandan-backed, ADFL rebels were bent on seizing it. The capture of Bunia would also give the ADFL rebels a strategic foothold in the Haut Zaire province and facilitate their campaign to seize the eastern capital of Kisangani, which, in turn, would open the way to Kinshasa. A month earlier, after the Kivu provinces had first been attacked in October, Kengo wa Dondo had met with military leaders in Kinshasa to plan a counter-offensive. At this meeting, the army chief of staff, general Eluki Monga Aundu, stated that the Zairean Armed Forces (FAZ) could not stage an effective counter-offensive against the Ugandan and Rwandan-backed rebels – who were reportedly 'equipped, trained and led on American lines'[20] – without the assistance of foreign mercenaries. Eluki asked Kengo to approve a plan to hire mercenaries, and from this time on the Kengo government and President Mobutu were agreed on the use of mercenaries.[21]

Despite this early decision, which precipitated a flurry of negotiations with military specialists, Kinshasa failed to implement an immediate plan. At first it looked to its traditional European and African allies for official support, but this was not forthcoming. Not even France, one of Mobutu's closest allies, would commit itself to intervene as it was unwilling to bear the risks and costs of such a potentially disastrous operation. After the end of the Cold War era and so many years of Mobutu's kleptocratic rule, foreign powers had little incentive to provide official support to Zaire and a greater incentive to unofficially favour the coming of a new government. Kinshasa was therefore diplomatically isolated and its only means of survival lay with the private military industry.

War Bonanza

As the Tutsi-led forces pushed the frontline deeper into eastern Zaire, a fever of commercial anticipation spread among international arms dealers and foreign military companies. Zaire – one of the largest, richest, and most corrupt countries in Africa – would provide the private military complex with its greatest opportunity to make a financial killing since the end of the Cold War. This was for several reasons. In the first place, mercenaries were not new to Zaire. For three decades the Mobutu government had allowed a wide range of foreign military and mercenary forces to either train or multiply its armed forces. During this time, pro-Western foreign forces had also been allowed to conduct their own covert offensive intelligence and military operations in Zaire in support of a number of armed groups, key of which were Holden Roberto's Frente National de Libertecao de Angola (FNLA) and Jonas Savimbi's UNITA forces in Angola. The most active of these Western-armed forces included the external wing of the American Central Intelligence Agency (CIA), the Special Tasks Directorate of the South African Defence Force (SADF) and a range of special forces from the French, Belgian and Israeli armies.

The post-Cold War expansion of the private military industry into Africa is rooted in the former external operations of these foreign national defence forces. Today, the private military companies that operate in African countries are all managed, advised or staffed by people who formerly participated in military-related support operations for their national governments in these same countries. For example, when the South African-based firm, Executive Outcomes (EO) first started out as a private military company operating in Angola, it was military officers who had links with the MPLA government and its FAA forces as well as officers who had formerly conducted military operations in Angola who were instrumental in clinching the mercenary deal with Luanda. Likewise, when EO first began negotiations to secure a contract in Sierra Leone, its executives used military and other specialists, who knew the country and were linked to the local political system, to broker EO's business deals. Therefore, a wide range of retired officers, who had formerly established good relations with the Zairean military leaders on behalf of their own governments, could expect to tender for similar business – but this time, on a private and far more lucrative basis. As a result, there were an unusually large number of French, Belgian, South African, American, British and even Israeli security companies that competed for Kinshasa's business.[22]

Zaire's great size and lack of infrastructure meant that its war purchases were likely to be considerable. It would require a large cache of small arms, significant air defence systems, massive logistical support, updated communication systems and a multi-skilled mercenary force. Such a large, various and costly contract attracted much attention not only because of the immediate profits it would create, but because it was expected to provide the means to develop a military-mineral empire similar to that initiated by the EO group in Angola between 1993 and 1996. A few of EO's former pilots and ground commanders, as well as a several lesser-known South African military companies, saw this opportunity and, believing that they could capitalise on EO's successful formula, they began negotiating contracts with a range of potential resource sector business partners and Zairean military leaders in late 1996.[23]

Lastly, the private military industry in Africa is economically dependent on quick cash and the hard currency of diamonds. Mineral and oil concessions also play a key role. Zaire's vast resources of diamonds, precious metals, cobalt, exotic woods and other natural resources were sufficient to attract a wide and ardent interest from the private military purveyors. In addition, such a military-mineral deal was expected to be an easy strike. The Mobutu government had been mortgaging state resources in exchange for patrimonial power and personal wealth for the past 32 years. Under Mobutu, formal state behaviour had collapsed and had been replaced by an informal, decentralised system that consisted of a patronage network of strongmen and clients who personally controlled Zaire's multiple centres of accumulation.[24] (These centres include the mining of diamonds, copper, cobalt, gold and other minerals as well as the trading of coffee, timber, diamonds and oil.) Many of the strongmen were of Mobutu's Ngbandi clan and were powerful figures in the military. It was not unusual for them and their wives to transport large parcels of diamonds and briefcases packed with hard currency (US dollars) between Zaire and Europe.[25]

Mobutu and many of his closest associates owned foreign bank accounts into which nine-figure sums of cash, derived from the sale of state resources, had habitually been deposited.[26] In addition, Mobutu had encouraged the development of a clandestine economy, that thrives on the illicit trading of diamonds, cobalt, arms, exotic hard woods and contraband as well as criminal rackets that manipulate state regulatory authority such as passport sales, money laundering and drug trafficking.[27] The kleptocratic political system as well as the total collapse of law and order in Zaire appealed to the private military dealers, who could expect to trade violence for hard cash in an unaccountable environment. For all the above reasons, Zaire was, at the outset, an ideal country in which the military-mineral industrial complex could do business.

Dozens of military brokers from around the world began consultations with Mobutu's foreign representatives. In South Africa, the Zairean chargé d'affaires, Isa Ghanda, was approached by several dealers. Out of these consultations arose a tender to the associated Ngbandi generals, Kpama Baramoto Kata (the commander of the Civil Guard) and Nzimbi Ngbale (the commander of the Presidential Guard) by the group of former EO employees. The tender was facilitated by several top-ranking former SADF officers who had long-standing military ties to the Zairean generals as a result of former SADF-FAZ cooperative programmes in Kinshasa and Pretoria. Not wanting to miss out on the opportunity, EO also approached Zairean representatives in Pretoria, Johannesburg and Morocco.[28]

Executive Outcomes

The EO tender was in direct conflict with the interests of their original client, the MPLA Angolan government, with whom EO still conducted military-mineral business through their associated companies in Luanda and Lunda Norte. The Angolan government favoured the ousting of Mobutu and his Ngbandi generals who were politically and economically close to its enemy, Savimbi, and who had facilitated the survival of UNITA for decades. The MPLA hoped that with the demise of UNITA's long-term ally, the incoming rebel (AFDL) government would help to destroy UNITA. This would be done by means of a cooperative agreement with the new Kinshasa government which would help to sever UNITA's logistical supply lines and rear bases in Zaire and stem its supply of funds from the trade of diamonds and arms in Kasai. With UNITA economically and logistically weakened, Luanda hoped it would then be able to launch a military assault to achieve the military victory that had not occurred in the 1994 war. The Angolan armed forces (FAA) therefore committed troops to Kabila's rebellion and opened a second front in western Zaire to clinch the success of the AFDL military campaign.

Despite their conflict of interest, EO executives Eeben Barlow and Lafras Luitingh conducted several communications with representatives of the Mobutu government and were said to have tendered a contract valued at about US$50 million to provide a fighting battalion in support of Mobutu in late 1996.[29] They were also linked to two French officers who were recruiting a multinational mercenary force for Mobutu, according to French newspaper reports in January 1997.[30] The reports said that EO had been contracted to supply logistic, communication and recruitment services to former gendarme, Robert Montoya, and former colonel, Alain le Carro. The latter had been hired by Kinshasa to recruit a force of about 200 to 300 men that was assembling in Kisangani to stem advances by the Rwandan-backed rebels.[31] Montoya and Le Caro both have private security operations in Francophone Africa (Togo, Ivory Coast, Burkina Faso and Congo-Brazzaville). Both also formerly served at the Elysée Palace under the late President François Mitterrand.[32] At the time of these reports, Barlow denied that EO had any personnel in Zaire and also denied knowing either of the Frenchmen. He stated that he had given the Angolan government an undertaking not to support 'the opposition', and 'If we started to work for the Zairean government against the wishes of the Angolan government then we would lose credibility and probably existing contracts we have with other governments.'

In spite of these credible denials, it is now clear that Barlow and Luitingh did consult Mobutu representatives and did tender their military services to a power that was hostile to their Angolan client government. Not only was this fact reported by EO personnel who worked closely with Barlow and Luitingh at the time, but a South African security firm, which successfully competed for the contract, reported that they had been present when Barlow emerged from a consultation with the Zairean envoy prior to their own appointment with him to negotiate military matters.[33] At the time, US intelligence sources also leaked information to the press that Barlow and EO were negotiating a deal with Mobutu.[34] The most conclusive proof emerged when a letter was discovered in Mobutu's residence in northern Zaire after it had fallen to the rebels.

This letter showed that in March 1997, as the AFDL rebels readied to seize the eastern capital of Kisangani, the Zairean consul in France called the EO offices in Pretoria in a last-ditch effort to secure mercenary support. As a result of this call, Barlow telephoned Professor W. H. Dixon, a representative for Mobutu in Monaco, who then wrote to the consul in France. Dixon asked the consul to set-up an urgent 'face-to-face meeting' with Mobutu, Barlow and Luitingh.[35] This meeting was 'necessary because of the urgency of the situation in Zaire and because His Excellency Marshal Mobutu [had] to make the final decision to implement the proposed cooperation'. In his letter Dixon explained that both Barlow and Luitingh would be prepared to 'change their plans and schedules if Mobutu 'so directs' and that arrangements had to be made quickly 'so that Executive Outcomes can do for Zaire what it has already done for Sierra Leone and Angola'. At this time EO was not only conducting business with the MPLA government in Luanda through its associated business entities, it was also reported to be assisting the Ugandan-backed AFDL through its close ties with Ugandan military leaders and affiliated military-mineral businesses in Uganda and Rwanda.[36] This was the reason, according to security experts in South Africa, that the Mobutu government did not hire EO's services in the first place. While in Luanda in early 1997, both Barlow and a British partner were taken to one side and given firm warning by the Angolan army chief, João De Matos, not to deal with Kinshasa or they would face serious economic losses in Angola.[37] It is clear that if the EO-group ever planned to be involved in Zaire, they could only have done so through a front company that had no apparent links to EO – or else the group would have lost significant clients and their diamond concessions in Angola.

Confusion in Kinshasa

In the last months of 1996, a stream of private military dealers visited Kinshasa to tender their services and hopefully arrange a face-to-face audience with Mobutu, but few were successful. Although Zairean officials expressed interest and many discussions took place, few deals were conclusively struck. The government lacked cohesion and dithered over an emergency plan. There was also an apparent shortage of hard cash. Ironically, while the state's resources had gone into the personal accounts of military leaders and politicians, few were willing to dip into their fortunes to finance the war effort. Mobutu's terminal illness and a paralysing distrust among politicians and military leaders in Kinshasa impeded the government from executing a quick and sensible plan. In this last hour of its existence, Zaire was effectively without a government and its prime minister, Kengo wa Dondo, was suspect in the eyes of many Zaireans because his mother had Tutsi origins, which later caused many to believe that he had sided with the Tutsi-led rebels. Frustrated by the delays, Army General Chief of Staff, Eluki Monga, accused Kengo of deliberately responding too slowly to the crisis.[38] In response, Kengo suspended Monga on November 20 and appointed Baramoto to take his place.

One of Zaire's most powerful strongmen, General Kpama Baramoto commanded his own force of some 10,000 soldiers which had often been used to put down civilian disturbances in the past. Baramoto had also accrued a vast personal fortune from his role in clandestine trade – mostly gained from gold and diamond mining operations he ran in Kivu and Kasai.[39] As a result, Baramoto kept his Civil Guard soldiers loyal by paying them well, unlike the Zairean Armed Forces (FAZ) who were unruly, untrained, unpaid and who looted for a living.

In 1996 Baramoto entered into a joint venture deal with US owned Barrick Gold Corporation to mine Bunia.[40] Barrick provided funds to refurbish the Bunia airport, which Baramoto used to facilitate his diamond mining operations and increase his stakes in private air cargo companies.[41] The airport was also useful to external allies like the Sudanese government who shipped weapons to Ugandan insurgents through Bunia. The illicit trading of diamonds and arms through Bunia was facilitated by a network of foreign arms dealers and chartered air companies, some of which were based in South Africa and Belgium and had close ties to Baramoto and the Mobutu clan. This relationship had been facilitated by Baramoto's close personal links with both the UNITA military leaders and UNITA-run diamond operations in Zaire.

Although he was not an esteemed military figure, Baramoto was not necessarily a bad choice to head the army. He could secure military support from the still powerful UNITA forces, he had a strong incentive to protect his property and financial interests in Bunia and he commanded his own forces that could be used to defend strategic assets in the region. In addition, he owned several homes in South Africa and had already entered into negotiations with a group of South African former SADF officers to mount a military intervention in Zaire. Baramoto was therefore expected to be able to raise, fund and execute a prompt counter-offensive. But Baramoto's appointment to army chief failed to produce immediate results, and a general lack of cohesion and decisive planning still prevailed in Kinshasa.[42]

By December 1996, the options had been considered and formal proposals for military support services had been tendered by about ten companies. These included three South African companies, several French and Belgian companies, an Israeli group, an American firm[43] and several companies with eastern European or Russian connections.[44] On 17 December 1996, after the rebels had seized a swathe of territory in the east, the terminally ill Mobutu finally returned to Zaire where a decision had been taken. A deal had been struck between the Mobutu government under Kengo wa Dondo, several wealthy Zairean businessmen and old allies in the French political and military networks. The mercenary force was to be led by a retired Belgian colonel, Christian Tavernier, who was close to Mobutu and who had been involved in military-related deals in Zaire since the 1960s.[45] Several associated deals with other companies were struck for the supply of arms, fighter and cargo aircraft and consumable rations.

The belated Kinshasa contracts of 1996 resulted in the hiring of a multinational mercenary force that was first deployed in a counter-offensive in late December – several days after the strategic Bunia airport and gold mines of Kilo Moto had been seized by the AFDL.

Mobutu had revived past tactics and restored old mercenary ties to solve a crisis brought by the new wind of change blowing through central Africa. Although many of the lead characters driving the new gale were old enemies who had been conquered before, Kinshasa was doomed to fall. Time was against the Mobutu clan. First, Mobutu was widely hated by his Zairean people who wanted him overthrown. Secondly, the giant FAZ army had very limited capabilities and only a handful of its estimated 60,000 soldiers had any combat skills. The FAZ were feared by the local populations and could not rely on civilian support. Since Mobutu was now diplomatically isolated, he stood little chance of winning the war. Thirdly, the mercenary force had arrived too late and ill-prepared to properly defend Kisangani against the AFDL two-pincer advance through Haut-Zaire. Lastly, Mobutu's terminal illness and the likely demise of the Kinshasa government had precipitated a power struggle at top levels of the armed forces which neutralised what little effective military action was possible.

Mobutu had no clear successor and increasingly a palace coup seemed likely. The primary candidates were the military leaders, Baramoto and Nzimbi Ngbale, who commanded the most efficient forces in Zaire – and who were the only figures deemed capable of posing a serious threat to Mobutu.[46] Although Mobutu retained the loyalty of his presidential guard (commanded by Nzimbi) and was still close to both Baramoto and

Nzimbi, the latter were nonetheless tempted to seize power in the event that Mobutu should suddenly die.[47] In September 1996 the Zairean Minister of Defence and Security, General Mavua Madima had 'called an urgent meeting of the general staff to discuss the situation and reassure the Zairean population that the army did not intend taking power'.[48] But French, South African and American intelligence sources continued to report that Baramoto had plans for a coup.[49] Mobutu had been informed of this but he treated Baramoto lightly and simply placed him under house arrest.[50]

In this respect, despite Mobutu's approval of a deal with French-Belgian and Serb mercenary forces, bidding on military contracts continued in Kinshasa and military dealers met with a variety of Zairean political figures who wanted to further their own plans. Both Baramoto and Nzimbi continued to meet with a clique of French and South African clients that included mercenaries, businessmen and former SADF officers.[51]

Foreign mercenary forces in aid of the Kinshasa government assembled in December 1996. They were the last of Mobutu's 'wild geese', and consisted of four main components:[52]

- a small, Western European group of 20–30 (mostly French) soldiers led by the Belgian ex-colonel, Christian Tavernier, who had dealt with paramilitary business in Francophone central Africa for 30 years;
- a larger group of about 80–100 Serb soldiers led by a Serb mercenary known as 'Colonel Dominic Yugo', who had previously worked with French security services in Bosnia;
- an unknown number of Ukrainian pilots who were hired by French recruiters to fly four Russian-built attack helicopters;
- a small team of South African security advisors and mercenary pilots who were recruited through a Pretoria-based company called Stabilco.

Western European Forces under Christian Tavernier

About 30 French-speaking soldiers were recruited through 'old-boy' French military and mercenary networks in Europe.[53] Many of these recruits had formerly served with Bob Denard, the infamous doyen of Franco-mercenarism, whose activities in Africa span almost four decades.[54] After commanding hired troops in the Congo mercenary wars of the 1960s, Denard conducted several operations in the Comoros Islands of the Indian ocean, where he toppled two governments in 1978 and 1989 and was finally arrested by French forces on his third attempt in 1995. Denard was imprisoned in Paris, tried and found guilty of mercenarism. Although France has strict anti-mercenary laws that allow for a five-year stay in prison, Denard was given a suspended sentence and a court order not to leave France. When the Mobutu contract surfaced in late 1996, neither Denard nor his second-in-command, Captain Dominique Malacrino ('Marquez') could therefore be directly involved in mercenary operations. However Denard's old military networks with their right-wing political backers were covertly involved in the Mobutu mercenary deal, according to French experts.[55] *Jane's Intelligence Review* also identified that a senior member of the right-wing National Party with close links to Denard played a crucial role in the recruitment process.[56]

While the ageing Denard kept out of sight, lesser known veterans of his group were deployed as senior commanders.[57] Chief among these was Tavernier, who was appointed the overall commander of Mobutu's mercenaries.[58] Tavernier had formerly served with Denard in the Congo mercenary wars when Denard led 100 white mercenaries and three Katangese battalions in a three-pincer formation to seize the eastern capital of Kisangani (then called Stanleyville). The operation was supposed to bring support to Belgian mercenary Jacques 'Black Jack' Schramme's forces in the capital but the plan went awry. Tavernier led his battalion, the 14th Commando, from their base at Watsa in north-eastern Congo. Denard led the 6th Commando from the south and a Belgian mercenary, Antoine de Clerq, led a third, the 8th commando, from their base at Isiro.[59] The mission collapsed under mysterious circumstances, leaving Denard with a bullet wound to his head that caused a life-long limp.[60]

Tavernier continued to operate in the region after the mercenary wars and secured close ties with both Mobutu and unofficial brokers of French foreign policy under the *éminence grise*, Jacques Foccart. One of the most powerful French figures in Africa, Foccart had approved the use of mercenaries to assist France's foreign interests as early as 1961.[61] Before his death in March 1997, Foccart was closely assisted by his deputy, Fernand Wibaux, who had seen an active diplomatic career as the ambassador to Mali, Chad, Senegal and Lebanon during the 1980s. Wibaux had supported foreign military intervention in Zaire and had maintained regular contact with Tavernier.[62]

A year before the ADFL war started, Tavernier had resurfaced in many of his old haunts in southern Sudan and eastern Zaire where he was involved in military provisioning and retraining projects.[63] At this time, he had also made several public statements about the importance of 'reorganising' the Zairean Armed Forces to curtail indiscipline.[64] He complained in particular about troops in north Kivu, along the Rwandan border, that were attacking Zairean Hutus and pillaging and selling weapons to rival (Tutsi-led) militias. Such a 'reorganisation' programme, if successfully launched, might have produced a more cohesive resistance to the ADFL rebels' advance in Kivu. It is interesting that a year before the Tutsi-led rebellion started, Tavernier had correctly identified the growing threat to the Mobutu regime.[65] His foresight is less surprising when one considers his activities in the region at the time. He was an influential agent for Mobutu who, in addition to other groups, met with a variety of anti-Ugandan and anti-Tutsi rebel forces in eastern Zaire. He also served as an agent of influence to senior Belgian officials and played a role in furthering Francophone interests in central Africa.[66] Tavernier had close ties to the exiled commander of the former Rwandan Armed Forces (FAR), General Augustin Bizimungu, whom he provided support to at the Mugunga refugee camp in eastern Zaire. [67] Bizimungu was a chief enemy of the new Rwandan government and was wanted in Kigali for genocidal war crimes. In 1994 he had fled to Mugunga with his forces, the former FAR and Interahamwe militia, who had carried out the Rwandan pogrom that saw over 500,000 people slaughtered in just three months. These forces – estimated to number about 30,000 – moved freely between Goma and Uvira along the Rwandan border from where they had direct access to arms supplies brought through Goma and Uvira. More than 150 tons of arms were shipped here by air between February and May 1996, according to a leaked UN report.[68]

Tavernier led an initiative to relocate the FAR based in Mugung to the Kamina air base in southern Zaire in 1996.[69] He attempted to draw several foreign diplomats into the scheme, including the former Belgian Premier, Leo Tindemans, and the US ambassador in Kinshasa, Dan Simpson. The move was supposed to defuse tension between Rwanda and Zaire, but it would also have strategically assisted Bizimungu's exiled army. The army would have been removed from the immediate reach of the Tutsi-led forces who were training and rearming for the October campaign and would have been relocated to a military base of strategic importance to the Ugandan and Rwandan plan to establish a buffer security zone in Zaire.[70] At Kamina the FAR troops could then have undergone their own military 'reorganisation'. But Tavernier's scheme did not come to pass, and instead he was forced to watch his predictions come true. His communications at this time also show that he had contact with Zairean-based Ugandan rebels he called 'Amin Dada militias', who had launched attacks on Uganda from Bunia.[71] In addition to these ongoing activities in Zaire, Tavernier also had military support projects in neighbouring Francophone countries such as Gabon, the Republic of Congo, Central African Republic and Cameroon.

In all of his projects, Tavernier was more an agent of influence for Francophone political circles than he was mercenary commander. His chief role was as an unofficial intermediary for Mobutu in Brussels and Paris. He had no significant reputation as a military figure and there is little mention of him in the standard texts on the Congo wars and Denard's operations in Africa.[72] Elderly and without much of a fighting portfolio, Tavernier was therefore an unlikely choice as commander-in-chief of the Mobutu hired forces. According to *Jane's Review*, the Paris recruiters had identified a more suitable Frenchman 'with long military experience and a right-wing background going back to the time of the Organisation Armée Secrète (OAS) in the early 60's', but he had declined their offer as the salaries were too low.[73] Surprisingly, the Mobutu project was troubled by a shortage of funds and salaries were not as high as expected. As a result, large numbers of 'high-calibre recruits' were not attracted to the project.[74] The question of money raised controversy both before the contract was effected and after the mercenaries had returned home. While the French fighters were paid average salaries and the Serbs received very low wages – about a fifth of the French rate, Tavernier and Denard are rumoured to have taken the lion's share of the budget. By contrast, the South African pilot-mercenaries with Stabilco were offered much higher salaries than their French and Serb colleagues – but these were not always promptly paid by their Zairean contractors. A further irony lay in the fact that the military generals and influential businessmen, who negotiated the contracts, had access to vast personal wealth and could have purchased the right military support for Kinshasa, but few were willing to self-finance the operations. The French and Belgian soldiers signed three-month renewable contracts at a monthly salary of FFR 30,000 (about US$ 5,000).[75] The members of the larger Serb force were each paid about US$ 1,000 per month.[76] The total budget was reportedly about US$ 5 million and was insufficient to raise a large mercenary force, as had been done in the Congo in the 1960s.[77]

In addition, the Elysée was unsure as to how it should continue its foreign policy in Zaire and Central Africa. It had recently approached Zaire's enemies, Uganda and Rwanda, in an attempt to initiate a new axis to their African policy and had adopted a cooler attitude to Mobutu, who was clearly the least favourite horse in the race for power in Kinshasa. These shifts in French policy made the operation less attractive to established experts and delayed effective support operations. French mercenaries deployed in Kinshasa at this time said that their operations were continually stalled as a result of a confused policy that arose out of a division between Africa experts in the Elysée.[78]

Tavernier's small force of western mercenaries was deployed in the last days of December 1996. They arrived from Europe by commercial flight in Kinshasa, and were relocated to the eastern base of Watsa via Kisangani by an Antonov Russian cargo-carrier flown by Ukrainian pilots. From Watsa, they split up into smaller groups and conducted various weapons training, intelligence and combat operations with FAZ and Katangan Angolan forces for almost three months. The Belgian-led forces took their commands from Tavernier who moved between Kinshasa, Watsa and other locations in the east. Their mission was to organise and lead the FAZ ground troops against the advancing northern front rebels while the Serb-Ukrainian forces were to provide air cover. In reality, the French commanders were often deserted by their FAZ troops – who fled into the bush – and the Serb air wing often failed to deploy at all. While the Serbs had access to three Mi-24 Hind choppers and three single-seat Aermacchi MB-326 K Light-attack aircraft, they often failed to understand the orders given by Zairean headquarters and did not therefore deploy air cover operations promptly or correctly.[79] A privately-leased Andover, flown by an ex-Vietnam American pilot, proved to be more effective to the beleagured French forces on the ground, who were rescued on more than one occasion by the humble cargo carrier.[80]

Eastern-European Mercenary Forces under Colonel 'Dominic Yugo'

For three decades, the former Zaire had witnessed the activities of French, Belgian and South African mercenaries, but when a group of about 80 to 100 Serb fighters was deployed at Kisangani it marked a new and disturbing trend in the growing private military industry.

The Serb forces constituted an air wing and a company-sized combat infantry unit. They were recruited in the Republic of Yugoslavia under the command of a senior Serb officer known as 'Colonel Dominic Yugo' who has close ties to the French internal security service, Direction de la Surveillance du Territoire (DST).[81] Most of the recruits were Bosnian Serbs from the Drina valley or refugees from the Krajina region of Croatia, and they are suspected of having been involved in Serbian 'ethnic cleansing' programmes. They were assembled by Lieutenant Milorad 'Misa' Palemis, who commanded an 80-man commando unit during the Yugoslav Civil War which has been accused of conducting the brutal massacre of civilians in Srebrenica in 1995. Yugo, who comes from the Kosovo region and was based at Lukavica barracks on the southern outskirts of Sarajevo during its siege in the civil war, was one of Palemis's senior officers.[82] He is accused of having committing atrocities in Srebrenica and elsewhere.[83] He is also accused of having personally executed and tortured dozens of Congolese civilians in Kisangani, whom he suspected of being rebel spies, and in another incident he was seen to shoot and kill two Protestant missionaries holding bibles.[84]

In December 1995, Yugo's close links with senior French officials in the internal security services (DST) were revealed when he helped the former French police commissioner, Jean-Charles Marchiani, successfully negotiate the release of two French airforce pilots who had been captured in Bosnia.[85] Marchiani was then a key aide to the former Interior Minister, Charles Pasqua, whose security operations in territories such in Algeria, Angola and Sudan had caused a deep rivalry to develop between the DST and the external military security service, Direction Générale de la Sécurité Extérieure (DGSE).[86] Signs of this rivalry were again visible in the Zaire intervention of 1996 and 1997 when the DSGE and the DST disagreed on the use of mercenaries to back Mobutu.[87] Marchiani is a powerful figure in the web of French officials who are involved in security and economic deals in both Bosnia and the mineral and oil-rich states of central Africa. He has been described as a 'weaver of politico-economic networks', who has brokered deals on behalf of French interests with both government and rebel leaders in central Africa.[88] He has also been linked to the French petroleum transnational, Elf, the French arms development group, Thompson-Armement and Brenco Trading, which is one of the largest suppliers of consumable goods in Angola.[89]

In addition to his contacts with Marchiani and French security, Yugo had direct contact with a Paris-based satellite and radio communications firm, GeoLink, which operated out of offices in Belgrade as well as Kinshasa.[90] GeoLink played a crucial role in bringing the three nationalities involved in the mercenary deal together. According to French newsletters, Yugo represented GeoLink in the sale of Thomson radar and other equipment in Serbia. GeoLink also had business relationships with both Mobutu and his close ally, Savimbi, as well as with the French secret services. GeoLink marketed specialised equipment for military intelligence requirements for sale in Africa and Europe. They had sold private satellite telephone services to several African heads of state, including Mobutu and Savimbi.[91] Through these commercial military activities, GeoLink also had close contact with the French DST security services.[92] French press reports in 1997 revealed that GeoLink was associated with the Serb mercenary contract and that its marketing executive, Philippe Perette, had been directly involved in the funding, provisioning and recruiting of the controversial Serb group.[93]

In 1996 Perette ran the GeoLink offices in Kinshasa, where the company had holdings of over FFR6 million and had plans to invest FFR25 million in the establishment of satellite telephone infrastructure.[94] In late 1996 a deal was made between Perette and several prominent Zairean businessmen who represented Kengo wa Dondo's government. Key among them were GeoLink's Zairean representative, Jean Mirow, and one of Mobutu's most trusted private advisers, Jean Seti Yale, who had earned a fortune through the trade of coffee and diamonds in Zaire and who owns multi-million dollar businesses in Antwerp, Portugal and South Africa. French and Zairean businessmen with close links to powerful security figures in their respective governments, were therefore responsible for the payment and recruitment of the Serb mercenaries.

In turn, powerful figures in the Serb government and security structures were also involved in the Zaire mercenary deal. The 100 Serb soldiers were hired through a special recruiting office at the Turist hotel in Belgrade.[95] Here they received their passports on the orders of Jovica Stanisic, head of the Department for State Security (SDB) of the Serbian Ministry of Internal Affairs.[96] The passports were then stamped with visas issued by the Zairean embassy in Paris.[97] In Belgrade, the mercenaries loaded up their military equipment and flew to Kinshasa, via Cairo for refuelling. A former logistics chief for the Yugoslav army, Jovan Cekovic, arranged permission with officials in Cairo to allow the planes carrying mercenaries and weapons to refuel.[98] The assembly, deployment and arming of the controversial Serb fighting group shows that a number of powerful business and security figures in both Paris and Bosnia converged to make the deployment of this controversial Serb fighting unit possible. According to a report in SOF there may also have been some joint involvement with South African-based mercenary recruiters who have military operations in Bosnia.[99]

The Serb mercenary force was deployed at Kisangani on 14 January 1997, where their mission was to protect the civilian and military airports from falling into rebel hands and to provide air support with four Mi-24 attack helicopters and three Aermacchi fighter jets. They also instructed members of the elite Zairean units, especially the SARM military intelligence force, on unarmed combat and the use of small arms.[100] They gave instruction on how to use the AK-47 assault rifle, the M53 general purpose machine gun and the Dragunov 7.62 mm sniping rifle that was widely used by Serb forces in the Sarajevo siege. A large cache of weapons had been imported from Yugoslavia and shipped via Egypt and Serbia, for which training was also provided. These included the M53 machine gun, 60 mm mortars and the RPG-7 rocket launcher. But the Serbs had little appetite for the Zairean 'fighting conditions'. They soon fell ill with malaria and dysentery and showed a hostile distrust of their own Zairean forces, who had little soldiering capabilities. After taking three casualties on their first reconnaissance of the frontline at Walikale in a shoot-out with rebel forces, the Serbs were unwilling to conduct combat or rescue operations.[101] Their operations in Kisangani did not impress either the more experienced French mercenaries or the South African mercenary pilots who were waiting in Kinshasa to secure an air wing contract.

South African Mercenary Forces under Former EO Commanders

Three South African mercenaries working for EO in the new (and short-lived) democracy of Sierra Leone saw the opportunity to initiate their own military business in Zaire when war broke out in October 1996. The three men were all contemporaries who had all fought in the Angolan bush war with the former SADF. They included a fighter chopper pilot of some repute, Major Neil Ellis, who had been decorated several times and who had been awarded an Honorus Crux – one of South Africa's highest awards for courage during combat. After the ANC came into power in 1994, an SADF general advised Ellis to resign. His controversial career as a combat pilot with the SADF's special forces would not assist him in the future, he was advised. Ellis resigned and first tried a number of failed farming and fishing ventures before turning to mercenarism in 1995. His first job was in Bosnia, where he was given a Soviet Mi-17 chopper to fly without any prior training. After taking a direct hit from ground forces on this first flight to deliver supplies, he was forced to limp back to base. Since his cargo had not been delivered, Ellis was not paid. He had to find his own way back to South Africa, where he joined EO and took up work in Sierra Leone, flying small fixed-wing aircraft and a converted Mi-17 chopper.

Here he met two of EO's ground commanders who were based in the Sierra Leone interior, Colonel Roelf van Heerden, who headed EO's operations in the diamondiferous region of Kono, near the Liberian border, and a man known only as 'P.P.', whose real name is Renier Hugo, who headed up a military base outside Freetown. Both van Heerden and Hugo commanded EO's successful ground campaign to recapture the diamond-producing centre of Kono in 1995. It was a jungle combat victory that they and their South African-based recruiters believed would stand the Zairean army in good stead. In late 1996 they entered into negotiations with a man called Mauritz Le Roux, who had close contact with the powerful general, Kpama Baramoto. A former SADF sapper, Le Roux had been involved in covert military resupply operations to UNITA forces in Jamba during the late eighties. He had later resigned from the SADF and served a brief stint with EO in their first operation at Soyo, in Angola in 1993. Thereafter Le Roux ran a small private security business, called SafeNet, in Pretoria and one of his main clients was Baramoto – a regular visitor to South Africa and the owner of several properties in Johannesburg.

Through his dealings with Baramoto, Le Roux hoped to secure a greater contract with the Mobutu government. Under the guidance of several military-related experts and more experienced senior militarists in Pretoria, Mauritz came up with a plan. He put together a team of mercenary commanders who were expected to clinch the deal with Baramoto, and set up a new mercenary firm under the name of Stability Control Agencies (Stabilco), which was registered in the Isle of Man. Stabilco had six South African directors, all of whom had experience with former SADF covert military and intelligence operations. The directors included Le Roux, Neil Ellis (Stabilco's air wing commander), Harold Miller (a military legal expert who had formerly advised the apartheid secret services), van Heerden (Stabilco's infantry commander who had mechanised infantry experience), a former SADF special forces operator, Johannes Lintveldt (Stabilco's base commander) and Hugo (Stabilco's second infantry commander).

In Kinshasa, Stabilco's military negotiations with FAZ leaders were led by van Heerden, Hugo and Ellis. The firm's financial negotiations were handled by Le Roux and Pretoria-based advisers. The group was assisted by the fact that Hugo had helped train Mobutu's Civil Guard in the early 1990s when he still served the former SADF's 5 Reconnaissance special forces regiment. Hugo was therefore known by General Baramoto. In early December 1996, as ADFL rebels advanced toward the gold fields of Bunia in the north-east, Stabilco concluded a deal with Baramoto, who been appointed FAZ Chief of Staff. The deal was facilitated by the former Zairean chargé d'affaires in Pretoria, Isa Ghanda.[102] But the contract was aborted when Mobutu returned to Zaire from France and appointed General Marc Mahele Liego Bokungu as the new chief of staff of the Zairean army. Baramoto and Stabilco's plans were nipped in the bud, and the South Africans were forced to open new negotiations with Mahele and the Defence Minister, Likulia Bolongo.

Baramoto's ousting was precipitated by the fact that the Mobutu government strongly suspected he would stage a coup d'état with the help of General Nzimbi and the South African mercenaries. Baramoto remained in contact with Le Roux throughout the next five months. For much of this time, the Stabilco team stayed in Kinshasa at the Intercontinental Hotel, at the state's expense. In January 1997, Stabilco deployed a small group of pilots who conducted air surveillance missions over Kisangani and the north-eastern provinces. Baramoto is said to have part-funded the Stabilco project during this time and been paid a fee of between US$140,000 and US$300,000.[103]

Although Stabilco provided the Zairean military leadership with repeated briefs, counter-attack proposals and some assessments of the ground situation, they were kept waiting on a final decision for most of their stay in Kinshasa. To further their chances, the group also negotiated with Zairean and Kinshasa-based businessmen as well as with French mercenaries in Kinshasa. In March, Stabilco's protracted negotiations with FAZ leaders, Mahele, Likulia and the latter's defence aide, looked more promising after the Serb-led group of mercenaries hired by Prime Minister Kengo wa Dondo failed to defend Kisangani and withdrew from the region. In Pretoria, Stabilco placed an advance force of 35 men on standby for deployment to Kisangani and recruited over 450 other troops who were to travel to Kinshasa in a chartered Russian Antonov 12. The salaries that Stabilco offered at this time were unusually high – mercenary pilots were offered almost double the fee that EO had offered them in 1994–97. Many were paid in advance, to maintain combat-ready forces on standby. In some cases, the standby salaries were paid for three months. This was an unusual development in the mercenary business and cost both Maritz and Stabilco dearly. It indicates, firstly, that the Mobutuist's situation in Kinshasa was critical and any counter-offensive Stabilco was likely to raise looked increasingly less likely to succeed. In addition, mercenaries approached by Stabilco expected to be paid handsomely for two reasons: they would back an internationally unpopular figure whose days were clearly numbered, and their Mobutuist financiers were known to be extremely rich. Yet once again, the Stabilco deal failed to be finally secured.

Desperate for effective intervention, Mobutu's representatives reopened the door to bidding for new military assistance contracts. As described earlier in this chapter, they made contact with EO's offices in Pretoria. The possibility that a formal link could have existed between Stabilco and EO remains questionable. Not only were former EO personnel key to Stabilco's operations, but all three former EO employees (Ellis, van Heerden and Hugo) returned to EO-related contracts thereafter in 1998 and 1999. In addition, the Stabilco team, by their own admission, maintained contact with EO executives in Pretoria. The company claimed that this was done in order to maintain friendly relations with their former employees. The Stabilco contract could well have disrupted the EO group's own business in Uganda and Angola, since both companies would have recruited from the same pool of South African fighters. EO executives Eeben Barlow and Lafras Luitingh repeatedly denied that EO had anything to do with Stabilco. Likewise, Stabilco repeatedly denied that they had any formal relationship with EO. However, EO clearly was interested in the Zaire contract in March. Yet – for a number of reasons – EO would not have been able to secure an open contract with Kinshasa. Perhaps key among them was the risk that Kinshasa would loose UNITA's support. UNITA was ill-disposed towards EO after their offensive military support of the MPLA government in the 1993–96 Angolan war.

In May 1997, a few days before the ADFL rebels seized Kinshasa, Stabilco secured a last-minute extraction contract with the Zairean military. It recruited a crisis air wing composed of South African helicopter and fighter jet pilots to extract valuable military assets from Gbadolite, as well as a small task force to escort Mobutuists to safety. The asset extraction plan went awry when an advance crew of pilots in Gbadolite was attacked by looting Zairean soldiers before they could extract a serviceable Mi-24 gunship. Five MiG-21 and three Jastreb Galeb (ex-Yugoslavian) fighter jets that were intended to be extracted had to abandoned. Stabilco director, Neil Ellis, and Joup Joubert, a top former SADF attack helicopter pilot who had worked for EO since 1994, as well as a French colleague were forced to abandon their aircraft and flee on foot from Gbadolite, across the Zairean border into Central African Republic. They were picked up by a French Puma helicopter carrying French special forces and were treated for shrapnel wounds at a military base in Bangui. Mobutu and his family fled from Gbadolite to Chad, while Baramoto and Nzimbi fled to South Africa with Le Roux. Likulia managed to find refuge at the French Embassy before being evacuated to France.

The Failure of the Zairean Mercenary Counter-offensives at Kisangani

At the time that the pro-Mobutu mercenary forces were fully deployed in early January 1997, the ADFL rebels had seized Bunia and had launched the second phase of their campaign to capture Kisangani. The capture of Bunia started a general pillaging retreat westward by FAZ troops and Baramoto's civil guard, ahead of the advancing rebels. From Bunia, the rebel troops divided into two groups: one opened up a northern front that threatened Watsa and Isiro and the other opened up a central front toward Kisangani via Bafwasende and Mambasa.[104] In the south-east, rebels had also captured Walikali and opened up a second axis on the central front that started a ripple of panic in Kisangani more than 300 km away. Further south, from Uvira and the shores of Lake Tanganyika, a third front had emerged facing Fizi and Kalemie.[105] But this southern front moved more slowly as a result of resistance from the Babembe, who had fought against the Tutsi in the Congo war of 1967. Rebel alliance forces had therefore fanned out from the Rwandan border region and were advancing from at least five main directions into eastern and southern Zaire. The Rwandan-led (proxy) war had turned into a broader-based, internationally-backed civil war to seize Zaire.

In Kinshasa, there was an urgent call for a counter-offensive which was launched from Kisangani by two FAZ battalions (supported by both former FAR and UNITA troops) on 20 January 1997. They moved out on the Bafwasende and Walikali routes to meet the advancing northern and central rebel fronts. In the south, a third battalion attempted to repel the advance toward Kalemie, but was soon stopped. Five days later, the counter-offensive was aborted and the FAZ troops fled back toward Kisangani. Mobutu was left with mainly his mercenary-led forces and former FAR troops to constitute any further threat to his enemies. The strategic objective of the mercenary-led Zairean forces was to defend the eastern capital, Kisangani, and its outerlying bases in order to prevent the ADFL rebels from seizing full control of eastern Zaire. Kisangani was the focal point because its capture would lead the way to Kinshasa (via the Congo river) and Mobutu's stronghold in the north.

In the event of Kinshasa's collapse, Mobutu, his Ngbandi generals and their army units were expected to retreat north into Equator, their native province from where they would wage an armed struggle against the successor regime.[106] Kisangani would give the rebels a crucial base (equipped with both military and civilian airports) from which they could launch attacks into Equator to drive the Mobutuists out of Zaire. As the third largest city in the former Zaire, Kisangani is also a crucial port and economic capital and its seizure by the rebels would cripple the morale and defensive capabilities of the pro-Mobutu forces. For these reasons, Kisangani was the main target of the ADFL campaign and the strategic key to controlling the rest of Zaire.

But the foreign mercenary forces failed to defend Kisangani and they were hastily extracted at the last minute, shortly before the AFDL troops seized Kinshasa on the 15 March 1997. The outraged Kinshasa government accused the mercenaries of betrayal and promptly arrested Christian Tavernier on charges that he had stolen millions of French francs from the original funds provided to hire mercenaries and had failed to honour his military contract with the Zairean government to defend Kisangani and other key eastern Zairean towns.[107] Tavernier was later released, but his Zairean military colleague, General Mahele – also accused of betrayal – was later shot and killed, allegedly by one of Mobutu's sons. In the ensuing months before Kinshasa's fall, most of the Mobutuist regime had fled across the river to Brazzaville and on to other African capitals, Europe, or to their luxury homes in South Africa. By the time they had all scuttled out of Zaire, the Ugandan and Rwandan-backed ADFL had seized Kinshasa with very little opposition and their leader, Laurent Kabila, had announced the coming of a new era in the history of central Africa.

Conclusion

The 1996 wind of change that was driven by Museveni's regional economic vision, the victorious Rwandan-backed ADFL military campaign and Kabila's rise to power, would not bring an era of stability and prosperity to the new Democratic Republic of Congo. Instead, the giant country would slip into further decline and political unrest, dragging with it neighbouring states such as Angola, Congo (Brazzaville), the Central African Republic, southern Sudan and Zambia. In the years to come, the DRC would continue to be plagued by anarchy and destabilised by a range of external and internal armed forces, each aided by foreign mercenaries and arms suppliers. Key among these forces would be:

- hostile FAZ troops in the Congo who were still loyal to Mobutuist leaders
- exiled former FAZ and special military forces under the command of former Zairean Generals, Baramoto and Nzimbi, who retreated

to South Africa to plan a counter-campaign with former SADF private military specialists

- Hutu Interahamwe militia who had retreated deep into the interior ahead of the ADFL advance in 1997
- remaining UNITA troops who had defended Kinshasa against the ADFL advance in the west and north-west of the country
- remaining Rwandan and Ugandan armed forces who had formerly constituted a significant part of the ADFL army
- warring military forces in nearby Brazzaville, where President Pascale Lissouba would soon be ousted by his nemesis, Denis Sassou Nguesso, with the assistance of Angolan troops
- Rwandan-backed rebel groups in the eastern provinces
- new deployments of Angolan armoured and infantry forces in western Congo.

Propped up by his backers, Kabila had successfully assumed control of a country that was potentially the richest state in sub-Saharan Africa but, awash with weapons and militia, it was teetering on the verge of dissolution. As such, the 1997 ADFL victory was only a brief respite in what would become a great battle for control of the capital of Kinshasa that would see over a dozen African countries drawn into the fray by late 1998.

In the period after his Inauguration, Kabila found himself pressured by several opposing political and military forces that threatened to usurp his control. He had been brought to power by the military might of foreigners, by the western media who had hailed him and the ADFL revolution as the key to peace in central Africa, and by Zairean popular sentiment, which was harshly anti-Mobutu and which called for a new government. But Congolese people and especially the citizens of Kinshasa, did not easily accept the foreign influences of Rwandans and Tutsis in Kabila's new Congo government and its armed forces. These influences were strongly felt in towns where Socialist-based 'reeducation programmes' for Congolese people were established by ADFL officers to drive the new era of Tutsi-led political influence. Kabila was therefore pressured to expel Tutsi influences from his government. On the other hand, he was pressured by the Kigali government to honour his security pact with them and assist both Rwanda and Uganda to eliminate their enemies in the Congo and solve border insecurity problems in the Kivu provinces by both military and constitutional means. Kabila was also facing increasing international criticism of his dictatorial style of governance, especially his refusal to allow human rights-based researchers to properly investigate allegations of human rights abuses and massacres committed during the war.

Faced with these pressures, Kabila's choice of action was simple. He chose to follow in his unpopular predecessor's shoes and seized the reins of power more firmly in his own, personal grasp. He ordered his Tutsi allies out of office and the Rwandan military forces out of the Congo. He reneged on his former security pact with the Rwandan government and instead fanned the flames of ethnic hatred by inciting violence against Tutsis and by collaborating with their enemy – the exiled Hutu militia in the Congo. Kabila also reneged on deals with several large mineral corporations who had helped finance the ADFL war effort and summarily cancelled all signed contracts with these parties in 1998. This precipitated the start of renewed hostilities between Rwanda and the Congo and a new war broke out. It also led to an international condemnation of Kabila's style of governance and made him an international pariah.

The formula for the second rebellion was the same as that used by the ADFL in 1996. The new rebellion sprang up in Kivu, it was led by Congolese leaders with close links to Kampala and Kigali and the rebel army was once again militarily backed by Ugandan and Rwandan forces. The new rebellion was also represented by the same key figures who had formerly played a significant role in the ADFL rebellion and in Kabila's early government before they were ousted. But the war would not be as swiftly won this time, and Kabila would not be toppled as easily as Mobutu had been. Kabila was able to win effective military support from Angola, Namibia and Zimbabwe, as well as logistical support from the oil-producing states of Chad and Libya, which prevented the second rebellion from a quick military victory in the Congo in late 1998. In response to this development, Congolese rebels and both Uganda and Rwanda entered into new security deals with their former enemies – the ousted Mobutuist leaders and the powerful rebel army, UNITA. This strategic exchange of military assistance between UNITA and Kigali-backed Congolese rebel forces was expected to secure two military advantages: Rwandan military expertise and support given to UNITA would pose a serious threat to Luanda, and the Angolan army was expected to withdraw its forces from the Congo and deploy them against UNITA, thus leaving Kabila more vulnerable to attack. In turn, UNITA would lend assistance to the Congo rebel advances on Kinshasa and UNITA would gain strategic military and economic advantages once former Mobutuists had returned to Kinshasa in the wake of Kabila's ousting.

But these new strategic regional and military alliances are precarious and have introduced several key trends to contemporary conflict in central Africa and to modern politics in Africa. First, the alliance system is fairly chaotic, with little trust between players, and often the same backers will provide support to different sides of the conflict. There is also a tendency for the alliances to shift and restructure themselves depending on which state-to-state and group-to-group players are involved at any one time. Underpinning these strategic alliances are national struggles for regional dominance, with South Africa, Angola, Libya and Uganda-

Rwanda as the key competitors. This has given rise to a new African scramble for political influence and economic advantage. Secondly, the axes of power have shifted from national realities to regional ones and, as a result, civil conflicts in Angola, DRC, Congo-Brazzaville, Uganda, Rwanda and Sudan have tended to spill over state borders and merge together since 1996. The belt of civil and regional conflict is therefore spread across a great swathe of territory from southern Angola to the Red Sea. Thirdly, there is great dissension and confusion in the Congolese rebel ranks with over a dozen separate factions each claiming the right to lead the rebellion. The ousted former Mobutuists are still a part of this scenario and are expecting to be part of a future solution to the Congo's intractable problems.

A largely hidden aspect of the new disorder in central Africa and the merging of civil conflicts is the role played by arms and mercenary organisations. Since EO closed its visible doors in 1998, silence has fallen over the mercenary business in Africa. Although mercenaries, private military advisors and private security forces are doing a brisk trade throughout Africa, there is today very little publicity on the industry since the boom years of 1996–97. However the military complex is still experiencing a growth in business although less is known about it. Researching the private military industry has become harder for several reasons. First, large, visible mercenary corporations like EO are no longer part of the industry. Since 1998, mercenary executives have adopted a low profile and prefer to keep both their clients and their operations hidden. To this effect, the giant EO-Branch Energy group has unbundled and contracted out its corporate organisation. Military and security contracts are also smaller, more specific and usually shorter in duration. This has been useful since it breaks down the cost of services and allows separate companies within a group to handle different contracts – often with opposing sides of a conflict. It also allows the firm to conceal its clients and maintain business relationships in the constantly shifting system of alliances.

Many military personnel deployed by the military firms are either South African or African. The increasing use of African mercenaries is useful, since non-white mercenary forces are cheaper and harder to detect. Many of the companies who provide military services to the armies involved in civil and regional conflicts in Africa are linked to one another and to the former EO group. As such, South African, European and African mercenaries with links to the former EO group are presently in the service of both rebel and state armies in Angola, the DRC, Congo-Brazzaville and Sudan. Their presence in these conflicts is unreported and remains to be analysed in the broader context of conflict resolution and destabilisation in Africa since 1997.

In this sense, the new era heralded by Kabila, Kagame and Museveni in 1997 would deliver war rather than stability and development. And it would be an era of prosperity only for traders of war. After the 1996–97 war, the resource-rich Congo would become more impoverished and war-torn than ever before. Although it is one of the most fertile countries in the world, hunger and malnutrition are growing on a daily basis. Congolese people in both cities and rural areas can no longer afford to eat regular meals, according to journalists who frequently visit the region. Few can remember eating red meat in the past two years. Many can only afford to provide their families with a single meal once every two days.[108] At the time of writing, regional and civil conflict continues to tear central Africa apart and there is no sign of an African Renaissance. The only beneficiaries of the new era of chaos and widespread destabilisation are the arms dealers, illicit traders and the mercenary and private security forces that continue to find abundant employ in the central African maelstrom.

Notes Chapter 4

1. The Congo was explored by Henry Morton Stanley, the American journalist who took up David Livingstone's cause in Africa to try and open a 'path for commerce and Christianity'. Stanley's expeditions to the Congo in 1879 and the early 1880s were backed by the Belgian king, Leopold II, who wished to make the Congo a fully-fledged Belgian state that would reap a vast fortune out of its trading monopoly. To this effect, Stanley discovered Leopoldville, the capital of the Congo, today known as Kinshasa. His great rival, the French sailor Pierre Savorgnan de Brazza, had raced against Stanley and the Belgian King to secure the region for France. He established the capital of Brazzaville on the Congo river, just a short distance from the site that Stanley later chose for Leopoldville. The rivalry between the two explorers captured the French imagination and spurred the French government on to greater military involvement in western Sudan. Historians would later discern the race for the Congo as the start of the scramble for Africa. See T. Pakenham, *The Scramble for Africa* (Jonathan Ball, 1991), p. 162.
2. Ibid., p. xxiii, preface.
3. Ibid., p. 678.
4. S. J. G. Clarke, *The Congo Mercenary: A history and analyis* (The South African Institute of International Affairs, 1968), p. 19.
5. Ibid., p. 18.
6. Ibid., p. 21.
7. Pierre Savorgnan de Brazza's diary (see note 1 above), September 1881, in Pakenham, *The Scramble for Africa*, p. 143.
8. Contemporised from the passage written by Pakenham, *The Scramble for Africa*, p. 678. (With apologies.)
9. P. Gourevitch, 'Letter from the Congo', *The New Yorker*, New York, 4 August 1997, p. 48.
10. P. Gourevitch, 'Letter from the Congo'.
11. R. Lemarchand, 'Patterns Of State Collapse And Reconstruction In Central Africa: Reflections On The Crisis In The Great Lakes'. Paper presented at

the XVIIth World Congress of the International Political Science Association, August 1997, Seoul. Published in *Journal of African Studies*, Vol.1, p. 13. <http://www.dir.ucar.edu/esig/ijas>

12. Notes taken at US hearing, 6 December 1996; 'Democratic Republic of the Congo: What Kabila is hiding', *Human Rights Watch Report*: October 1997, Vol. 9, no. 5.

13. W. Thom, 'Congo-Zaire's 1996–1997 Civil War in the context of Evolving Patterns of Military Conflict in the Era of Independence'. Paper delivered at the 40th Annual meeting of the African Studies Association, Ohio, 16 November 1997, p. 20.

14. W. Thom, 'Congo-Zaire's 1996–1997 Civil War', p. 19.

15. R. Lemarchand, 'Patterns of State Collapse And Reconstruction In Central Africa: Reflections On The Crisis In The Great Lakes'. Paper presented at the XVIIth World Congress of the International Political Science Association, August 1997, Seoul. Published in *Journal of African Studies*, Vol.1, p. 13. <http://www.dir.ucar.edu/esig/ijas> Accessed January 1999.

16. In a report it published in 1997, Médecins Sans Frontières accused the ADFL of pursuing a 'deliberate strategy of elimination of all Rwandan refugees, including women and children.'

17. R. Lemarchand, 'Patterns of State Collapse'.

18. Interview quoted in W. Thom. 'Congo-Zaire's 1996–1997 Civil War', p. 45

19. R. Copsen, *Zaire, Congressional Research Service Issue Brief*, Library of Congress, 5 December 1996, Washington, p. 10.

20. Comments quoted were made by a French general and were published in *Le Monde*, Paris, 7 January 1997, reprinted in S. Laufer, 'Report links SA mercenary group to Zaire conflict', *Businessday*, 8 January 1997.

21. Author's interview with former Prime Minister, Kengo wa Dondo, Johannesburg, March 1999.

22. Israel had shifted away from supporting Mobutu since the 1994 genocide in Rwanda and had increasingly lent assistance to Rwanda and Uganda. However, Israeli officers admitted that 'private security ventures' could still be helping Mobutu. *Africa Confidential*, Vol. 37. No. 24, 29 November 1996, p. 4.

23. Interviews with the company in Johannesburg, December 1996.

24. This analysis is borrowed from W. Reno, 'Sovereignty and Personal Rule in Zaire', *African Studies Quarterly*, University of Florida, Gainesville (no date shown), <http://web.africa.ufl.edu/asq/index.htm>

25. Interview with French mercenary in Paris, June 1997.

26. W. Reno, 'Sovereignty and personal rule', and interviews with South African mercenaries deployed in Zaire in 1996.

27. W. Reno, 'Sovereignty and personal rule', pp. 4–6.

28. Interviews with EO-affiliated personnel who either facilitated or knew of these meetings. Interviews were conducted in early 1997 and in 1998 with several informants in Johannesburg and Pretoria.

29. K. Pech, 'Former SADF soldiers earning small fortunes as mercenaries in white legion in Zaire.' *Independent on Sunday*, 12 January 1997.

30. M. Dejevsky, 'Former French officers recruit mercenaries for Zaire', the *Independent*, 8 January 1997. She quotes an article from *Le Monde*, 7 January 1997.

31. M. Dejevsky, ibid; W. Wallis, 'Mercenaries train soldiers in Zaire, sources say'. Reuters, 8 January 1997.
32. Le Caro headed the Presidential Guard until 1994 and is an adviser to west African heads of state in Cote d'Ivoire and Burkina Faso. M. Dejevsky, 'Former French officers'.
33. Interviews with security consultant and former EO employee, Johannesburg, 1997.
34. 'Congo-Kinshasa: Paid fighters – and their paymasters', *Africa Confidential*, London, Vol. 38, No. 13, 20 June 1997, p. 2.
35. Letter written by W. H. Dixon on 15 March 1997 and addressed to the 'diplomatic counsellor of Zaire' in Cap Martin, France. It was found in Mobutu's looted residence in Camp Tsha-Tshi in Zaire and was published by *Africa Confidential*, London, Vol. 38, No. 13, 20 June 1997. Dixon sent a copy of the letter to Mobutu's son, Nzanga.
36. A. Kotze, 'SA Soldate teen hul eie mense in Zaire' (South African Soldiers against their own men) *Naweek-Beeld*, Pretoria, 22 March 1997.
37. Interview with senior EO-affiliated informant in Johannesburg, 1997.
38. R. Copsen, *Zaire*, Congressional Research Service Issue Brief, Library of Congress, 5 December 1996, Washington, pp. 10, 11.
39. A close aide to Baramoto estimated that his hidden wealth made him one of the richest Ngbandi figures after Mobutu. The latter was estimated to have pocketed a personal wealth of US$4 billion. Interview with aide, Johannesburg, July 1997.
40. W. Reno, 'Sovereignty and personal rule', p. 4.
41. Ibid.
42. Interviews with mercenaries who were in Kinshasa at the time, conducted in Johannesburg and Pretoria, 1996 and 1997. South African private intelligence informants have suggested that Baramoto and Nzimbi deliberately stalled the process so that the deteriorating situation in the east would provide them with a plausible incentive to stage a coup d'état.
43. 'Central Africa: the Influx of Arms and the Continuation of Crisis', National Security News services, May 1998. This report says that MPRI offered assistance to Kinshasa. According to sources in Kinshasa at the time, an American firm connected to MPRI tendered but it used a different name.
44. Interview with informant who tendered in Kinshasa, Johannesburg, June 1997.
45. Tavernier is a Belgian citizen who was born in the former Congo colony, in the town of Lodja in Eastern Kasai, according to a French journalist, Francois Misser, who has compiled extensive research on Tavernier. Misser has pointed out that Tavernier's military title was not given to him by the Belgian armed forces and that he failed his entrance exam to the Belgian military school, the Ecole Royale Militaire. His military title was given by Mobutu in 1964 when he was hired to command a force of soldiers in the 'Armee Nationale Congolaise'. F. Misser, Email letter, 18 June 1999.
46. R. Copson, *Zaire*, p. 4.
47. Interviews with mercenaries deployed in Kinshasa in early 1997. Interviewed in Pretoria, June 1997.
48. *Zaire Situation Report: 25 February 1997*, p. 10. Report produced by Gauteng-based commercial intelligence company.

49. *Zaire Situation Report*, 25 February 1997. Also Copsen, *Zaire*, and interviews with informants in Paris and Gauteng in May and June 1997.
50. *Zaire Situation Report*, p. 11.
51. Interview with mercenary close to Baramoto who was in Kinshasa at the time. Interview, Johannesburg, June 1997.
52. This assessment of foreign hired forces that were deployed on behalf of the Zairean government in 1996 and 1997 does not include the estimated 2,000 armoured UNITA troops that were deployed in the north-west of Zaire in order to slow down the ADFL in the last months of the war.
53. This group consisted of about 16 Frenchmen, several Belgians, an Italian, a Chilean and a Portuguese, according to Thierry Charlier, who spent three months in the Congo and was the only foreign journalist to travel with Mobutu's mercenaries. The initial size of the group was 24 and was later boosted to about 33 in total. T. Charlier, 'French Merc's "Rumble in the Jungle": Francophones' Last Stand in Zaire', *Soldier of Fortune Magazine Inc.*, September 1997, pp. 56–7.
54. In the late 1950s, Denard had been an NCO in the French Marines and later a gendarme in Algeria where he met Colonel Trinquier and Roger Faulques, who later headed the first French mercenary initiative in the Congo in 1961. A. Mockler *Mercenaries*, Macdonald and Company Ltd, (1970), pp. 161, 166.
55. T. Charlier, 'French Merc's', p. 56; F. Misser and P. Chapeleau, *Mercenaires S.A.*, *Desclee de Brouwer*, 1998. Interviews with French experts, June 1997 and January 1999.
56. S. Boyne, 'The White Legion: Mercenaries in Zaire', *Jane's Intelligence Review*, London, June 1997, p. 279.
57. T. Charlier, 'French Merc's', See his article for references to these ground commanders.
58. Ibid.
59. 'Zaire: The Tired Mercenaries', *Lettre du Continent*, Indigo Press, No. 275, 6 February 1997, p. 6 ; A. Mockler, *Mercenaries*, chapters 7–11.
60. Denard's forces were suspected of acting more in Mobutu's interests than Jacques Schramme's and Denard was thought to have been shot by a fellow mercenary. The mystery has never been fully explained.
61. 'There can be little doubt that most of the French mercenaries, from Trinquier onwards, took orders from Foccart – or if not their orders, at least their orientation. Minor incidents and private conversations have confirmed this; but the surest evidence is France's obvious interest in the Congo, by far the largest and richest of Africa's francophone states and France's open support for Biafra.' A. Mockler, *Mercenaries*, p. 209.
62. 'Zaire: Outside Agents', *Africa Confidential*, Vol. 38, No. 3, 31 January 1997, pointers.
63. In August 1995 Tavernier offered to purchase camouflage uniforms on behalf of the Zairean army. Faxed letter written to Jean-Pierre Kimbulu wa Moyanso, the Zairean ambassador to Brussels, 3 August 1995, Brussels.
64. 'Zaire: A Barbouze Returns', *Africa Confidential*, Vol. 37, No. 4, 16 February 1996, pointers.
65. In a letter addressed to Belgian politicians in 1997, Tavernier refers to an earlier communication when he had correctly predicted the cross-border invasion into Kivu.

66. Copies of several letters from Tavernier to Belgian and French officials between 1995 and 1997, recording events in eastern Zaire.
67. Copy of Tavernier's hand-written note recording a meeting on 5 January 1996 in Mugenga with Bizimungu, an ex-FAR officer, and a foreign envoy, Jean Divoy. In his minutes of the meeting, Tavernier had agreed to arrange a future meeting with Mobutu, Simpson, Tindemans and the Rwandan delegation of the ex-FAR' in order to 'find an acceptable solution to the refugee problem'.
68. W. Thom, 'Congo-Zaire's 1996–1997 Civil War', p. 38.
69. *Jane's Intelligence Review*, June 1997, p. 280.
70. In a letter Tavernier wrote in 1995 to Belgian officials he identified Kamina as the southernmost strategic point in the Tutsi plan to extend their influence into Zaire.
71. See Tavernier's 1995 letter.
72. Interview with Francois Misser, co-editor with Chapeleau, *Mercenaires S.A.* Desclee de Brouwer, 1998
73. S. Boyne, 'The Recruiting of the White Legion', *Jane's Intelligence Review*, June 1997, p. 279.
74. Ibid.
75. Charlier, 'French Merc's', p. 56. Although this is not a bad fee, as has been pointed out by French writers, it was lower than many French mercenaries expected to receive for the Zaire contract.
76. Mercenary informant who was deployed in Zaire in 1996 and 1997. Corroborated by Charlier's report, 'French Merc's', in *Soldier of Fortune*.
77. Interview with mercenary (identity witheld) in Paris, June 1997.
78. Ibid.
79. Charlier, 'French Merc's', p. 59.
80. Ibid.
81. The officer's rank is not certain and it has been assumed by a number of writers that Dominic Yugo is a nom de guerre, an apparent reference to former Yugoslavia.
82. *Jane's Intelligence Review*, June 1997, p. 279.
83. Ibid., *Human Rights Watch Report*, October 1997, Vol. 9, No. 5.
84. According to testimony from local Congolese aid workers, countless journalists, and international humanitarian workers, Yugo committed numerous abuses and violations of international humanitarian law. He also allegedly 'claimed responsibility for air attacks on Walikale and Bukavu, incidents which resulted in numerous civilian deaths and casualties.' 'Democratic Republic of the Congo: What Kabila is hiding', *Human Rights Watch Report*, October 1997, Vol. 9, No. 5 (a).
85. 'La France a-t-elle procuré des Serbes a Mobutu', *Marianne*, Paris, 12 May 1997, p. 20.
86. Pasqua was revealed to have been involved in the donation of US$300 million to Unita in 1997 and had to account for his African policy to Chirac's government in Paris in early 1997 (*Lettre du Continent*, No. 273, Paris, 27 March 1997, p. 1). He also developed contacts with the Sudanese government and its internal security services – contrary to official foreign policy. 'Pasqua's Shadow', *Africa Confidential*, Vol. 36, No. 5, 3 March 1995, p. 7.
87. According to several writers, the DST played a role in the recruitment and provisioning of the mercenaries while the DGSE opposed the plan.

Interviews with French mercenaries in Paris, June 1997 and with French experts in Brussels and Paris, January 1999. Also several articles on Zaire in *Lettre du Continent* produced between 1996 and 1997.

88. 'Pasqua's Shadow', *Africa Confidential*, p. 7.
89. 'Pasqua's Shadow', *Angola: Brenco Trading, Lettre du Continent*, No. 279, 9 January 1997 (Indigo Press).
90. Interviews with French expert and the editor of Indigo Press, Antoine Glaser, January 1999. Also 'Zaire: L'Affaire Geolink', *Lettre du Continent*, 3 April 1997, No. 279 (Indigo Press), p. 6.
91. 'Comoros islands – Geolink gunning for new contracts', *Indian Ocean Newsletter*, No. 729 (Indigo Press), p. 7.
92. Personal comment made by Antoine Glaser of Indigo Press.
93. *Marianne* 'La France a-telle'.
94. *Lettre du Continent*, No. 279.
95. Report in *Sunday Times*, 9 March 1997 and in *Jane's Intelligence Review*, June 1997.
96. 'France-Zaire-Congo 1960–1997, Echec aux Mercenaires', *Dossiers Noirs de la Politique Africaine de la France No. 9*, L'Harmattan, 1997.
97. Ibid.
98. Ibid.
99. Al Venter, 'Today's Dogs Of War', *Soldier of Fortune*, August 1998, p. 56.
100. Service d'Action et de Renseignment Militaire. *Jane's Intelligence Review*, June 1997, p. 280.
101. Charlier, 'French Merc's'.
102. The author conducted several interviews with mercenaries close to Stabilco between the period of December 1996 and June 1997.
103. Interviews with mercenaries.
104. W. Thom, 'Congo-Zaire's 1996–1997 Civil War', p. 27.
105. Ibid.
106. Copsen, *Zaire*, p. 11.
107. E. Allio, 'Zaire locks up mercenary chie', *New Vision*, 24 March 1997.
108. Interview with Belgian-based journalist Marie-France Cros in Brussels, May 1999.

5

Mining for Serious Trouble: Jean-Raymond Boulle and his Corporate Empire Project

Johan Peleman

A news release by the America Mineral Fields corporation, dated 20 May, 1997 read: 'America Mineral Fields Inc. wishes to recognise the triumphant liberation of the people of Zaire by President Laurent Kabila and the Democratic Liberation Forces of the Congo.'[1] That very evening rebel leader Laurent Kabila was to enter Kinshasa as the new President of Zaire and instantly rename the vast central African state the Democratic Republic of Congo. The mining corporation must have been the first entity to 'recognise' Kabila as the new leader of the Democratic Republic of Congo. One week before the US government officially recognised Kabila as Zaire's new leader, America Mineral Fields had flown a group of investors to meet him in Goma, in the area controlled by the anti-Mobutu forces – Alliance des Forces Démocratiques pour la Liberation (AFDL). Aboard the corporate jet were American investors, Democratic Party congresswoman Cynthia McKinney and Robin Sanders, Director of African Affairs at the United States National Security Council.

The corporation had every reason to congratulate Congo's new leader. On 16 April, America Mineral Fields had been awarded Kolwezi, a copper and cobalt tailings recovery project situated in the 'liberated zone', by the President-in-waiting, Laurent Kabila. Although dictator Mobutu Sese Seko was still in office in the yet-to-be-renamed Zaire the company had stated, in an 16 April 1997 communiqué, that it had been awarded the project by 'the Democratic Republic of Congo'.[2] A further announcement on the deal released by the company on 23 April 1997 stated that Kabila's brother, Kambale Kabila Mututulo, and another official of the ADFL rebels had endorsed several agreements between the Democratic Republic of Congo, the state-owned mining company Gecamines and America Mineral Fields.[3] Although an internationally recognised government was still in power in Zaire's capital, the country's assets were already up for offer by the self-declared representatives of a rebel force, signing mining agreements on behalf of state-owned companies and on behalf of the Democratic Republic of Congo.

This chapter traces the emergence of Jean-Raymond Boulle, a key figure behind the America Mineral Fields incursion into Africa. The analysis will attempt to throw up the complicated corporate arsenal at the disposal of some transnationals which provides them with the flexibility and options to manipulate weak governments better than even some powerful Western states.

Le Jeu de Boulle

AMF's principal shareholder, Jean-Raymond Boulle, had created another company, America Diamond Buyers, in February 1997. He had also paid rebel leader Kabila US$25,000 for a diamond-buying licence and started running the only licensed diamond-buying office in Kisangani, the eastern Zairean town that had been under the control of the AFDL since 15 March, purchasing about US$100,000 in gems a day from April 1997.[4] At that time, lacking an airforce of his own, Kabila was flying around the rebel-controlled areas in Boulle's chartered Lear jet. When De Beers, the transnational diamond giant, refused to deal directly with the rebels in Kisangani, Boulle had quickly jumped in to inject the cash the AFDL leaders needed to lead their army into the capital Kinshasa, still thousands of kilometres away at that time. In return, America Diamond Buyers was awarded the wartime diamond-purchasing monopoly.

In October 1997 America Mineral Fields issued another communiqué in which the company proudly claimed to have been 'the first mining company to support President Kabila and his government' and 'the first to be awarded a mining contract by President Kabila'.[5] This, as we have seen, was no exaggeration since Boulle's dealings with Kabila began long before the latter became president. The contract referred to in the latest communiqué was for the Kolwezi tailings project, which made America Mineral Fields 'a partner with the people of the Democratic Republic of the Congo'. This 'partnership' conferred the ownership of the project on AMF, giving the company 'exploration rights to a resource of 600 million pounds of cobalt and 3 billion pounds of copper with a gross in-situ value in excess of US$14 billion at current metal prices'. That makes the deal worth about 42 times AMF's original capital investment of US$350 million.[6] Another 'exclusive agreement' on the world's largest zinc mine in Kipushi gave AMF a combined area the size of Switzerland to explore and develop in southeastern Zaire. The initial negotiations for the Kipushi deal had taken place between Mobutu's Prime Minister Kengo wa Dondo and Anglo-American, in the latter's capacity as AMF's partner in June

1996, but it was Kabila who eventually signed it.[7] AMF's capitalisation value at the time of the signature was no more than US$37 million, with mineral properties worth only a few million dollars in Brazil and a few exploration agreements in Zambia, Liberia and Angola.[8] When the deal was struck, the stock price of the company rose in one week from CAN$3 a share to CAN$7. When Laurent Kabila took over the country a few weeks later, the price jumped up to nearly CAN$10.[9]

America Mineral Fields

America Mineral Fields was created in 1995 by Jean-Raymond Boulle. He is the principal shareholder of the company and controls roughly 42 per cent of its stock. Except for the name and the location of its headquarters, America Mineral Fields can hardly be considered an American company operationally. It is listed on the Vancouver and Toronto stock exchanges but based in the United States. Its corporate headquarters are today in Dallas, Texas, but at the time AMF entered into its unique partnership with Kabila, a remote building in Hope, Arkansas, was the official seat of this small two-year-old exploration company.

The location of America Mineral Fields in the hometown of American president Bill Clinton did cast some doubts on the private nature of the company's relationship to the new Congo leadership. 'Congogate', wrote one Congolese newspaper poignantly, not amused by Kabila's rise to power. 'Friends in high places' was the title of a *Forbes* Magazine article on AMF's owner, Boulle. It was of course no secret at the time that the American government had already given up on Mobutu, even officially, after the fall of Kisangani to the rebels. Did AMF have any advance knowledge of the way the military balance would change in Zaire or of the fact that the US was about to get rid of Mobutu, its long-time ally in Africa?[10]

Jean-Raymond Boulle

J.-R. Boulle is a French-speaking, Mauritius-born, British citizen living in Monaco. His mining operations are run through 100 per cent owned Luxembourg-based investment company, MIL Investments, and his principal portfolio manager is based in a small apartment in the Belgian city of Antwerp, the world's main diamond trading centre.[11] Boulle started his career as a buyer for De Beers Consolidated Mines but made his fortune when he began a partnership with another controversial mining investor, Robert Friedland. Friedland is the financial wizard and Vancouver stock exchange guru behind DiamondWorks, the holding company of the Branch Mining and Branch Energy enterprises that pop up whenever and wherever the mercenary companies, Sandline International and Executive Outcomes, are active. His investment company, Ivanhoe Capital Investments, helped DiamondWorks to raise capital for its mining operations. Friedland's brother, Eric, was the chairman and chief executive of DiamondWorks until July 1997.[12]

Boulle, in one way or the other, usually heads for the same trouble spots as Robert Friedland. He became seriously rich when he and Friedland started prospecting for diamonds in Canada's Voisey Bay. Their company, Diamond Field Resources, never produced a single carat of diamonds but metaphorically struck gold when it discovered one of the world's richest nickel deposits. In 1995 Diamond Field Resources was sold to Inco, a giant nickel producer, in Canada's biggest ever corporate takeover worth CAN$3.1 billion. Both Boulle's and Friedland's shares were suddenly worth hundreds of millions of dollars.[13]

Jean-Raymond Boulle is also a controlling shareholder of the American public company Nord Resources, which has its corporate offices in Dayton, Ohio.[14] Nord owns 50 per cent of one of the world's rare titanium oxide mines, Sierra Rutile Ltd, in Sierra Leone. Boulle bought a considerable part of Nord Resources' shares in early 1996, at a time that the Sierra Rutile mine had already been overrun and was under the control of the rebels of the Revolutionary United Front. The mine has been inactive since it was captured in 1994.

When the then military junta in Sierra Leone, the National Provisional Ruling Council, contacted a number of private military companies to provide assistance, the British company J&S Franklin came up with a strong proposal to train the Sierra Leone military and subcontracted the operation to Jersey-based Gurkha Security Guards.[15] GSG arrived in January 1995 under the command of an American and two British veteran military officers. The American, Bob Mackenzie, was in charge of the operation. However, despite his long experience, first in Vietnam and then as a commander in the Rhodesian Special Air Service and some of the crack units of the apartheid South African Defence Force in Mozambique, he was killed shortly after GSG's arrival in Sierra Leone.

J&S Franklin's role in the arrival of Executive Outcomes in Sierra Leone has not been documented, but the company did remain present. J&S Franklin, or the Franklin Group, are British suppliers of military equipment. The company produces non-lethal items, such as tents and uniforms, but also carries out a brokering role for most of Britain's and Northern Ireland's main weapons producers. The Franklin Group co-operates closely with the Foreign Office, a service that was generously rewarded with the Queen's Award for Exports in 1996. According to the company's website, J&S Franklin provides sophisticated military equipment and the necessary support and advise, together with training packages, at specialised British centres or at the client's own facilities.

The group acts as an agent for several leading defence manufacturers and counts the British Ministry of Defence, as well as the UN and several relief agencies, among its clients. A page on the marketing abilities of the company at its website in 1997 announced, among other achievements, 'the supply of a complete Mobile Air Traffic Control system to Central Africa'.[16]

J&S Franklin's presence in Sierra Leone deserves some attention. First, the company's instrumental role in the recruitment of a private military company to train the armed forces of Sierra Leone, is interesting. Robert Mackenzie, the American commander in charge of the private Gurkha contingent, had a notorious past, helping to prop up white minority regimes in southern Africa. He was one of the leading commanders of the Rhodesian Special Air Service, the expatriate mercenary force that spearheaded the armed forces of Ian Smith's Rhodesia. After the Rhodesian regime collapsed, he conducted operations with the South African Special Forces in Southern Africa. He was in charge of the South African- and Rhodesian-led destabilisation and terror campaigns in Mozambique in support of RENAMO.[17] When he returned to the United States, he remained one of the key supporters of the RENAMO rebels even after the publication in 1988 of an official American State Department report documenting RENAMO's atrocities. He was also known for his articles in the mercenary monthly *Soldier of Fortune*.[18]

Mackenzie was part of the small far-right community of private groups, businessmen and individuals within the American intelligence community that supported RENAMO all the way through. While most conservative groups in the US started distancing themselves from RENAMO because of its atrocities, former CIA Deputy Director Ray Cline and his son-in-law Major Bob Mackenzie were totally committed to the cause and kept exerting influence, however marginal, in the corridors of power in Reagan's Washington.[19] Mackenzie's role as the commander of the Gurkha Security Guards for the Sierra Leone operation gives some credence to the reports that the Gurkhas were not hired just for training purposes. Having served as a commander in South Africa's destabilisation units, Mackenzie was of course no stranger to the Executive Outcomes operatives who arrived in late 1995 to replace the demoralised Gurkhas.

The signing of the GSG contract by Captain Valentine Strasser's NPRC junta took place at a time when the rebel Revolutionary United Front was taking over the mining areas and had targeted Sierra Rutile and the iron ore producer, Sieromco – the two most important sources of Sierra Leonean revenue. However, the British government, confronted with the task of securing the release of British citizens taken hostage by the RUF in January 1995, distanced itself from the Gurkha force in Sierra Leone. In March 1995, the Foreign and Commonwealth Office officially denied 'giving any military assistance to the government of Sierra

Leone', but said it could not be held accountable for 'people who were no longer in the British army and who decided to sell their services elsewhere'. The RUF accused Britain of supplying weapons but, while London may have condoned the operations of the mercenaries, it cannot be denied that private individuals and corporate players were largely autonomous in their relations with the Sierra Leone junta. In this regard, J&S Franklin's relationship to Boulle and the deployment of private armies is worth analysing.

Jean-Raymond Boulle had been active in Sierra Leone before the 1990s. In 1989, at the time when Charles Taylor was preparing for his invasion of Liberia, Boulle had offered to come to the rescue of the then Sierra Leonean President, Joseph Momoh, who was at the time in desperate need of foreign investors. Boulle arrived with his company, Sunshine Boulle, based in Dallas, Texas. Sunshine Boulle was interested in deep kimberlite diamond deposits. The exploitation costs would have run close to US$100 million, but the concession would have secured the company a virtual kimberlite monopoly in the diamond-rich area of Kono. However, due to the massive scale of informal and illicit diamond exploitation and trade in the mining area, the company pulled out of the deal. The analysis William Reno made of the power and influence of informal politico-commercial networks in Sierra Leone at the time showed that the government had virtually relinquished central authority and state collapse was already irreversibly imminent, even before the rebellion started in 1991.[20]

It seems that Sunshine Boulle's concern over the chaotic and insecure environment in the mining areas sparked a presidential order to sweep the diamond areas in April 1990 and expel thousands of mine labourers involved in illegal prospecting. The Sierra Leonean Special Security Department and loyal army units who were deployed in the subsequent operation apparently met armed resistance from local strongmen and police personnel involved in the informal circuits. A year later, and after repeated military forays into the diamond fields, Sunshine Boulle eventually signed a contract promising a rapid US$70 million investment. The company brought in its own private force of foreign military experts, but abandoned the project nevertheless shortly thereafter, in 1991.[21] What is known about the kimberlite contract is that the company had brokered an agreement allowing it to pay no more than 22.5 per cent tax on profits, 'a rate even less then what Sierra Leone Selection Trust, a De Beers subsidiary, had enjoyed in 1933'.[22]

Fall-Out and Rivalry

Despite Sunshine Boulle's alleged departure from Sierra Leone at the height of the civil war, it was rumoured in March 1997 that the Sunshine

Mining Company still held claims to certain diamond properties in the country. A report in the Canadian *Globe and Mail* caused quite a storm on the Vancouver Stock Exchange. DiamondWorks, led by Boulle's former ally Robert Friedland, had to issue a news release stating its legal rights to one of Sierra Leone's richest diamond properties, the Branch Energy-owned Koidu lease.[23] The paper wrote that Sunshine Mining and Boulle had acquired the Koidu concession in July 1994 from the government of Valentine Strasser. A year after that, however, a new mines minister in the Strasser government had offered the rights to the same concession to Branch Energy Ltd in a deal allegedly connected with the Executive Outcomes mercenary contract.[24]

The Sierra Rutile mine also seems to have constituted a bargaining chip in the EO contract with the NPRC junta. The regime had committed itself to allocating 50 per cent of its tax intake from the Sierra Rutile mine to meet its obligations to the mercenaries, with whom it had signed a US$20 million contract.[25] The Sierra Rutile mines held little prospect for EO, which had an eye for quick returns, since the enclave surrounding the mine remained too volatile for any mining to resume. Consequently, EO seems to have focused its attention on the diamond areas so as to generate fast tax returns for the government, which would then be in a position to meet its financial commitments to the mercenary outfit. In July 1995 Branch Energy, which had close links with EO, acquired the Koidu lease to which Sunshine Mining and Jean-Raymond Boulle also laid claim. The former allies, it seemed, had fallen out.

In spite of the apparent animosity between Friedman and Boulle each kept a lively interest in the other's moves in the investment market. At the time that Boulle was buying up large quantities of shares in Nord Resources, the holding company of Sierra Rutile, Branch Mining had already secured some of Sierra Leone's finest diamond fields. The control of natural resources is the major factor propelling the civil war in Sierra Leone. Consequently, virtually all the major mining sites are located in zones of fierce warfare. The unique alliance with the South African mercenary force ensured that Branch could outgun all other rival non-state or private strongmen and thus confer a monopoly status on Branch Mining. No unarmed company could compete with it.

Equally Armed and Dangerous

Given the reconfiguration of power in the mineral sector, Jean-Raymond Boulle did not return to Sierra Leone unprepared. As a controlling shareholder of almost one third of the shares of Nord Resources, Boulle's company MIL Investments was entitled to nominate its own voting directors to the company's Board. Among the nominees appointed to the Nord Resources Board were Allan McKerron who, until 1996, served as an Anglo-American executive, and Jean-Raymond Boulle's brother, Max. Interestingly, Mark Franklin of J&S Franklin also became a director of the Board. The significance of Marc Franklin on the Board of Nord Resources was not lost on keen observers. That one of Britain's key brokers of military supplies and training packages holds shares in Nord Resources and is a voting director on behalf of Boulle's operations in Sierra Leone, mirrored the modus operandi of the Branch Mining-EO/Sandline alliance, whatever their corporate links.[26]

Friedland's rivalry with his former ally Boulle did not stop at the Koidu lease. In 1996, Sandline International signed a contract with the government of Papua New Guinea to deploy Executive Outcomes to quell a secessionist rebellion on the island of Bougainville. This contract sparked a military revolt when the commander of the armed forces of PNG, Brigadier-General Singirok, exposed the presence of the mercenaries and publicly accused prominent cabinet members of corruption. The Australian Broadcasting Corporation, focusing on the shady background of the deal, zoomed in on Robert Friedland and his Singapore-based investment company. ABC also connected Friedland and British businessman Anthony Buckingham to the Sandline contract and especially to the BCL copper mine in Bougainville, the most important target of the operation.[27] In July 1997, a few days before Papua New Guinea's major donor, Australia, officially announced it would sign the UN Convention against Mercenaries, Brigadier-General Singirok acknowledged that he had received a significant amount of money on a private account from a British arms manufacturer before leading the revolt against Sandline. The British company that had made the payment was J&S Franklin.[28] Was the company just trying to undermine a rival brokering company or was Boulle settling a score with the mercenaries connected to Friedland?

EO's operations in Angola have been widely documented, but Boulle's operations in Angola are worth mentioning as well. In May 1996, at a time when America Mineral Fields had started negotiations with the Kengo Government in Zaire, AMF concluded an agreement with International Defence and Security (IDAS) Resources NV, a private Netherlands Antilles company, to establish a 50/50 joint venture to develop diamond concessions and other important mineral properties held by IDAS in Angola. IDAS in turn had formed a joint venture with the Angolan State mining company Endiama.[29] IDAS was originally established in Belgium under the name International Danger and Disaster Assistance (IDDA), with offices in Antwerp, Brussels and Dubaï (UAE). It was founded by Michel Jean-Pierre, a Belgian veteran of the French Foreign Legion. After Jean-Pierre's sudden death in 1993, the company apparently moved its headquarters to Beverwijk in the Netherlands, although a Belgium-based office still remained in place. Documents of the Dutch Trade Register show that IDAS (the

Netherlands) was established in May 1993 and registered in August of that same year. It was integrated into a Holding structure based in the Netherlands Antilles. Documents of the Curacao Commercial Register list this holding company and a controlling trust[30] with the company IDAS Resources as part of the structure established in June 1995.

IDAS had gained a very good international reputation for its demining and explosive ordnance disposal expertise. During the Gulf War, IDAS was a subcontractor to the American company Brown & Root for demilitarisation and clearance of the US contingent's ammunition and missiles stocks in Saudi Arabia. It had also done a very good job in demining El Salvador in 1993. Praise for high professionalism was showered on the explosives expert personnel of IDAS in letters of recommendation from embassy officials and military commanders of US, Belgium, El Salvador, Saudi Arabia, Kuwait and Cambodia.[31]

IDAS in Angola

In Angola, IDAS had an ongoing demining operation. According to *The Economist*, the Belgium-based IDAS subsidiary was well connected in Washington and played an instrumental role in the MPLA's war strategy against UNITA in 1996 in Angola. At the time, the hard core of Jonas Savimbi's UNITA rebels were reported to have withdrawn to strongholds in the Cuango Valley and across the border in Zaire. IDAS was hired to guard the route from Zaire, thus blocking rebel incursions against government forces. In payment, the Angolan government granted IDAS a diamond concession near the Angolan-Zairean border in August 1996.[32] According to Paul Beaver of the British-based Jane's Information Group, IDAS's Angola contract was an incentive contract: 'Clear out the rebels, and a share of the diamonds is yours.'[33]

IDAS's Belgian Chief Executive Michel Jean-Pierre had approached Gurkha Security Guards in mid-1994 and planned to deploy up to 2,000 Gurkha soldiers to pacify the Lunda Norte area which, at that time, had fallen under UNITA control. GSG had already been active in Angola under a contract with Odebrecht Mining Services for the protection of the Brazilian company's assets in the province in 1991, but had to evacuate its staff when UNITA overran the mine in 1992.[34] From this perspective, Boulle's interests in the IDAS properties and his America Mineral Fields operations across the border in Zaire seem to overlap. The MPLA's decisive support for Kabila in 1997 and IDAS's military presence in the border region reveal a remarkable alliance between private military companies, Angolan strategic objectives, a rebel army in Zaire and a private sponsor, Jean-Raymond Boulle. It is not clear whether J&S Franklin had any involvement in the Gurkha contract in Angola. IDAS's activities during the Kabila rebellion have not been documented, but the company was certainly active in Angola.

On 4 March 1998 America Mineral Fields provided the 'friends and investors of AMF' with a copy of a report from Maedel's *Mini-Cap Analyst*, an investment newsletter published in Zürich, Switzerland. In the newsletter's coverage of America Mineral Fields, a section dedicated to the security situation in Angola's diamond region read:

Taking no chances, Endiama (Angola's state mining company) also hired a company associated with IDAS to bring about a thousand Gurkha troops from Nepal into the concession under the supervision of American and European officers.

The report, promoting America Mineral Fields' shares, added:

The Gurkhas are legendary for their prowess in hand to hand combat, a necessary skill for any soldier operating in the jungle. During the Falklands war, for example, the Gurkhas spearheaded the British assault ... once they hit the beach, they were estimated to have severed an Argentinian head every 30 seconds.[35]

At the time this report was published, AMF had already purchased 100 per cent of IDAS Resources,[36] a few days after the Angolan government announced it had taken full control of the mining area in the northern Cuango Valley.[37]

The clearout of UNITA strongholds in northern Angola coincided with a fall out between America Mineral Fields and Laurent Kabila. In late December 1997 the Congo authorities cancelled the contract awarded to America Mineral Fields in Kolwezi, as 'AMF had not satisfied the financial conditions for granting the contract'.[38] AMF subsequently accused South African mining giant Anglo-American of bribery and deliberate sabotage of AMF's relations with the Kabila government. On 7 January 1998 AMF issued a press statement announcing legal proceedings in the State District Court of Dallas, Texas, against Anglo-American for punitive damages in excess of US$ 3 billion.[39] The company appointed economist Robert Stewart as a senior advisor and chairman to renegotiate the Kolwezi project with Gecamines and the Congo government. Immediately prior to that, Stewart had been the senior African adviser to Bechtel and had been involved in the American construction group's development plan for the Democratic Republic of the Congo. Stewart's appointment was terminated soon after however, and in May 1998 he showed up in Brussels at a peculiar press conference, together with retired Colonel Willy Mallants, announcing a plan to oust Kabila. Colonel Mallants had been a Belgian mercenary officer who worked as a 'volunteer' for the AFDL rebel alliance of Laurent Kabila during his 1996–97 rebellion, but had now turned against him.

AMF's claim against Anglo-American was eventually withdrawn in March 1998, and on 4 August 1998, the company announced that it had formed a joint venture with Anglo-American and had entered into final negotiations with the Democratic Republic of Congo concerning the Kolwezi project. This announcement came soon after a new rebellion, this time against Kabila by his former allies in the AFDL, had broken out in the east of Congo. America Mineral Fields was back in business.

Jean-Raymond Boulle's position in war-torn regions mirrors that of the financiers behind Sandline International and Executive Outcomes. Be it in Sierra Leone, Angola, Congo or Papua New Guinea, mercenary troubleseekers and their private financiers show up whenever a shifting military balance can be manipulated. In the case of Sandline/EO, the close relationship with Northern governments and intelligence agencies has become obvious after the embarrassing Arms-for-Africa scandal in Britain caused some serious trouble for the Blair government.

When the American Defense Intelligence Agency organised its Africa Symposium on 'privatisation of security in Sub-Saharan Africa' in June 1997, a month before Sandline chief executive Tim Spicer went negotiating the Sierra Leone contract with a fugitive banker in Canada, Boulle's people were present as well. Next to the representatives from Sandline, EO, Military Professional Resource Increment and several American intelligence agencies, the participants list of the DIA conference showed a Benny Bray of Sierra Rutile America, one of the few mining representatives present.[40] The J&S Franklin link to the British Foreign Office, the IDAS link to Washington or GSG's late commander Mackenzie's links to a right-wing intelligence milieu in Washington add to the suggestion that Boulle is pretty well-connected. Boulle is known to have met Bill Clinton, then governor of Arkansas, on several occasions. Clinton signed a bill in 1987 allowing another Boulle-controlled company, along with a few others, to start digging for diamonds in the Crater of Diamonds State Park in Arkansas. The project was finally shut down in 1996, when it turned out that the State Park kimberlite was not worth mining, in spite of Boulle's earlier claims that the site held a mining potential of US$5 billion worth of diamonds.[41]

Conclusion: Treading on Minefields Where States Dare Not

Notwithstanding his network of connections in high places, Boulle is not necessarily mining for trouble on behalf of any state or its agency. A small but well organised investor, he operates a corporate empire with all the necessary attributes and distinctive features that a state needs in the international environment. More flexible and rationally organised then any state or state agency, Boulle can manoeuvre as a privileged and powerful player in several regions and markets. In environments of chaos, state collapse and economic decay, a powerful private player with access to significant amounts of money, raw materials, military technology and even a private rapid reaction force can decisively pursue his own private agenda. This is what Boulle does.

In Sierra Leone, the International Monetary Fund prefers a corporate entity like Sierra Rutile as an interlocutor, over the debt-ridden and beleaguered institutions of a collapsed government.[42] Press releases from Nord Resources, America Mineral Fields or most other of the companies that Boulle controls read like diplomatic cables from the war zone, boosting the morale of shareholders on the home front in Europe or North America.

Indeed, Boulle coordinates his geo-economic agenda in the corporate boardrooms, military headquarters or presidential palaces of any interested ally just as any other 'sovereign' actor on the international scene. When and where a confrontational approach backed by military muscle is required for the desired result, his corporate empire can either deploy its own private army or openly sponsor the local rebellion or the armed forces of the government of the day. In Angola, some international aid agencies and donors may even find that they have subcontracted demining operations and devoted public money to a private army that has become an active participant in combat operations on behalf of a public company and its shareholders. IDAS Belgium's liquidation in late 1998 has thrown up questions about the financial and other activities of Belgium's former demining experts. In 1999, judicial authorities in Brussels started probing IDAS's liquidation and mercenary recruitment practices, forbidden under Belgian law since 1979. Jean-Raymond Boulle will no doubt send in a battalion of legal experts to the rescue. If not the Gurkhas.

Notes Chapter 5

1. 'Kabila and Liberation Forces Triumphant'. News Release, America Mineral Fields Inc., 20 May 1997.
2. 'America Mineral Fields Awarded Kolwezi'. Press release on Canada News Wire, 16 April 1997.
3. 'America Mineral Fields Inc. Kolwezi Update'. Canada News Wire, 23 April 1997.
4. R. Block, 'Taking Sides: As Zaire's War Wages, Foreign Businesses Scramble for Inroads'. *Wall Street Journal*, 14 April 1997.
5. 'America Mineral Fields Inc. Completes Drilling of Kolwezi Tailings Katanga in the Democratic Republic of Congo'. America Mineral Fields, 8 October 1997.
6. C. M. Oliver, an adviser in securities, in his report on America Mineral Fields Inc., 'Central African Copper Belt Properties', 7 October 1997.
7. America Mineral Fields Inc. *Annual Report* 1996.

8. Ibid.
9. Source, Bloomberg Business News.
10. Richard C. Morais, 'Friends in high places', *Forbes* Magazine, 10 August 1998.
11. Analysis of corporate documents of Nord Resources Corporation, a publicly traded company in the US.
12. Branch had a tiny exploration agreement in Zaire when Boulle started partnering with Kabila. Friedland's corporate links to Sandline/EO and the Branch assets were documented in the Australian daily *Sydney Morning Herald* (7 April 1997).
13. *Forbes*, 10 August 1998.
14. At the end of 1998 MIL Investments and Boulle held 28 per cent of the shares and were still purchasing large quantities of stock, according to MIL's filings for the US Securities and Exchange Commission.
15. Jim Hooper, 'Sierra Leone – the war continues'. *Jane's Intelligence Review*, January 1996, p. 41.
16. <http://www.franklin.co.uk/marketing.html>, downloaded on 24 September 1997.
17. Some of these operations are described in detail in Anthony Rogers', *Someone Else's War: Mercenaries from 1960 to the Present* (Harper Collins, 1998). Also Tim Ripley, *Mercenaries: Soldiers of Fortune* (Parragon, 1997).
18. On Mackenzie's connections in Washington see: Alex Vines, *Renamo: From Terrorism to Democracy in Mozambique?* (Centre for Southern African Studies, University of York/James Currey Ltd, 1991, revised edition 1996), pp. 42–50.
19. William Minter, *Apartheid's Contra's: An Inquiry into the Roots of War in Angola and Mozambique* (Witwatersrand University Press/Zed Books, 1994). MacKenzie married Cline's daughter in 1989.
20. William Reno, *Corruption and State Politics in Sierra Leone.* (Cambridge University Press, 1995). See also Chapter 3 of this book.
21. Reno, *Corruption*, pp. 164–5. Also: 'Sunshine Boulle Mining Company Told to Pay Damages'. *Wall Street Journal*, 23 January 1992.
22. Reno, *Corruption*, p. 166. SLST, Sierra Leone Selection Trust, was a De Beers subsidiary that acquired a 99-year monopoly on the British colony's diamonds in 1932–33.
23. 'DiamondWorks Confirms Legal Rights to Koidu Diamond Property'. Press Release, Canada News Wire, Vancouver, 3 March 1997. Eric Friedland, Robert Friedland's brother, was chairman and CEO of DiamondWorks until mid-1997.
24. K. Howlett and M. Drohan, 'Canadian Miners living dangerously'. The *Globe and Mail*, 26 July 1997.
25. Letter received from David Isenberg containing corrections in his monograph *Soldiers of Fortune Ltd.: a Profile of Private Sector Corporate Mercenary Firms*, Washington D.C., Center for Defense Information, November 1997. The corrections were added after Isenberg received some comments from a source who worked directly with EO.
26. Annual meeting of stockholders of Nord Resources Corporation, June 1997, as reported to the US Securities and Exchange Commission.
27. 'Robert Friedland: King of the Canadian Juniors'. Background Briefing, ABC-Radio, Stan Correy, 6 April 1997.

28. Luc Palmer, 'Singirok Admits Taking Arms Cash'. *Sydney Morning Herald*, July 14 1997. Also: D. Leppard and M. Ashworth, 'Arms firm made secret payments'. *Sunday Times*, 3 August 1997. We contacted the offices of J&S Franklin repeatedly after the company's name had showed up several times during an official hearing on the mercenary contract, but representatives were not available for any comments.

29. 'America Mineral Fields concludes 50/50 joint venture on important diamond properties in Angola'. AMF press release on Canada News Wire, 21 May 1996.

30. I.R.M & O. Holdings registered in March 1993 with Pietermaai Trust NV as the controlling entity.

31. A Belgian section of an international relief organisation provided us with a copy of a glossy IDAS prospectus including several of these letters of recommendation.

32. 'The diamond cut: Angola's struggle is about wealth and power'. *The Economist*, 13 July 1993.

33. Quoted in 'Friends in high places', *Forbes*, 10 August 1998.

34. Alex Vines, 'Gurkhas in Africa: the Works of Gurkha Security Guards Ltd', draft paper presented at the ISS 'Profit and Plunder Round Table' in Johannesburg, South Africa, 1998.

35. Excerpt from Maedel's *Mini-Cap Analyst*, 27 February 1998. Reprinted by America Mineral Fields with the permission of Maedel's *Mini-Cap Analyst*. Document received on 3 April 1998 from Early Young, Sr., Vice President, Corporate Relations, AMF.

36. 'America Mineral Fields Purchases 100% of IDAS Resources'. AMF News Release, Vancouver, 20 January 1998.

37. 'America Mineral Fields and IDAS' Cuango Basin Diamond Project Accessible'. E-mail received from America Mineral Fields on 14 January 1998.

38. 'Democratic Congo's Gecamines challenges AMF claim'. Reuters, Kinshasa, 20 January 1998.

39. 'America Mineral Fields Launches US$3 Billion Texas Legal Action Against Anglo American/De Beers/Minorco Group'. AMF News Release, 7 January 1998.

40. 1997 DIO Africa Symposium Attendees list. Washington D.C., Defense Intelligence Agency, 24 July 1997.

41. 'Friends in High Places'. *Forbes*, 10 August 1998.

42. Pratap Chatterjee, 'Afrika: Geruchten over een nieuwe huurlingenoperatie in Sierra Leone', IPS, 17 August 1997.

6

Mercenaries, Human Rights and Legality

Alex Vines

Africa in the 1990s has seen a significant growth in the private security sector. This has been driven by increased perceptions of insecurity caused by terrorism, kidnapping, random acts of violence, urban unrest, increasing general crime, corporate crime and weakened and poorly resourced and trained state law enforcement agencies. The rapid expansion of this industry has given rise to many thousands of providers of security-related services and products worldwide. In South Africa alone there are 5,939 companies regulated by the Security Officers Board.[1] Many of these companies remain relatively small in global terms (average size is under US$5 million in annual revenue), highly specialised yet undercapitalised and serving finite geographic regions. The combined US and international security market had estimated revenues of US$55.6 billion in 1990. Revenues in this industry are expected to increase to US$202 billion by 2010, a compounded annual growth of 7 per cent from 1990.[2]

Private security firms operating in Africa in the 1990s can be divided into three general categories: classic mercenary firms, private military companies that engage in mercenary-like activities, and private security firms. The divisions between these categories are grey and what these firms engage in is fluid, often depending on profit margins and commercial opportunity. A number of them exhibit a corporate nature, including ongoing intelligence gathering and concern for public relations. These firms handpick their employees and enjoy large pools of qualified applicants, a spin-off from the end of the Cold War and the political realignments and defence cutbacks that followed. Many of these firms have close relations with multinational oil and mineral companies, which provide them with additional funding and political contacts.[3]

Is this privatisation of security a good thing in Africa, and will regulation work? This paper examines some ethical and practical problems generated by the growth of this industry, and assesses where it will be going in the next decade.

LEGAL ASPECTS OF THE MERCENARY TRADE

A number of firms operating in Africa have been called mercenary. The traditional notion of a mercenary is 'a soldier willing to sell his military skills to the highest bidder, no matter what the cause'.[4] The established definition of a mercenary in international law is set out in Additional Protocol 1 to Article 47 of the Geneva Convention (1949). This classifies a mercenary according to the following criteria:

- Specially recruited locally or abroad to fight in an armed conflict
- Does, in fact, take part in hostilities
- Motivated to take part in the hostilities essentially by the desire for private gain
- Neither a national of a party to the conflict nor a resident of a territory controlled by a party to the conflict
- Not a member of the armed forces of a party to the conflict
- Has not been sent by a state which is not a party to the conflict on official duty as a member of its armed forces.

Both the OAU and international conventions on mercenaries contain definitions that focus on acts aimed at overthrowing or undermining the constitutional order and territorial integrity of the state. By January 1999 only 16 states had ratified the international convention; 22 are required to before it comes into force.[5] Many of the private military companies mentioned below also claim they work only for recognised governments and are therefore exempt from the terms of these conventions. The UN Special Rapporteur on the role of mercenaries understands this, noting in his 1997 report that security companies 'cannot be strictly considered as coming within the legal scope of mercenary status'.[6]

Domestic Legislation

The weakness of international law has placed greater responsibility on domestic laws. In the US, the Neutrality Act of 1937 prohibits the recruitment of mercenaries in the US. Australia has a similar law. Under the Australian Crimes (Foreign Incursions and Recruitment) Act of 1978, it is an offence to recruit mercenaries in Australia. However, in both countries it is not illegal to be a mercenary. The UK's Foreign

Enlistment Act of 1870 makes enlistment of mercenaries both within and outside the UK an offence, but this has not stopped British citizens playing important roles in mercenary-like activities. The British government is now considering updating the legislation to take into account firms like Sandline International,[7] but also believes that the International Convention is too weak to be of any use.

South Africa's parliament passed a tough new amendment to its Foreign Military Assistance Act in February 1998 in which anyone convicted of an offence involving mercenary activity would forfeit property to the state.[8] The new prohibition simply says: 'No person may within the Republic or elsewhere recruit, use, train, finance or engage in mercenary activity', and adds a new definition of mercenary activity as 'direct participation in armed conflict for private gain'. It is also necessary that South African companies providing military assistance abroad seek prior government approval.[9] However, this Act does not provide adequate mechanisms for public and parliamentary scrutiny and accountability of the activities of private security companies.[10]

The UN Special Rapporteur recommended in January 1998 that:

> Given the legal gaps and inadequacies which permit the existence of mercenaries whose activities can pass as normal, it is recommended that the Commission on Human Rights should propose that the States Members of the United Nations should consider adopting legislation to prohibit mercenary activity and the use of national territory for such unlawful acts.[11]

He also concluded that:

> In what appears to be a new international trend, legally registered companies are providing security and military advisory and training services to the armed forces of legitimate Governments. There have been complaints that some of these companies go beyond advisory and instruction work, becoming involved in military combat taking over political, economic and financial matters in the country served. It is therefore recommended that the evolution of these companies, the relevant legislation of States and the conditions under which States agreed to conclude contracts with such companies should be monitored closely. It needs to be assessed whether the security and internal order of a State, which has lost part or all its capacity to keep order, should henceforth be left to the action of specialised companies which will take charge of its security.[12]

In his January 1999 report Ballesteros added that:

Mercenaries base their comparative advantage and greater efficiency on the fact that they do not regard themselves as being bound to respect human rights or the rules of international humanitarian law. Greater disdain for human dignity and greater cruelty are considered efficient instruments for winning the fight. The participation of mercenaries in armed conflicts and in any other situation in which their services are lawful may jeopardize the self-determination of peoples and always hampers the enjoyment of the human rights of those on whom their presence is inflicted.[13]

A number of the private military companies that Ballesteros mentioned have operated in Africa. The following case studies assess their success on the ground, and question how beneficial they have been to the countries in which they have operated.[14]

PRIVATE MILITARY COMPANIES

Executive Outcomes (EO) in Angola[15]

The most widely known firm is Executive Outcomes. Its involvement in Angola began in 1993 when it was contracted by the Angolan State oil company, Sonangol, to provide security for the Soyo oil installations against UNITA attack. Although EO lost these installations to UNITA, the Angolan government gave EO a one-year contract in September 1993 (US$20 million for military supplies and US$20 million directly to EO). The Angolan government then renewed the contract every year till mid-1996.[16] EO acted in Angola as a 'force multiplier', offering a small group of individuals who trained up the effectiveness of a larger fighting force. EO fielded some 550 men and trained over 5,000 troops and 30 pilots in this period.[17]

A watershed event for EO was the successful capture from UNITA rebels of Ndlatando city by the 16th Brigade in May 1994. EO had trained the Brigade and fought alongside it. The success increased the Angolan government's confidence in the firm. EO personnel also helped recapture the diamond areas around Cafunfo in mid-July and the oil installations at Soyo in November. They were also active in Uige and Huambo.[18] Pilots belonging to Ibis Air – in which EO was a significant shareholder – flew combat missions in Mi-8, Mi-17 and MiG-23 fighters and Pilatus planes. EO activities in Angola cost US$60 million, with 20 fatalities.

EO wanted to stay on and train the new Angolan army, but the Angolan army of the time came under increasing pressure to withdraw its support of EO and replace it with the US firm, MPRI. In January 1996 EO officially left Angola. A number of EO's personnel stayed on, redeployed to companies linked to EO such as Branch Mining, Shibata

Security, Stuart Mills International, Saracen and Alpha 5. EO and its partners were rewarded with a number of economic concessions across the country, including diamond mines. Many former personnel remained on EO's books and could be mobilised for new contract work with the firm.

EO's involvement in Angola's war did not create peace or stability. Five years later, the Lusaka peace process is dead and the country is back at war with UNITA rebels. According to the government, UNITA is employing Israeli, Serb, Ukrainian and South African mercenaries, while the government has been beefing up its foreign military specialists, including Portuguese, South Africans and possibly Cubans.[19]

What is evident, however, is that EO's activities went far beyond providing training and assistance. An EO member admitted on South African television in 1995 that his organisation did not just train soldiers but engaged in aggressive military operations:

> A team went ahead to clean up Cafunfo. We followed up later and, on the way to Cafunfo, we killed about 300 enemy soldiers. Executive Outcomes was engaged in attacks all the time. It did give some training as well, but the successes of the MPLA could be directly attributed to Executive Outcomes' involvement.[20]

A former EO employee described his work to Human Rights Watch in June 1995:

> We are professionals. We don't engage in unnecessary violence. We are specialists in counter-insurgency operations. Most of our work was training but we also engaged directly in operations that had direct commercial consequence. Some civilians get hurt in these operations but they were not our target. In the Cafunfo operation we did encounter problems and things went out of control for a short time. That was unusual.[21]

UNITA claimed to have killed over 125 'foreign mercenaries' during 1993–94. EO's Eeben Barlow admitted that 15 of his employees had died in Angola from malaria, training accidents and UNITA attacks in the same period. In September 1994, another EO representative said that 14 employees had died in Angola in the past year.[22]

UNITA took EO employees prisoner and, according to a UNITA official, executed them. Two EO personnel were captured in eastern Angola in March 1994. Two more were captured on 15 July 1994 when their PC-7 plane was shot down by UNITA shortly before government forces recaptured the strategic diamond town of Cafunfo.[23] One of these was identified as D. C. O'Connell, a South African. Since January 1993 UNITA had publicly threatened to execute any captured mercenaries.

Although South African Deputy President Thabo Mbeki made an appeal for clemency in July 1994, a UNITA official told Human Rights Watch that they were executed shortly after capture. He justified the executions by saying,

> These are mercenaries. They fight for money not ideology. They have no rights in UNITA eyes. Both black and white mercenaries did not have the right to live. They designed and conducted operations that killed our people. We have no responsibility to them under international law. They earned thousands of US [dollars] a month trying to kill us.[24]

The July/August 1994 edition of UNITA's journal *Terra Angolana* included photos of three white soldiers killed by UNITA; two were identified from the photos as EO employees by their families.

There is no doubt that EO quickened the pace of the war in Angola and added pressure on UNITA to sign the Lusaka Protocol in November 1994.[25] However, the tide had already turned against UNITA and EO's claims to have won the war are inflated.

Herb Howe calls EO an 'apparently stabilising force' in Angola, and discounts the number of human rights abuses that EO personnel were involved in as 'generally correct treatment of the civilian populations'. Howe is supported by David Shearer, who argues that 'private military forces cannot be defined in absolute terms: they occupy a grey area that challenges the liberal conscience'. Moral judgements on the use of mercenaries are usually passed at a distance from the situations in which these forces are involved, he argues: those facing conflict and defeat have less moral compunction.

What Shearer and Howe fail to acknowledge in their writings are the serious human rights abuses that occurred in Angola, and were documented by groups like Human Rights Watch.[26] The looting of a town by EO is even captured on a video diary.[27] Howe himself said in public that EO introduced an indiscriminate weapon into Angola – fuel air explosive.[28] Indeed, EO pilots have boasted about such weapons. Jim Hooper recorded an EO pilot describing the use of weapons in an indiscriminate fashion as documented by Human Rights Watch during the 1992–94 war,[29] saying:

> As far as armament [is concerned], the MiG's twin-barrelled 23 mm cannon is an excellent piece of kit. Very accurate, very effective against ground troops. Our most common weapons loads were 250kg or 500kg bombs, but we occasionally carried napalm and rockets. Interestingly enough, we also had some MK 82 bombs kindly provided by the Israelis, who had modified the American kit to fit the hardpoints on Soviet aircraft. The most effective bomb we used was the Russian RBK

SWAB, a 500kg cluster bomb. Once we'd pulled off and looked back at that target you could see hundreds of explosions going off [within] at least a 300m radius.[30]

With a new war in Angola heating up in 1999 and increasing reports of a proliferation of mercenary use in it by both the government and UNITA rebels, the need to seek ways to stop mercenaries from fuelling and encouraging conflict is urgent. Angola's rich past of human rights abuses contains a significant number by the hands of mercenaries.

Executive Outcomes in Sierra Leone[31]

In May 1995, Sierra Leone contracted with EO to help its faltering four-year campaign against the RUF. The Sierra Leonean government signed three separate security contracts with EO over the 21 months they were in the country for a total of $35 million – an average of more than US$1.5 million a month, to be paid in cash.[32] Because the government was cash-starved, EO negotiated mining concessions in return for its services.[33]

EO's military progress was rapid. Again the company acted as a force multiplier providing technical services, combat forces and limited training. By late January 1996, EO-backed forces had retaken the southern coastal rutile and bauxite mines, belonging to Sierra Rutile and Sieromco. EO claims that only two of its personnel were killed during its operations, which lasted a year and a half. As in Angola, a cease-fire followed in November 1996. With the signing of the November 1996 peace agreement, the rebels demanded that all EO personnel should leave. EO technically withdrew at the end of its contract in January 1997, although around 100 of the 285 EO personnel stayed on in different companies, some of them, such as the security firm LifeGuard, linked to EO.

There are fewer reports of EO abuses in Sierra Leone, possibly because EO learnt from its mistakes in Angola. EO appears to have used fuel-air explosive in Kono district. Their pilots had also complained that they could not differentiate between civilians and rebels due to the thickness of the jungle.[34] The Strasser government responded by ordering them to fire indiscriminately.[35] Human Rights Watch has also obtained evidence of EO soldiers having kept photo albums of people they had killed and there is anecdotal evidence of minelaying around the diamond mines.[36]

EO's claim to have returned stability to Sierra Leone was short-lived. In May 1997 the newly elected government was overthrown in a coup led by Commander Johnny Paul Koroma and a junta remained in control of Freetown until a counter-coup in March 1998 restored the government of President Kabbah. The country has been engulfed by violence from the time of the 1997 coup.

The UN Special Rapporteur on Mercenarism also challenges the hypothesis that EO brought stability to Sierra Leone:[37] his view is that 'the presence of the private company which was partly responsible for the security of Sierra Leone created an illusion of governability, but left untouched some substantive problems which could never be solved by a service company'.

This danger was flagged in paragraphs 64 and 65 of the Special Rapporteur's previous report to the Commission on Human Rights (E/CN.4/1996/27), in which he noted that EO was involved in such delicate activities as 'training of officers and other ranks; reconnaissance and aerial photography; strategic planning; training in the use of new military equipment; advising on arms purchases; devising psychological campaigns aimed at creating panic among the civilian population and discrediting the leaders of the RUF etc, etc' (paragraph 65). Moreover, EO was responsible for overall security and was directly active in the Kono and Koidu districts, Kangari Hills, and Camp Charlie at Mile 91.

The issue was also dealt with in the Special Rapporteur's 1996 report to the General Assembly (AC/51/392, § 33), in which he described the precarious situation in Sierra Leone, pointing out that the presence there of a company that worked in security matters but employed mercenaries was a debilitating factor which at some point would impair the stability of the legal government. Prompted by an enlightened sense of unity, discipline and subordination to civilian rule, Sierra Leone should accord priority to organising its police and armed forces, which should assume sole responsibility for security. Retaining the company for that function until late 1996 was a mistake and a waste of valuable time, and according to the thesis developed, it weakened the legal government of President Tejan Kabbah.

EO defended its record in Sierra Leone. It argued that 'Forces employed to restore a legitimate democratically elected government, displaced by a coup, should be supported and praised and not shot down in flames by "ethical" journalists taking the moral high ground.' EO then warned that 'Coup plotters should take cognisance of the fact that if the international community refrain [sic] from taking action against them, there are private forces who may just do so.'[38]

EO was sensitive about being classified as a mercenary group, although Ballesteros calls it 'a private security company that works with mercenaries'. It also claimed that as it possessed no military equipment and had no military infrastructure such as bases or barracks it could not be described as a private army. Yet EO possessed shares in Ibis Air and during the Sandline International adventure in Papua New Guinea, purchased, rather than hired the heavy military equipment.[39]

EO claimed to have refused to work for regimes such as Sudan, the military dictatorship in Nigeria, and President Mobutu, but other sources suggest that Mobutu found EO too expensive.[40] In October 1997 EO also claimed that President Kabila had approached it and there is evidence that in 1998 EO worked for Kinshasa. EO also denied that it had ever been involved with non-state actors, although there is some evidence that EO worked with UNITA in early 1993 and was employed briefly for intelligence-gathering purposes by the RUF in Sierra Leone.[41] EO has strongly denied allegations made by Human Rights Watch that it provided military services to CNDD forces in Zaire in 1996.

Sandline International in Papua New Guinea

Sandline is the sister company to EO and offers the same sort of services.[42] It has been operational since 2 July 1993, when Sandline International (originally Castle Engineering) was incorporated in the British Virgin Islands and acted as a parent administrative company.

Sandline, which offers a range of products from systems procurement to combat operations, admits to having undertaken six international operations since 1993. One of these was located in Papua New Guinea, where Sandline was hired by the government in March 1997 to import Russian small and large arms, including four helicopter gunships (see the tables at the end of this chapter), and 44 mercenaries supplied by Executive Outcomes. The objective was to defeat the Bougainville Resistance Army (BRA) and recapture the Panguna copper mine, one of the largest copper mines in the world, and set up as a joint venture between the Papua New Guinea (PNG) government and the mining giant, Rio Tinto Zinc. The mine has been inactive for the last eight years because of the internal conflict.[43]

The mission was explicitly defined in their contract with the PNG government. Sandline was to:[44]

- train the state's Special Forces Unit (SFU) in tactical skills specific to the objective
- gather intelligence to support effective deployment and operations
- conduct offensive operations with PNG defence forces to render the BRA military ineffective and repossess the Panguna mine
- provide follow-up operational support, to be further specified and agreed between the parties and subject to separate service provision levels and free negotiations.

In order to fulfil the objectives outlined above, Sandline agreed to:

- send a 16-man command, administration and training team in the first week of implementation of the contract to establish the necessary liaison with the Papua New Guinea Defence Forces; develop a logistical and communications infrastructure; prepare for the safe arrival of the military and aeronautical equipment contracted for; initiate the information-gathering and intelligence operations and begin training the Special Forces Unit
- set up bases at Jackson Airport and the La Selva training centre in Wewac
- send and deploy throughout the territory of PNG within ten days of the arrival of the command, administration and training team, the following: Special Forces officers and troops, aircraft and helicopter crews, engineers, intelligence agents, special teams of operatives, mission troops, medical and paramedical personnel, etc.
- send arms, ammunition and equipment, including aircraft, helicopters, electronic warfare equipment and communications systems, as well as any personnel necessary for their maintenance and for training in their use
- ensure that personnel sent to the country have appropriate identity papers, and assume responsibility for any expense caused by loss of its personnel, unless that loss was the result of negligence by the state.

The contract was worth US$36 million, and part of the payment was a stake in the Panguna mine. Sandline's staff were also granted tax exemptions, facilities and privileges in connection with their import of goods and their entry into and departure from the country. The government also undertook to instruct its civil servants and members of the Defence Forces to recognise the military ranks of Sandline personnel and to obey their orders. The government was forced to cut budgets in health and education drastically to raise money for the Sandline venture.

The project fell apart when the commander of the armed forces publicised the plan, which caused a military revolt.[45] At the end of March 1997, the 44 Sandline mercenaries were forced to leave as the result of widespread protests within the Defence Forces. Two commissions of inquiry investigated the matter, and while Sandline was cleared of illegality, their armaments were impounded in Australia,[46] and US$18 million in payments was withheld although it was finally paid in October 1998 after a long legal battle with the Port Moresby government.[47] The commission of inquiry revealed that the head of the armed forces 'blew the whistle' in part because he had been accepting payments from a rival organisation, the UK-based J&S Franklin Corporation, referred to as 'The Franklins',[48] but also probably out of concern for the long-term dangers to the state of using foreign mercenaries.

The UN Special Rapporteur concluded that:

The crisis in Papua New Guinea suggests that it was triggered by the implementation of the contract between the Government of Papua New Guinea and the private security company Sandline International. For the armed forces and the population, the presence of foreign soldiers recruited by a foreign company, which was responsible not only for training but also commanding military operations in an internal armed conflict, thereby subjecting the country's military leaders to its orders, and which received for the initial period of the contract (three months) $36 million as well as the promise of participation in the company Bougainville Copper Limited, was considered to be an act that violated sovereignty and the right to self-determination.[49]

Sandline's behaviour in Papua New Guinea was aggressive, pushing for business and advocating the use of force. The prices it quoted were above market rates for the equipment. The consultancy fees were high, such as US$1,165 an hour and a cost price of US$35,714 per mercenary per month. Spicer was poorly informed about Papua New Guinean politics and culture, and Sandline did not appear to understand that this was a country where institutions still worked, suggesting that Sandline's predatory style of business stands a chance only in collapsed or very weak states.[50]

Subsequent events on Bougainville illustrate just how wrong Sandline's hard military recipe for pacification was. Singirok's actions to terminate the Sandline deal provided the impetus for new mediation attempts, through the good offices of New Zealand government official John Hayes. A round of contacts and talks starting in July 1997 resulted in the Lincoln Agreement of January 1998, which set out a process that would lead to the formation of a Reconciliation Government for Bougainville. In April 1998 a permanent cease-fire was signed and a multi-nation Peace Monitoring Group made up of unarmed soldiers and civilians from Australia, New Zealand, Fiji and Vanuatu were put on the island as observers.[51] The paradox was that Sandline's abject failure has been its greatest ever contribution to the enhancement of peace and stability.

Sandline in Sierra Leone

Sandline was also involved in another scandal regarding an arms shipment to Sierra Leone in February 1996. As discussed in depth in Chapter 3, Sandline was contacted by the deposed government of President Kabbah (which was overthrown in a military coup) in order to help the democratically elected government return to power through a counter-coup. The mission was scaled down because of the limited resources of the government-in-exile, and primarily consisted of importing weapons for ECOMOG forces and local Kamajor militia members. The scandal broke when approximately 35 tons of Bulgarian arms were reported to have been imported into the country in violation of the UN arms embargo on Sierra Leone and British law emplaced by the Ordering Council. Sandline claims that the British government was fully aware of, and acquiesced, in the operation and that the US State Department knew of the operation. Two inquiries have followed in the UK over this.[52]

Since this scandal erupted, Sandline's activities have been curtailed, although it still appears to be operational in Sierra Leone with both government and ECOMOG forces. The company had also been involved with former President Lissouba in Congo-Brazzaville but had fallen out with him over non-payment. The company was warned off by the British government from breaking an arms embargo on the former Yugoslavia in 1998 to help the Kosovo Liberation Army, although a feasibility study was carried out.

DiamondWorks

DiamondWorks, founded in 1996, is a Vancouver-based mining firm that aspires to be a major player in the international diamond market. It acquired Branch Energy, a mining and oil interest firm with strong links to Sandline and Executive Outcomes, in that year. Part of the payment negotiated for the activities of Branch Energy's private army has been lucrative mining concessions in countries like Angola and Sierra Leone. Consequently, DiamondWorks is developing mining concessions in Angola and Sierra Leone obtained as payments for the activities of EO. In 1997 it became Canada's largest producer of diamonds. Its diamond concessions (primarily in Angola and Sierra Leone) are estimated at US$3.7 billion. In Sierra Leone, it holds three major properties: the Koidu diamond mine, the Sewa diamond concession, and the Sierra Rutile mine, the world's largest titanium mine.

In its 1997 Annual Report, DiamondWorks stated that:

> The only disappointment that we experienced in 1997 occurred in Sierra Leone. As our shareholders are aware, our development activities in Sierra Leone were suspended after a military coup in May 1997. While the coup did not directly impact our operations, we decided to stop work in the country pending the restoration of peace. On March 10, Sierra Leone's democratically elected government was returned to power, paving the way for us to resume our development work at the Koidu Mine in eastern Sierra Leone.

The company's board includes individuals with well-known links to Sandline/EO. Bruce Walsham, the CEO of DiamondWorks, has denied many times that his firm has any links to EO or Sandline. This denial is made despite the presence of Michael Grunberg on its board of directors; Simon Mann as its Southern Africa director; Tim Spicer being asked to liaise with the British Foreign Office in November over the kidnapping by UNITA of a British employee; and despite the offices of DiamondWorks and Sandline being in the same building in London (King's Road, Chelsea). This contradiction is also to be found on the DiamondWorks website, where in one section the company denies any link to Sandline and in another publishes a press release about Tim Spicer's acting on its behalf in seeking the release of the British hostage. [53]

The White Legion

It is easy to recall images of the late 1960s in relation to mercenary activities in the Congo. In the last months of President Mobutu's reign in early 1997, mercenary forces once again featured in Mobutu's final attempt to stop rebel forces from ousting him. The mercenary force that was hired did not make much impact, at the most delaying the fall of Kisangani by a number of weeks.

There were two distinct elements in the mercenary force which began assembling in late November and December 1996. One was a small, Western European element headed by Christian Tavernier, who had overall operational command of the force. [54] The other, larger element (280 strong) was the East European group, consisting of former members of Bosnian Serb forces. Their pay ranged from US$3,000 to US$10,000. The Yugoslav authorities, which authorised the sale of military equipment, also assisted with recruitment and the transfer of war material via Luxor Airport in Egypt. The mercenary force hired four Mi-24 'Hind' helicopters manned by Ukrainians, and also used six French-made Puma and Gazelle helicopters and some Yugoslav-built SOKO tactical strike jets. [55]

According to the UN Special Rapporteur on Mercenarism, the Mobutu government spent some US$50 million in public funds to pay and arm the mercenaries. He also reported that: [56]

> The mercenaries hired to defend Mobutu came principally from Angola; the Federal Republic of Yugoslavia (Serbia and Montenegro); Bosnia and Herzegovina; South Africa; and France, where two former members of the presidential security teams, Alain le Carro and Robert Montoya, were alleged to be in charge of recruiting between 200 and 300 mercenaries. There were also, but in smaller numbers, Belgians, Britons and Mozambicans.

The mercenaries gave training to the Força Armada da Zaire (FAZ), and especially to elite units at Kisangani airport. Although they enjoyed a success on 17 February 1997, the fall of Kindu on 1 March reversed their fortunes and caused increasingly low morale amongst the regular forces and poor discipline amongst the Serbs. The Alliance of Democratic Forces for the Liberation of Congo-Zaire (ADFL) rebels in support of Laurent Kabila continued to make progress toward Kisangani, despite the mining of roads and the airport. The mercenary unit and the FAZ 48th Regiment were able to hold off the rebel advance on the city for some five days before the rebels broke through to capture the city. As the rebels closed in on Kisangani airport, the mercenaries tried to stop Zairean soldiers fleeing, which led to an exchange of fire. Finally, the remaining few dozen Serb fighters fled aboard their helicopters after blowing up their HQ. A number of these Serbs moved to Gbadolite, but were overrun and killed by looting troops a few months later.

According to *Africa Confidential*, some of Tavernier's men were working for Sandline in Sierra Leone in 1998.[57] As in Angola, it has emerged that mercenaries may be fighting with the rebels, with some 300 Ukrainian mercenaries reported in the north of the country.[58]

Stabilco

Although the Belgian-led Serb and Croat mercenary forces withdrew from Congo, a South African firm, Stability Control Agencies (Stabilco), owned by Mauritz Le Roux, attempted to win a government contract from Mobutu to turn the tide of the war. In January 1997 Stabilco deployed a small group of pilots to Kinshasa, and conducted air surveillance missions over Kisangani and the north-eastern provinces. Kpama Baramoto Kata, at the time Chief of Staff of the Zairean army, funded this operation to the amount of US$300,000. In March, Stabilco placed an advance force of 35 men on standby for deployment to Kisangani, and recruited 450 other troops to travel to Kinshasa by air. A significant number of those employed by Stabilco were former EO personnel. Stabilco was also commissioned to extract assets and people from Gbadolite, but the plan failed when looting Zairean soldiers attacked an advance party and the operation was aborted. Stabilco has closed down and its personnel now work for Sandline in Sierra Leone and for ECOMOG.[59]

Mercenary Intervention in Congo-Brazzaville

In mid-1997, another African country, the Republic of Congo-Brazzaville, was afflicted by an armed conflict. This civil war, fought mainly by militias, again saw the intervention of mercenaries. Paradoxically, it was the legal government of President Lissouba that had hired the mercenaries, from Israel and South Africa, to provide military training. On 10 October 1997 several mercenaries returned to South Africa and Namibia. President Lissouba, as his fortunes declined further, also hired Ukrainian helicopters and crews for MiG-21 planes in addition to getting assistance from UNITA rebels. He also appears to have used the services of Sandline International, but fell out with them over non-payment. His government fell to former dictator Denis Sassou Nguesso's forces, backed up by Angolan government troops. Nguesso reportedly captured and imprisoned several mercenaries of Russian nationality.

Unrest continued in Congo-Brazzaville in 1999, and there may have been a fresh influx of mercenaries in this conflict. The Vatican's news agency reported that 100 Cubans arrived in January 1999 to fight for dictator Sassou Nguesso.[60]

Gurkha Security Guards (GSG)

GSG is a privately owned British company offering security services. It was formed in late 1989, and was for much of its existence run as a private security company specialising in the recruitment and deployment of Gurkhas from Nepal. GSG has concentrated its work in Africa, protecting Lonrho's estates in Mozambique from 1990–92, and in 1991 providing a fully integrated security team of Gurkhas for Oderbrecht Mining Services in Angola. In 1994 GSG had an internal management split and faced financial ruin. In an attempt to reverse the decline, in late 1994 this small firm accepted an approach by J&S Franklin Ltd to train the Sierra Leone Military Force. GSG sent 58 Gurkhas and three European managers to Sierra Leone to train the military, and found themselves in active military operations to secure their training base.[61]

Disaster struck on 24 February when two of the European trainers, five Gurkhas and a platoon of RSLMF infantry walked into a rebel training camp by mistake, while on a reconnaissance mission. A firefight followed, in which at least 21 were killed, including the two GSG European trainers. By late March, GSG had pulled out and EO had begun preparing its operations. The negative publicity generated by GSG's mercenary adventure in Sierra Leone has ensured that this company has been unable to attract further lucrative contracts, and has since 1994 remained little more than a letterhead.

PRIVATE SECURITY COMPANIES

The growing number of private security firms operational in Africa form a category distinct from the private military companies and mercenary groups described above. The 1990s have been a boom period for such firms, although no audit has yet been conducted. What is less well known is the mushrooming of local private security firms.[62] Many of these firms provide only guards, but some are increasingly used by cash-strapped governments for selected tasks. For example, in Zambia, the Movement for Multiparty Democracy (MMD) government appears to use private security firms for surveillance of opposition politicians and some types of crowd control when civil unrest is feared. One important development in recent years is joint venture agreements with international firms, which could assist the transfer of international norms and encourage higher standards and best practice conduct in this industry. International private security companies will come under increasing pressure to adopt codes of conduct that are internationally recognised and monitored. The recent fortunes of Defence Systems Limited demonstrate the opportunities and challenges for a private security firm.

Defence Systems Limited (DSL)

Defence Systems Ltd was founded in 1981 by Alastair Morrison, a former British SAS officer who had experience operating in hostile environments, and realised the demand for such services.[63] Today the company has over 130 contracts for 115 clients in 22 countries. It is composed mostly of former British special forces officers and was originally associated with Hambros Bank. DSL claims that it 'never gets involved in other people's wars. It's simply not an aspect of our business, and business is good. We want to establish clear blue water between us and mercenary firms.' However the UN Special Rapporteur on Mercenarism, during his report to the 54th UN General Assembly, referred to DSL's activities as 'mercenary'.[64]

DSL began providing security and logistical personnel to the UN mission in former Yugoslavia in 1992, becoming the largest such contractor there to the UN with at least 430 personnel by February 1995. The firm was approached by the Papua New Guinea (PNG) government in the mid-1990s, to help establish a paramilitary police force for the country. Although the contract was never concluded, due to lack of funds on the part of the PNG government, DSL Chairman Alastair Morrison did recommend Sandline to PNG.[65]

In April 1997, a US firm, Armor Holdings, bought out DSL.[66] The takeover transaction was accounted for as a pooling of interest for US$7.6 million in cash for all preferred shares and US$10.9 million in stock (Armor also assumed US$7.5 million in debt). DSL's net revenues for 1996 were reported at US$31.1 million in annual pro forma revenue; it had more than 5,000 employees in 22 countries throughout South

America, Africa and South-east Asia, of which 100 only have access to 'firearms'.[67]

DSL's core business is 'devising and then implementing solutions to complex security problems'. DSL is contracted to the US State Department for the provision of security to high-risk embassies, as well as to other diplomatic clients such as the British High Commission in Uganda. DSL currently offers comprehensive services in eight main areas:[68]

- Mining and oilfield security: security consultancy, security audits and the provision of security management, training, personnel and equipment to the petrochemical and mineral extraction industries worldwide
- Specialist manpower: provision of qualified and experienced personnel for all levels of security, and for specialist training and project management
- Guard force management: selection, training, deployment and management of local guard forces for key installations including embassies in high-risk areas
- Humanitarian mine clearance operation: training of indigenous personnel in demining, mine awareness training to NGO staff, UXO clearance, and supply of demining equipment
- Security of communication routes: extensive experience of airline, airport and port security. Security of rail, sea and overland routes including cash and high value Goods in Transit
- Threat assessment: analysis of risk and exposure and recommendations of appropriate countermeasures
- Crisis management: work with the client to consider various potentially damaging scenarios and cogently plan the responses
- Technical security equipment: recommendations of cost-effective security equipment and systems complementing the security manpower.

DSL mishandled its operation in Angola badly, but was also a victim of the domination of the Angolan private security business by a number of senior MPLA military and security officials. There is fierce competition in the security market in Angola, with a number of security firms concluding joint ventures with Angolan partners – usually senior Angolan officials who are paid hefty dollar salaries to sit on the local companies' boards of directors.[69] Although there are some 90 registered private security firms in Angola, two dominate the market and control most of the business – Teleservices along the coast and oil areas, and Alpha 5 in the diamond areas.[70] Teleservices' main shareholders are the Chief of Staff of the armed forces, the commander of ground forces, the head of intelligence and the current Angolan ambassador to Washington. In January, apparently as a reward for his help in evicting DSL, the Interior Minister was given a 25 per cent stake in the company. Gray Security enjoys a management contract with both Teleservices and Alpha-5. This entails placing Gray personnel in key management positions and training positions in these companies.[71]

The division of responsibilities between Teleservices and Alpha 5 was developed by government officials in 1992 and has ensured a tight operation. However, DSL's success threatened the domination of the market by these two Angolan firms, which made its eviction necessary.

The domination of security firms tied into the state is a source of concern. A senior Angolan police commissioner complained in September 1998 that, 'these private security firms erode the State further. They are dangerous, we cannot regulate them as they are politically controlled by senior government officials'. The overrunning of Yetwane diamond mine in November by armed assailants was facilitated by lax security. Teleservices' personnel responsible for the mine's security responded to the attack by looting the mine themselves.[72]

DSL in Colombia

Angola is not the only country in which DSL has caused controversy. Its operations in Colombia have come under media scrutiny, and its conduct has been questioned by human rights organisations. British Petroleum contracted DSL in 1992 to run their security operation in Colombia. The BP contract is handled by a DSL subsidiary, Defence Systems Colombia (DSC). BP employs DSC to co-ordinate the defence of oil rigs and staff with the Colombia army and police. The oil field is situated in Casanare, a conflict zone where one of Colombia's strongest guerrilla forces, the Castroite National Liberation Army (ELN) is active.[73]

The area is the focus of a dirty counter-insurgency war by the Colombian army in which many human rights abuses have been reported. Although DSC was not directly involved, investigative journalists suggest that two ex-SAS DSC security advisers trained Colombian police with BP approval. On 30 April 1996, BP signed a contract with the Colombian police to create and dispatch a unit of policemen to Casanare to protect their oil installations. The contract was worth some US$5 million a year. Following an attack on a BP rig in May, BP tasked a team of DSC to train the police. The trainers wore police uniforms too. In two open letters released in April 1998, Human Rights Watch criticised the contractual relationship between Colombian security forces and two international consortia of oil companies operating the principal oil fields and pipelines in the nation.[74] The letters detailed terms of the multimillion-dollar security contracts and reports of killings, beatings, and arrests committed by those forces responsible for protecting the companies' installations. Human Rights Watch called on the companies to implement contractual and procedural structures to ensure respect for human rights in their security arrangements.

In Casanare department, the location of the Cusiana-Cupiagua oil fields developed by British Petroleum, ECOPETROL, Total and Triton, contracts came up for renewal in June 1998 as military, paramilitary, guerrilla and criminal activity increased in the area. The renegotiated contracts between the companies and the Ministry of Defence restructured the flows of funds to avoid direct company payments to state security forces. Payments for security were to be made to the state-owned ECOPETROL as a conduit to the Defence Ministry, instead of directly from the companies to the army. At the time of writing, the oil companies still made direct payments to the national police.

There were also some substantive changes in the contracts. BP, the only consortium member with a human rights policy, reported that human rights clauses were included in the new contract; an auditing mechanism was implemented to monitor the flow of funds; and a committee was established to monitor the performance of the military units providing security for the companies. Human Rights Watch could not assess the effectiveness of these programmes because the contract was not available to third parties. No mechanism to ensure that the personnel guarding these installations would be screened for human rights violations was apparent.

The conduct of private security providers for the BP-led consortium continued to be a problem in 1998. Following allegations in 1997 that DSC had imported arms into the country and trained Colombian National Police (PONAL) in counter-insurgency techniques, a government inquiry was launched to determine the role of this company and the police. DSC refused to co-operate with the investigation. In September 1998, BP reported that it had formed an oversight committee to monitor its private security providers, was developing a code of conduct for DSC, and had urged the company to cooperate fully with the government. Despite the allegations BP renewed its contract with DSC for one more year.

In October, new allegations were made that DSC and an Israeli private security firm, Silver Shadow, intended providing arms and intelligence services for the Colombian military while they were security contractors for the Ocensa pipeline. Reports also alleged that DSC had set up intelligence networks to monitor individuals opposed to the company. BP steadfastly denied these claims, but suspended a senior security official while investigating the allegations.

AirScan

The US Florida-based company AirScan has been under contract to protect the oil installations in Angola's oil-rich enclave Cabinda since 1995. Chevron is the main operator there.[75] AirScan also carries out day/night airborne operations and security missions, such as protecting US military space launch sites and classified US government assets. At present AirScan has deployed in Cabinda a modified Cessna E337 aircraft equipped with five tactical radios and a Global Positioning System (GPS) receiver. These conduct 24-hour searches for insurgents in the vicinity of the oil fields. Aerial equipment includes the Northrop Grumman WF-360TL and WF-160DS infrared/television systems, which provide x10 magnification and allow for a stand-off surveillance range of between 8,000 ft and 12,000 ft.

AirScan has made a number of proposals to the Angolan government to protect the oilfields and fisheries. It has offered a 'maritime surveillance and security system', which would include training crews, and providing surveillance aircraft, patrol boats and operational command centres.

AirScan has also been engaged in a joint venture project to train the security forces for the oil town of Malongo in conjunction with Alerta, a private security firm founded by the previous governor of Cabinda. Alerta is to take over complete responsibility for Malongo's security in 1999.

AirScan is itself not without controversy. It has been alleged to have been involved in the Angolan-Congolese intervention in Brazzaville in October 1997 to remove the democratically elected Pascal Lissouba in favour of former dictator Denis Sassou Nguesso. AirScan has also reportedly been involved in arms-trafficking from Uganda to southern Sudan to support the Sudanese People's Liberation Army (SPLA).

Conclusion

As we have seen in the case studies above, private military companies and classic mercenary firms have not been the silver bullet they market themselves as providing. In the case studies of EO, Sandline, the White Legion, Stabilco and GSG, none of these firms has shown the ability to provide anything but short and localised respites from conflict. They certainly have not enhanced stability or encouraged business confidence. Indeed, their poor human rights record, their lack of transparency, their engagement in arms transfers, their training in psychological warfare against civilians, their erosion of national self-determination and sovereignty in situations of crisis and their use of people with track records of human rights abuse does not bode well for the upholding of international law.[76] One has to question the legitimacy of these firms.

What gives them the right to choose a client or to infringe the established rights of sovereign states to non-intervention, as enshrined in the UN charter? Also, they do not offer an integrated approach to conflict resolution, although they later invest in protecting assets they have gained through the initial contract.

There is also increasing evidence of private military companies engaging in hostile actions to undermine one another in their efforts to obtain new clients. In this way, the foreseeable rivalry between oil and mineral companies and their accompanying private security companies could signify a dangerous step towards the 'privatisation of warfare',[77] which also raises the question of whether peace is in the interest of such companies, who would otherwise find themselves out of business. Sandline's aggressive push for a strong military response in Papua New Guinea is an example of this.

All of the case studies above focus on firms working with governments or governments in exile. However, at the end of the 1990s there is an increasing incidence of classic mercenary operations conducted with non-state parties, in Angola, in both Congos and in Sierra Leone. Some of these mercenaries seem to have worked for 'responsible' private military companies and then to have moved on to fighting for rebel forces to increase their wages. This trend has serious implications and urgently needs monitoring and tracking. But it is equally clear that it is only countries with lucrative mineral resources that can sustain this sort of enterprise. There is no sign of mercenary activity in countries that derive their wealth from tourism or agriculture.

Another problem is that private security firms may not be completely disassociated from the foreign policies of the countries they operate from. Security companies have often claimed that they operate with the tacit approval of their home governments and are a conduit for privatised intelligence-gathering. The controversy over Sandline International suggests that in this case British intelligence may have played a role in the operation. This in turn raises the issue of accountability, as private military companies are at present accountable only to their shareholders and clients.

It is for a number of these reasons that the days of private military companies like Sandline are numbered. The closure of EO on 1 January 1999 occurred not because it could not find a steady client base, but because it was too controversial and too high profile. Also, both Sandline and EO have been dogged by non-paying clients. EO pulled out of Sierra Leone with US$19.5 million still owed to it by the government. The debt was to be repaid at a rate of US$600,000 a month between March 1997 and the end of 1999, and was budgeted for in the country's yearly expenditures.[78] Following its efforts in Papua New Guinea, Sandline clawed back only some of what it was owed in 1998. Their failure to be paid cash up front and their need to tie their operations to mineral assets that will take time to generate cash has also been burdensome. In 1999 Heritage Oil & Gas and DiamondWorks both suffered from a weak market for oil and diamonds, and there was a rush to sell DiamondWorks shares in Canada in late January 1999 following a collapse in confidence in the company. As their assets are located in insecure zones, the production costs are also high, making these assets even less lucrative. In such a climate it is the large firms that have the institutional muscle to see them through the depression.

EO and Sandline have made an attempt to move into upmarket security, becoming more like Defence Systems Limited. Both EO and Sandline have argued for regulation.[79] In March 1998 Sandline published a paper called 'Private Military Companies – Independent or Regulated?' in which it supported a system of 'Registration, Approval, Project Authorisation and Operational Oversight', which it argued should be drawn up by 'governments and international institutions (such as the UN, EU and OAU)'.[80]

EO failed and Sandline will continue to find it difficult to get business that is paid for. Indeed, in 1999, the lack of business forced Sandline to change the criteria on who it will work with as an attempt to widen its potential client pool. Sandline now not only aims to work for recognised states but also says in its Corporate Overview that it will work with 'Genuine, internationally recognized and supported liberation movements.'[81] This raises the question over who decides who is a 'Genuine' non-state actor and who is not. Sandline's failed attempts to work with the KLA was the first tangible sign of this switch in strategy. The established private security companies are also vigilant and protective of their niche markets, and will increasingly ensure that there is 'clear blue water' between them and a Sandline type firm, especially as they will not want to face losing clients that include a growing number of NGOs. The flip side of the coin is that the flow of former Soviet and Eastern Europeans turning up to fight in Africa's wars is likely to continue. Although their numbers are still small, exposing their firms and the methods by which they are recruited is an urgent task.

Legislation regulating private military companies needs to be carefully considered. Alternatives to private security companies should also be considered. Countries can formally invite other governments to provide advice. The creation of a UN rapid reaction force would also mean that such firms would not be needed in situations of crisis. Another way would be to put firms under direct home state or UN control.[82]

The above are all responsive measures. However, the best way to avoid the use of such military firms is to invest in preventive action, dealing with core issues such as poverty and equitable, accountable use of resources and equal rights long before conflicts begin. The UN Special Rapporteur in his 1998 report recommended that:

The Commission on Human Rights should call for a study on ways of reinforcing international prevention, action and intervention machinery in order to strengthen the exercise of human rights and promote the rule of law in countries threatened or weakened by armed

Table 6.1: Sandline's Contract Breakdown

Item	Quantity	US$
Mi-24 helicopter	2@4,100.000	8,200,000
Mi-17 helicopter	2@1,500,000	3,000,000
Mi-24 ordnance	Table 6.2 below	2,500,000
Mi-17 ordnance	-	400,000
Night Vision equipment	18@33,880	610,000
Mi-24 aircrew	6	680,000
Mi-17 aircrew	6	860,000
Surveillance platform	1 (spotter aircraft)	2,400,000
On-board systems	1(?)	4,850,000
SP crew	4 (spotter pilots)	280,000
Ground System	1	600,000
Mission Operators	5	480,000
Ground Staff	5	270,000
Electronic warfare (EW)	Inc.	120,000
SP trainers	Inc.	120,000
Project co-ordinator	1	Inc.
Personnel equipment	30	250,000
Personnel movement	Pack	250,000
Insurance	Inc.	Inc.
Logistics support	Client	client
Asset positioning	4	1,200,000
Spares – helicopters	Pack	1,500,000
Spares – SP	Pack	600,000
Subtotal		29,170,000
Special Forces Team		
Manpower (40 plus doctors)	42	4,500,000
Equipment	Table 6.3 below	2,500,000
Positioning	Pack	100,000
Subtotal		7,100,000
Communications Equipment		
HF radio system	1+15	400,000
Hardened tactical radio sys	1+16	500,000
Satellite comms units	15	200,000
Subtotal		1,100,000
Contract Totals		37,370,000
Package price reduction		−1,370,000
Contract Fee to Client		36,000,000

conflicts, thereby ensuring that the purpose of hiring private companies of this nature, if indispensable, is solely to obtain technical and professional advice on military matters or police protection, within the legal framework expressly laid down.[83]

Unlike private military companies, private security companies are booming and their market is likely to grow. They will prosper as long as they avoid becoming involved in mercenary-type activities. DSL's problems in Colombia and Angola are good examples of some of the challenges this industry will face – how to keep clean in a hostile environment, and using mainly local staff.

Table 6.2: Mi-24 Ordinance

Item	Quantity
57mm rocket launcher pods	6
57mm high explosive rockets	
(for use against fixed installations, vehicles or boats)	1000
23mm ball	20,000
23mm tracer	5,000
23mm links	12,5000

Table 6.3: Special Forces Equipment

Item	Quantity
AK-47 assault rifle	100
PKM light machine guns	10
RPG-7 shoulder-held rocket propelled grenade launcher	10
Makarov Pistol	20
60mm mortar	10
82mm mortar	6
AGS-17 automatic 30mm grenade launcher	4
7.62x39 (for AK-47)	500,000
7.62x54 (for PKM machine gun)	250,000
12.7mm ball	100,000
12.7mm tracer	25,000
PG-7 (rounds for RPG-7)	1,000
40mm grenade	2,000
Illumination flare	200
smoke/frag grenade	800
AK-47 magazines	1,000
60mm HE (mortar rounds)	2,500
82mm HE (mortar rounds)	2,500
Ammo links	250,000
Personnel kit and uniforms	100

Source: Sandline site: <coombs.anu.edu.au/specialproj/png>

The private security industry requires regulation, as well as training in human rights, transparency and best practice. Multinational firms can encourage such practice by scrutinising who they hire and also building into their security arrangements independent auditing of operations by human rights and environmental rights specialists. As standards in the oil industry on Health, Safety and the Environment (HSE) became increasingly benchmark practice for a number of mineral and oil companies in the late 1980s, so will respect for human rights in the next millennium. This will make firms with links to EO-type operations very exposed in the market place.

Notes Chapter 6

1. As of 21 October 1998 there were 136,310 private security guards in South Africa operating with these firms.
2. Equitable Securities Corporation, *Equitable Securities Research*, 27 August 1997.
3. H. Howe, 'Global Order and Security Privatisation' in *Strategic Forum*, May 1998.
4. K. O'Brien, 'Military-Advisory Groups and African Security: Privatised Peacekeeping?' in *International Peacekeeping*, Autumn 1998, p. 81.
5. Countries that have satisfied the Convention include Angola, Belarus, Democratic Republic of Congo, Germany, Morocco, Nigeria, Poland, Romania, Uruguay, Yugoslavia.
6. Office of the UN High Commissioner for Human Rights, *Report on the Question of the Use of Mercenaries as a Means of Violating Human Rights and Impeding the Exercise of the Rights of Peoples to Self-determination* (E/CN/.4/1997/24), 20 February 1997, available at <www.unhchr.ch>.
7. D. Shearer, *Private Armies and Military Intervention*, Adelphi Paper 316, February 1998, p. 20.
8. Regulation of Foreign Military Assistance Bill (as amended by the Portfolio Committee on Defence (National Assembly) B 54B-97); see also M. Malan and J. Cilliers, *Mercenaries and Mischief: The Regulation of Foreign Military Assistance Bill*, ISS Papers, 25 September 1997.
9. This allows for advice or training, personnel, financial, logistical, intelligence or operational support, personnel recruitment, medical or paramedical services or procurement of equipment. Authorisation may be refused if they are 'in conflict with the Republic's obligations in terms of international law'. They can also be refused in the 'infringement of human rights and fundamental freedoms', where the assistance to be rendered may endanger peace by introducing destabilising factors into a region. The kind of operations Executive Outcomes conducted in Angola and Sierra Leone come under the bill's ban on 'direct participation as a combatant in armed conflict for private gains'.
10. See Kader Asmal, 'Restricting Mercenary Activity: the approach of the South African Government', Paper presented at The Hague Appeal for Peace Conference, 14 May 1999. Asmal wrote in this paper that, 'A kernel of the Act is to have an outright and explicit ban on direct military partici-

pation as combatants in armed conflict for private gain. This we wrote into the Bill when it was before Parliament as a result of the intervention of non-governmental organisations who wished this outright ban to be placed up-front in the legislation.'

11. *Report on the Question of the Use of Mercenaries*, Enrique Bernales Ballesteros, Special Rapporteur, pursuant to Commission resolution 1995/5 and Commission decision 1997/120, 27 January 1998, available at <www.unhchr.ch/html/menu4/chrrep/98chr31.tm>

12. Commission on Human Rights, Fifty-fifth session, 'Report on the question of the use of mercenaries as a means of violating human rights and impeding the exercise of the right of peoples to self-determination', submitted by Mr. Enrique Ballesteros (Peru), Special Rapporteur pursuant to Commission resolution 1998/6. E/CN.4/1999/11, 13 January 1999.

13. Ibid.

14. The US-based Military Professional Resources Incorporated (MPRI) is not covered in this chapter, although it attempted to operate in Angola with little success to date. However, MPRI has not been without controversy in its training of the Croatian and Bosnian armed forces and is reported to have had a significant planning role in the Croatian army's 1995 offensive known as Operation Storm, which according to the International Criminal Tribunal for Former Yugoslavia, is the largest single incident of ethnic cleansing during the Yugoslav civil war. In January 1997, peacekeeping forces impounded part of a US-MPRI weapons shipment of tank ammunition on the grounds that the amount had been under-reported to avoid a breach of the Dayton Peace Accord (see *Washington Post*, 29 January 1997).

15. For more information on EO's involvement in Angola see Chapter 2.

16. By this time EO was actually two companies, one based in Pretoria, South Africa, which originally had Eeben Barlow, Lafras Luitingh and Nico Palm as directors. In July 1997 Barlow resigned, leaving Luitingh and Nico Palm with 50 per cent of the shares each. The other, EO, was incorporated in England and Wales and its directors were Luther Eeben Barlow and his wife Susan; more on EO's structure in Chapter 2.

17. E. Rubin, 'An Army of their Own', *Harper's*, February 1997.

18. H. Howe, 'Private Security Forces and African Stability: The Case of Executive Outcomes', *Journal of Modern African Studies*, 36, 2, June 1998.

19. Ibid.; R. Silva, 'Angola Expulsa Jornalista do "DN"', *Publico*, 21 January 1999.

20. SABC TV (Johannesburg), in Afrikaans, 1800 GMT, 17 November 1995.

21. *Human Rights Watch* interview, 8 June 1995.

22. *Mail and Guardian*, 16–22 September 1994.

23. A.J. Venter, 'Executive Outcomes' Mercs & Migs Turn Tide in Angola', *Soldier of Fortune*, January 1996.

24. *Human Rights Watch* interview, Luanda, 14 March 1995.

25. A. Vines, 'La troisieme guerre angolaise', in C. Messiant (ed.), *L'Angola dans la guerre, Politique Africaine*, 57, March 1995.

26. 'Angola: Between War and Peace. Arms Trade and Human Rights Abuses since the Lusaka Protocol', *Human Rights Watch*, 8, 1 (A) pp. 17–21.

27. *Executive Outcomes. The War Business*, 1997, Journeyman Productions documentary film.

28. H. Howe, 'Executive Outcomes: A 999 Service for Africa'. Talk given at King's College, University of London, January 1997.

29. See *Angola, Arms Trade and Violations of the Laws of War since the 1992 elections, Human Rights Watch*, 1994.

30. J. Hooper, 'Air War in Angola', *World Air Power Journal*, 1997.

31. For more information on mercenary involvement in Sierra Leone see Chapter 3.

32. Shearer, 'Private Armies'.

33. J. Harding, 'The Mercenary Business: Executive Outcomes', *Review of African Political Economy*, March 1997.

34. Philip Trewhitt, *The Business of Killing* (Parliamentary Human Rights Group, London, July 1999), p. 15.

35. Elizabeth Rubin, 'An Army of One's Own', *Harper's* Magazine, February 1997.

36. International Campaign to Ban Landmines (ed.), 'Sierra Leone', *The Landmine Monitor*, (Human Rights Watch, 1999).

37. <www.unhchr.ch/html/menu4/chrrep/98chr31.htm>

38. <www.eo.com.> EO Press Release signed by Nick Van der Berg, EO's Managing Director.

39. A. J. Venter, 'Market Forces: How Hired Guns Succeeded where the United Nations Failed', *Jane's International Defence Review*, 1 March 1998.

40. Some sources claim EO tendered two contracts in March 1997, one worth US$36 million and another US$18 million.

41. *Human Rights Watch, Angola*.

42. See <www.sandline.com> and Chapter 2.

43. The best account of the Sandline debacle is by S. Dorney, *The Sandline Affair: Politics and Mercenaries and the Bougainville Crisis* (ABC Books, 1998).

44. The contract is published at <www.theage.com.au/special/asiaonline/sandline.htm.>

45. The Commander-in-Chief of the Defence Forces, Brigadier-General Jerry Singirok, was dismissed by the Prime Minister for criticising the contract. On 18 March 1997, the army protested against his dismissal, mutinied at the Murray barracks and marched on parliament. General Singirok later accepted his dismissal but called for a commission of inquiry. On 26 March an inquiry was set up and on 11 August 1997 it was expanded.

46. Australia agreed on 27 March 1997 to store at Tindal RAAF base in Australia's Northern Territory an Antonov 124 full of Sandline's military equipment until ownership was sorted out. A PNG delegation that inspected the store in April 1997 reported that the transport helicopters were earlier models than those PNG had been charged for and that the rockets were in poor condition.

47. 'Sierra Leone/Britain: Militias and Market Forces', *Africa Confidential*, Vol. 39, No. 21, 23 October 1998.

48. J&S Franklin placed £31,000 in a London account for the Papuan army chief. The firm, an agent abroad for Shorts of Belfast, GKN and other big defence contractors, has been accused by Amnesty International of offering to sell electric batons, which have been used by some regimes to torture dissidents. Franklin has a long history of supplying military equipment to Africa and introduced GSG to Sierra Leone. Sandline sources claim Franklin has been trying to erode their presence in that country by offering alternative bids. Franklin was also an agent of the Singapore government's

arms company, Unicorn International Pty, and organised the sale to the PNG government of small arms and heavy mortar launchers.

49. 'Report on the question of the use of mercenaries', 1998, Enrique Bernales Ballesteros, Special Rapporteur.
50. S. Dorney, *The Sandline Affair*.
51. Ibid.
52. Alex Vines, 'Mercenaries and the Privatisation of Security in Africa in the 1990s', Greg Mills and John Stremlau (eds), *Privatisation of Security in Africa* (South African Institute of International Affairs, March 1999), pp. 63–6.
53. <www.diamondworks.com.>
54. Tavernier's group consisted of 16 French, two Belgians, one Italian.
55. S. Boyne, 'The White Legion: Mercenaries in Zaire', *Jane's Intelligence Review*, June 1997.
56. <www.unhchr.ch/html/menu4/chrrep/98chr31.htm>
57. 'Sierra Leone/Britain', *Africa Confidential*.
58. S. Kiley, 'Freetown burns as rebels slaughter hundreds', *The Times*, 13 January 1999.
59. Personal communication from Sam Kiley, 11 December 1998.
60. SAPA news agency, 18 January 1999.
61. Vines, 'Gurkhas and the Private Security Business in Africa', in J. Cilliers and P. Mason (eds), *Peace, Profit or Plunder: The Privatisation of Security in War-Torn African Societies*. Institute of Security Studies and Canadian Council for International Peace and Security (1999).
62. See, for example, Network of Independent Monitors (NIM) (compiled in conjunction with S. Bletcher), *Safety in Security? A Report on the Private Security Industry and its Involvement in Violence* (NIM, 1997).
63. Morrison made his name leading the famous SAS rescue of a hijacked Lufthansa plane at Mogadishu Airport in 1977.
64. Enrique Ballesteros' presentation to the 54th session of the UN General Assembly, GA/SHL/3484, 23 October 1998.
65. *Daily Telegraph*, 3 April 1997.
66. Armor Holdings, Inc. is a leading provider of security products and services for law enforcement, governmental agencies and multinational corporations around the world. Founded in 1969 as American Body Armor & Equipment Inc, the company primarily manufactured armoured products until a controlling interest was acquired by Kanders Florida Holdings in January 1996. Since this time, Armor Holdings has made five strategic acquisitions which have broadened the company's product offering, expanded its distribution network, and facilitated entry into security services.
67. However, this has not always been the case in Mozambique, and in DSL's proposal for paramilitary training in Papua New Guinea, firearms featured prominently.
68. Note sent to Human Rights Watch, no date.
69. *Monthly Review Bulletin*, February 1998.
70. Alpha 5's main shareholder is the state-owned diamond mining company Endiama.
71. Gray's main market is South Africa. However, it also has through joint ventures expanded into Botswana, Mozambique, Namibia, Nigeria and Zimbabwe. It claims to operate under the code of practice under the International Standards Organisation, the body that sets standards for industrial

enterprises. There remains a suggestion that Gray has benefited from a close relationship with EO, something Gray denies.

72. *Sunday Times*, 15 November 1998.
73. 'Corporations and Human Rights' in *Human Rights Watch World Report 1999*. New York: Human Rights Watch, 1998, pp. 456–7.
74. A consortium composed of Occidental Petroleum, Royal Dutch/Shell, and the national oil company, ECOPETROL, which operates the Cano-Limon oil field in Arauca department, took no action to address reports of extrajudicial executions and a massacre committed by the state forces assigned to protect the consortium's facilities. Although the companies' response was that human rights violations were the responsibility of governments, and they did not announce any programs to ensure that their security providers did not commit human rights violations, Royal Dutch/Shell, the only member of this consortium with a human rights policy, announced its intention of selling its share of the project as part of an overall divestiture of its Colombian holdings.
75. Interview with Chevron, November 1998; A. Venter, 'Market Forces: How Hired Guns Succeeded where the United Nations Failed', *Jane's International Defense Review*, 31, 3, p. 24.
76. Both EO and Sandline claim they have human rights codes of conduct. EO claimed that 'As a rule, our training programmes emphasise the need for good manners which forms the foundation discipline and a high regard for universally accepted values and norms, based on the Universal Declaration of Human Rights (UN)', taken from EO Public Hearing Presentation to the Portfolio Committee, 13 October 1997.
77. O'Brien, 'Military-Advisory Groups'.
78. D. Shearer, 'Private Armies', p. 52.
79. EO stated at its public hearing to the Portfolio Committee on 13 October 1997 that, 'We are here today, not to object to the Regulation of Foreign Military Assistance (because we are not against regulation in principle), but in partnership with government to make an effort to ensure good legislation is drafted. We think the legislation should be objective, fair and practical – without denying people their basic and generally accepted rights.'
80. Sandline International, 'Private Military Companies – Independent or Regulated?'. Sandline International, 28 March 1998. Available at <www.sandline.com>
81. <www.sandline.com/company/index.html>
82. L. Nathan, 'Lethal Weapons: Why Africa needs Alternatives to Hired Guns', *Track Two*, August 1997.
83. <www.unhchr.ch/html/menu4/chrrep/98chr31.htm>

7

The OAU Convention for the Elimination of Mercenarism and Civil Conflicts

Kofi Oteng Kufuor

There is well-documented evidence of mercenary activity in a number of African conflicts since the collapse of the colonial system. In these instances, mercenaries have been recruited either by established governments trying to hold on to their authority, or by rebel movements committed either to overthrowing the government of the day, or radically altering the power configuration within a given state. For example, during the 1960s in Kinshasa, in the Congo, Moise Tshombe, the prime minister at the time, relied on the services of mercenaries for military support.[1] Mercenaries played a role in the abortive invasions of Guinea in 1970[2] and Benin in 1977,[3] and there was an attempt by mercenaries to overthrow the government of the Seychelles in 1981.[4] Furthermore, a coup d'état in the Comoros Islands in 1978 that resulted in the overthrow of the government was backed by French, German and Belgian mercenaries,[5] and another attempted coup in that country involved the use of mercenaries by the rebels.[6]

This chapter sets out to discuss the OAU Convention for the Elimination of Mercenarism (CEMA, or the Convention),[7] and particularly its capacity to prohibit incumbent governments in Africa from hiring the services of mercenaries[8] for the purpose of suppressing rebel uprisings.[9] The chapter critiques the Convention, highlighting the problems arising when its provisions are applied to the government of an OAU member state in an internal conflict. It examines the various approaches that could be used to interpret the Convention and the resulting uncertainty about whether its provisions bar governments from recruiting soldiers of fortune. The chapter concludes by suggesting means by which the uncertainty could be cleared up.

The OAU began to address the problem of mercenary activity in Africa when the events in the Congo in the 1960s led the OAU Council of Ministers to adopt a Resolution calling on the Government of the Congo among other things, to expel all mercenaries from its territory.[10] The crisis in the Congo also resulted in the OAU establishing a Special Committee[11] on mercenaries at its Kinshasa Conference in 1967, and the adoption of a Resolution by the Assembly of Heads of State and Government.[12] In response to the invasion of Guinea by mercenaries in 1970, the OAU Council of Ministers adopted another Resolution condemning this act.[13] In 1977, the OAU demonstrated its major concern about mercenary activities in the continent by drafting the Convention for the Elimination of Mercenarism in Africa.[14]

The Impact of Mercenaries on Armed Conflicts

Governments use mercenaries for the purpose of gaining the upper hand in a civil war. Closely linked to this is the fact that when mediation in a civil war has proven unsuccessful, the significance of an efficient fighting force becomes important as the parties to the conflict will tend to resolve the issue by force.[15] Moreover, current evidence points to the fact that the use of force to resolve a conflict is more likely to be the case than a peaceful outcome. Consequently, the better the military preparations that one side makes, the more likely it is to prevail.[16] Mercenary forces do pose problems. First, they can contribute to the increased flow of arms to a region that is already volatile. Observers of African security have pointed out that there is a link between private security firms, suppliers and governments searching for markets for a fiercely nationalist arms industry.[17] On a number of occasions, mercenaries have been used to subvert democratically elected governments, given that their main motivation is profit and not ideology. Finally, notwithstanding the hype about their expertise and efficiency, mercenary force is easily replaceable in conflicts. The effective use of force that mercenaries are credited with is not beyond the capability of regional or global multinational armies, given the right support and conditions. From the point of view of conflict management, multilateral forces are clearly more desirable. They do not have to do their paymasters' bidding as mercenaries necessarily do and are thus more likely to be non-partisan. Multilateral intervention can therefore be better geared toward creating an environment where a genuine political settlement can be achieved.[18] Moreover, private armies do not come cheap. The Sierra Leone government signed three separate contracts for the services of mercenaries over the 22 months they were in the country. They were for a total of US$35 million – an average of more than US$1.5 million a month to be paid in cash.[19]

Another perceived disadvantage inherent in mercenary deployment is that a private military force could easily upset the delicate balance between the country's political leaders and its military, as the civil war in Sierra Leone has clearly demonstrated. A decision by a government to recruit mercenaries may be construed as a sign of desperation and of lack of faith in its own armed forces. It is likely to cause severe loss of face for its military personnel. In Sierra Leone, for example, the military overthrew the government in May 1997 because it believed its position had been undermined. Arguably, the mercenary presence in Sierra Leone at the time was a contributing factor, as it highlighted the shortcomings of the Sierra Leone armed forces.[20]

CEMA and internal conflicts in Africa

In the late 1970s and early 1980s the problem of mercenaries attracted little interest at OAU summit meetings. The liberation struggle was virtually over and most states had consolidated their independence. As such, mercenary activity on the continent had declined. However, in the late 1980s the continent was forced to confront the issue again, due to the proliferation of mercenaries in a number of internal conflicts. This time, however, it was not the colonial or racist governments who hired the mercenaries but predominantly incumbent African regimes. A number of African governments, such as Sierra Leone,[21] and Angola,[22] relied on mercenaries to help repel rebel attacks.

Among the several mercenary outfits active in Africa, probably the most renowned is Executive Outcomes. EO claims to provide general security services, military training, infantry troops and air support. It assembles teams of military personnel for each contract it enters into from a list of over 2,000 former members of the apartheid-era South African Defence Force and the South African Police, as well as other southern African and European ex-servicemen. EO selectively recruits former members of the notorious and feared special force units like the 32nd Battalion, the Reconnaissance Commandos, the Parachute Brigade, and the Paramilitary Koevoet force. These were men used regularly in the apartheid era to covertly destabilise neighbouring countries and to frustrate internal opposition. Former members of the SADF are attracted to employment with Executive Outcomes because it offers high salaries, medical and life insurance benefits, and the opportunity to use their military training.[23]

EO's first major contract was in Angola in 1993, initially securing stability around mineral fields and later fighting on behalf of the Angolan government against Jonas Savimbi's UNITA forces. EO was also employed in Sierra Leone in 1995 in a bid by the military government to defeat the rebels of the Revolutionary United Front.[24] Although EO works as an independent security firm, it has close ties to mineral exploration companies. This combination has been likened to a modern version of the British East India Company, which employed its own security force to protect its economic interests in India during the eighteenth century. EO is a commercial venture motivated principally by profit and will charge what it feels the market will allow, and not merely to cover its costs. The expense may prompt a desperate regime to empty its coffers, thereby jeopardising future government programmes.[25] There is little incentive for a security company to prolong its stay if a country is unable to pay. The company may cut short its operations when the money runs out, regardless of the consequences. Alternatively, a lucrative contract may act as an incentive to prolong violence and justify a company's existence.[26] It is obvious from the above that EO, and similar mercenary outfits, play an important role in Africa's civil wars. What is the position of CEMA on this score? Does it prohibit the use of mercenaries under all circumstances? Or, crucially, does it make an exception, albeit implied, for established governments?

Any attempt to answer these questions must start from the definition that the Convention gives to the mercenary, as any person who:

- is specifically recruited locally or abroad in order to fight in an armed conflict
- does in fact take a direct part in the hostilities
- is motivated to take part in the hostilities essentially by the desire for private gain and in fact is promised by or on behalf of a party to the conflict material compensation
- is neither a national of a party to the conflict nor a resident of a territory controlled by a party to the conflict
- is not a member of the armed forces of a party to the conflict
- is not sent by a state other than a party to the conflict on official mission as a member of the armed forces of the said state.[27]

The first task that confronts us is to ascertain whether the mercenaries recruited by governments are participating directly in conflicts or are merely offering technical advice.[28] The wording of Article 1 (b), which defines a mercenary as a person recruited to 'take a direct part in hostilities', does, on the face of it, mean that a mercenary is someone who, for example, engages in actual exchange of fire with opposing forces, or flies fighter planes in combat, or is part of a crew of a naval vessel that engages enemy forces.

The problem is that this provision contains a loophole, the effect of which can possibly defeat the entire purpose of the Convention. This is because verifying the exact role of mercenaries fighting on the side of a government in civil conflict could be quite difficult. In the absence of a statement by the host government admitting the direct participation in combat of persons recruited within the definition of CEMA, or one or more of the mercenaries being captured during combat by rebel forces, there will be a lack of evidence to establish the fact that the host has violated Article 1 (b) of the Convention. Any allegation levelled at a particular government that it has breached the Convention could be met with the riposte that persons so recruited have only been contracted to train the government armed forces and that their activities do not violate Article 1 (b).

The above point assumes importance when juxtaposed with Article 1 (2). This Article, taken at face value, prevents governments from using mercenaries. Among those capable of committing the crime of mercenarism under this article are the state or its representative. The article prohibits individuals, groups, associations, states or representatives of states from conducting a range of affairs that facilitate the presence and activities of mercenaries on the continent. This paragraph of the CEMA provides greater guidance than the draft document on mercenaries produced by the OAU in 1972.[29] This draft Convention did not specifically mention the state or its representatives in connection with the offence of recruiting, financing, training or protecting mercenaries.[30] The argument here therefore is that this conscious decision to bring the State or its representatives within the scope of CEMA amounts to a specific intent to bar them from violating the Convention.

Under Article 6 there is a general obligation on all the contracting parties to adopt measures aimed at eradicating all mercenary activities in Africa. Thus the contracting parties are to prevent their nationals or foreigners in their respective territories from engaging in any of the acts set out in Article 1 of the CEMA. Furthermore, contracting parties are to prevent mercenaries from entering or passing through their territory and they are to also prohibit on their territory any activities by persons or organisations who use mercenaries against, inter alia, the people of Africa in their struggle for liberation.

This struggle for liberation might not necessarily be limited to the anti-colonial /anti-racist struggle. After all, with the collapse of the apartheid system in South Africa the colonial-racist context is no longer applicable. Hence this would either mean that the CEMA is irrelevant and obsolete, or it is still very valid, given the continuing struggle by the people of Africa against authoritarian and undemocratic regimes. If the contracting parties are under an obligation to eradicate mercenary activity, it will then seem odd that any of the same contracting parties can turn round and actively employ mercenaries. The latter action is inconsistent with the former undertaking. Thus, in our opinion, Article 6 makes no exception for governments caught up in civil wars. Of course, it could be argued that in internal conflicts there is no violation by force of arms of the territorial integrity of another state, or the stability of another state, or a process of classical self-determination (independence). Thus, governments can freely employ mercenaries in civil wars. As far as the current role of mercenaries is concerned, the focus is on their role in suppressing rebel uprisings. Such uprisings can easily be construed as a process of self-determination. Self-determination is no longer limited to the colonial context. It has been proven to be central to recent political developments in the former Soviet Union, Yugoslavia and to the Kurds, just to mention a few examples. [31] To labour the point, the African Charter on Human and Peoples' Rights states that resistance to oppressive rule is part of the process of self-determination. [32] If, for instance, a government noted for its human rights violations and its undemocratic nature uses mercenaries to crush a popular rebel uprising, this could arguably be a violation of Article 1 (2).

However, the failure of the CEMA to expressly bar governments from recruiting mercenaries does create the possibility that instead of relying on the provisions of the CEMA in isolation, we can broaden our analysis and rely on the context within which CEMA was drafted. It is this alternative standpoint that throws up what is possibly a different answer to the question of whether the CEMA applies to governments. The arguments marshalled in support of this view are as follows:

1. The customary maxim in international law is that if a particular action is not expressly required or banned under international law, then it is up to the states in the international system to decide whether to carry out or refuse to carry out a given act. [33] Applying this to CEMA, it could be argued that there is no specific provision that prohibits states from recruiting mercenaries; hence the operations of EO and other firms is not a violation of CEMA.

2. None of the governments that have relied on the use of mercenaries has been a colonising power or a racially oppressive state. This was the context or situation that underpinned the series of OAU Resolutions on mercenaries, culminating in the adoption of CEMA in 1977. It could therefore be argued that the use of mercenaries by a government against rebels in a civil war does not amount to the kind of threat envisaged in the Convention's Preamble. [34]

3. How do we reconcile the principle of non-interference in the affairs of a sovereign state under the OAU Charter [35] with the probable cause for concern by the member states of the OAU about the role of mercenaries in Africa? OAU members have generally resisted attempts at external interference in their affairs, upholding vigorously the doctrine of state sovereignty. Nothing was written in the Convention to suggest that this traditional doctrine did not apply. Thus, whom a government employs to help sustain it in power has never really formed part of the concerns of the OAU. Indeed, the history of Africa's civil wars is dotted with instances where member

states of the OAU have relied on other African governments and foreign forces to shore them up in the face of internal threats. Arguably, therefore, mercenaries fall into this grey but quasi-permissible category of forces that provide external assistance to African states. In effect, the right to conduct internal affairs in a manner the state thinks fit provides sufficient justification for the employment of firms such as EO.

In strong counterargument to the above, there have been clear signals since the end of the Cold War that the OAU is shifting away from the notions of 'territorial integrity' and 'non-interference in the internal affairs of other states'. Though this shift in position is not codified in law, the attitude of neighbouring states to internal conflicts has become more interventionist, rather than aloof. It is obvious that the impact of some domestic conflicts on their neighbouring regions is, in fact, transforming such conflicts from civil wars into international conflicts. Besides, the genocidal wars that have characterised Africa of late have brought to the fore gross violations of human rights. The questions of impunity and human rights abuses are no longer seen as 'internal' issues but matters that must provoke immediate international responses. Thus it is perhaps not so accurate to assert that a state may do whatever it so wishes within its borders.[36] A government in a civil war may still be in violation of the CEMA depending on the impact of the war and the construction of the Convention's provisions.

Conclusion

Mercenaries are in demand because of the perceived strategic advantage they give to the parties in a conflict that hire them. In Sierra Leone, by the time EO entered the civil war, the RUF had advanced to almost 20 km of Freetown, the capital. EO, responding to the crisis, achieved two key results. Firstly, the government was put in a much stronger strategic position; the intervention helped make elections possible and parts of the country were made safe for some of the more than one million displaced persons to return home. Secondly, there was a clear correlation between the outcome of EO's military operations and the willingness of the RUF to negotiate. A senior international diplomat, remarking on the influence of EO's military operations against the RUF, observed that military pressure always needed to be put on before negotiations could succeed.[37]

There remain, however, unanswered questions about the long-term impact of private military intervention in civil wars. For instance, how does a 'victorious' party in conflict, bolstered by such unaccountable military muscle, respond to the need for long-term stability? How does the 'victorious' party reconcile the desire to defend its territorial sovereignty with the fact that the instruments needed to maintain sovereignty are controlled by an external body accountable only to its commercial interests? The answers to these questions, based on the Sierra Leone experience, do not speak in favour of mercenary intervention in civil wars.

In the light of the perceived usefulness of mercenaries to vulnerable regimes, it is probable that governments in Africa, when faced with internal security threats, will continue to rely on the services of those willing to hire themselves out as soldiers of fortune. Accordingly, the question as to whether the CEMA applies to civil conflicts will continue to be of relevance. In particular, there is a need to redraft the Convention in order to clear the ambiguity in the text, or for the OAU to come out with a resolution specifying whether its members can or cannot use soldiers of fortune.

If the OAU chooses to expressly ban the employment of mercenaries in civil wars, then the following comments could be of some assistance in the redrafting of the CEMA. Firstly, there should be a clause inserted into the Convention expressly forbidding the recruitment of mercenaries by governments of the OAU. Secondly, to avoid possible exploitation of the loophole in Article 1 (b), the Convention could be amended to ban even the mere presence of mercenaries in the ranks of a party to a conflict even if they are used only as instructors and advisers. Thirdly, a means of investigating claims regarding the use of mercenaries in internal conflicts should be established with rules regarding states which have standing in any investigation or related action. Fourthly, if governments wish to retain the services of foreign advisers and technicians, it will be necessary to provide an exception for such personnel. It is submitted that their case may best be dealt with by administrative arrangements annexed to the CEMA. A request for such personnel would have to be made by one state to another. It would be up to the state providing such specialists to decide whether to accede to the request or not. If the providing state decides to honour such a request, it will be its responsibility to organise the assistance and answer, jointly with the soliciting state, for the future conduct of the specialists. It would be a defence for a person charged with a CEMA offence to show that he was recruited under such an inter-state agreement and the requested state would have a defence against a charge of intervention.[38]

Notes Chapter 7

1. See F. Scott Bobb, *Historical Dictionary of Zaire* (Scarecrow, 1980) at p. 215 under the heading 'Tshombe', and p. 143 under the heading 'Mercenaries'. See also Hoare, *Congo Mercenary* (Robert Hale, 1967). Tshombe's actions led to the OAU Council of Minister's adopting a resolution in the matter.

The essence of this resolution was that the Council of Ministers of the Organization of African Unity:

- is deeply concerned by the deteriorating situation in the Democratic Republic of Congo resulting from foreign intervention as well as the use of mercenaries principally recruited from the racist countries of South Africa and Southern Rhodesia
- reaffirms the resolutions of the Organization of African Unity inviting all African states to abstain from any relationship whatsoever with the Government of South Africa because of its policy of apartheid
- considers that foreign intervention and use of mercenaries has unfortunate effects on the neighbouring independent states as well as on the struggle for national liberation in Angola, Southern Rhodesia, Mozambique and the other territories in the region which are still under colonial domination, and constitutes a serious threat to peace in the African Continent
- appeals to the Government of the Democratic Republic of the Congo to stop immediately the recruitment of mercenaries and to expel as soon as possible all mercenaries of whatever origin who are already in the Congo so as to facilitate an African solution. ECM/Res.5 (III) 10 September 1964 (Addis Ababa).

2. See 'Guinea, Attempted Invasion', *Africa Research Bulletin: Political Social and Cultural Series*, 7, 1–30 November 1970, p. 1933. This led to the OAU adopting a resolution condemning the attempted invasion. See ECM/Res. 17 (VII) (December) 1970 (Lagos).

3. See 'Benin: Abortive Coup', *Africa Research Bulletin: Political, Social and Cultural Series*, 14, 1–31 January 1977, pp. 4288–9.

4. See *Africa Research Bulletin: Political, Social and Cultural Series*, 18, No.11, 1981, pp. 6254–6.

5. See 'Comoros Islands, President Overthrown,' *Africa Research Bulletin: Political, Social and Cultural Series*, 15, 1–31 May 1978, p. 4849.

6. For reports of mercenary involvement in this coup attempt see the following accounts: British Broadcasting Corporation (BBC) *Summary of World Broadcasts*, Africa, Latin America and the Caribbean, 'Comoros, coup attempt led by French mercenary Bob Denard, president held by coup leaders', 29 September 1995, AL/2421 A/6; BBC *Summary of World Broadcasts*, Africa, Latin America and the Caribbean, 'Comoros, Coup leaders set up transitional military committee; French troops placed on alert; Head of transitional military committee named as Captain Kombo; Interview with coup leader Captain Kombo; Coup attempt said successful; new committee says political leaders to be tried'. AL/2422 A/5; 32 *Africa Research Bulletin: Political, Social and Cultural Series*, 'Comoros Islands, French End Coup', 1–31 October 1995, p. 12021.

7. The text of the Convention is reprinted in Gino J. Naldi (ed.) *Documents of the Organization of African Unity*, 58 (Mansell, 1992), and is also reprinted in the Appendix to this article. The Convention entered into force on 22 April 1985. See Report of the Secretary-General on the Status of OAU Treaties, *African Journal of International and Comparative Law*, 10, No.3, 1998, pp. 522, 528–9.

8. For an introduction to the issue of mercenaries in international law, see the following: Marie-France Major, 'Mercenaries and International Law', *Georgia Journal of International and Comparative Law*, 22, 103 (1992); Edward KwaKwa, 'The Current Status of Mercenaries in the Law of Armed Conflict', *Hastings Journal of International and Comparative Law*, 14, 67 (1990); James L. Taulbee, 'Myths, Mercenaries and Contemporary International Law', *California Western International Law Journal*, 15, 339 (1985); P. W. Mourning, 'Leashing the Dogs of War: Outlawing the Recruitment and Use of Mercenaries', *Virginia Journal of International Law*, 22, 589 (1982); Henry W. Van Deventer, 'Mercenaries at Geneva', *American Journal of International Law*, 70, 811 (1976); N. C. Burmester, 'The Recruitment and Use of Mercenaries in Armed Conflicts', *American Journal of International Law*, 72, 37 (1978); Juan Carlos Zarate, 'The Emergence of a New Dog of War: Private International Security Companies and the New World Disorder', *Stanford Journal of International Law*, 34, No.1 (1998); Report on the Question of the Use of Mercenaries as a Means of Impeding the Exercise of the Right of Self-Determination, UN ESCOR, 44th Session, UN Doc. E/CN.4/1988/14 (1988); Report of the Ad Hoc Committee on the Drafting of an International Convention Against The Recruitment, Use, Financing, and Training of Mercenaries, UN GAOR, 40th Session, Supplement No.43, UN. Doc A/40/43 (1988).

9. Historically, mercenaries have always been relied upon by governments in times of war. See for instance Charles Clavert Bayley, *Mercenaries for the Crimea: The German, Swiss and Italian Legions in British Service, 1854–56*, (McGill-Queens University Press, 1977). For an account of mercenaries in Africa, see S. J. G. Clarke, *The Congo Mercenary: a History and Analysis*, (South African Institute of International Affairs, 1968); and Mike Hoare, *Congo Mercenary* (Robert Hale 1967).

10. ECM/Res. 5 (III) 10 September (1964) (Addis Ababa). Reprinted in Robert E. Cesner and John W. Brant, 'Law of the Mercenary: An International Dilemma', *Capital University Law Review*, 6 (1977), 328, Appendix II, p. 363.

11. See *Africa Research Bulletin: Political, Social and Cultural Series*, 4, 1967, No. 9, p. 856 A.

12. See AHG/Res. 49 (IV) September (1967) (Kinshasa) in Cesner and Brant, *Law of the Mercenary*, p. 364.

13. See ECM/Res. 17 (VII) December (1970) (Lagos). See Cesner and Brant, *Law of the Mercenary*, p. 365.

14. Also, as a response to the overthrow of the Government of the Comoro Islands in 1978, the OAU Ministerial Council expelled the delegation from that country on the grounds that it had been installed by mercenaries. See 'Emergency Powers for the OAU', *West Africa*, 17 July (1978), 1375.

15. David Shearer, 'Dial an Army', *The World Today*, 53, Nos. 8–9 August/September 1997, pp. 203–5, 205.

16. Ibid., p. 205.

17. Ibid., p. 205.

18. See Funmi Olonisakin, 'Mercenaries Fill the Vacuum', *The World Today*, 54, No.6, June 1998, pp. 146–7.

19. Shearer, 'Dial an Army', p. 205.

20. Ibid., p. 205.

21. Events in Sierra Leone at the height of that country's civil war demonstrate the role of mercenaries in Africa's civil conflicts. The Revolutionary United Front (RUF) is a rebel organisation that commenced its military activities in 1991 with the aim of overthrowing the government of General Joseph Momoh in Sierra Leone.
    Despite a change in Government, and the determination of a new military junta at the time, the rebels continued to wage war against the new military authorities with a considerable degree of success. For example see 'Sierra Leone: Freetown Within Sight?', *Africa Research Bulletin: Political Series*, 32, April 1–30, 1995, p. 11832. As a result, the new administration engaged the services of Executive Outcomes as part of its war effort with the hope that EO could train its soldiers and improve their fighting capabilities.

22. See *Africa Research Bulletin: Political Series*, 30, No.2, p. 10932, 1–28 February 1993.

23. See Alan Rake, 'Dangerous Dogs of War', *New African*, November (1995), 10. See also, 'Sierra Leone, Rent an Army – At a Price', *Africa Research Bulletin: Political, Social and Cultural Series*, 32, 1–30 June 1995, 11894; 'Mercenaries under Attack', *West Africa*, 5–11 June 1995, p. 882.

24. Shearer, 'Dial an Army', p. 203.

25. Ibid.

26. Ibid., p. 205.

27. Convention for the Elimination of Mercenaries (CEMA), Article 1 (1), (a–f).

28. For instance, in the Sierra Leone civil war this was not very clear. This is because on the one hand it was reported that EO was only to train the Sierra Leone army. See *Africa Research Bulletin*, 32, p. 11894; and also Rake, 'Dangerous Dogs of War,' *New African*, 12. However, since then there have been reports that they have participated directly in combat, for example by flying helicopter gunships for the NPRC. See Christopher Johnson, 'Troops Foil Coup in Sierra Leone', *Independent*, 4 October 1995, 'Mercenaries at Work', *West Africa*, 15, 23–29 October 1995, p. 1633.

29. OAU Convention for the Elimination of Mercenaries in Africa, OAU Doc CM/433/Rev.L, Annex 1 (1972) reprinted in P.W. Mourning, 'Leashing the Dogs of War', in *Virginia Journal of International Law*, 22, p. 613.

30. See ibid., Article 2.

31. See Robert McCorquodale, 'Self-Determination Beyond the Colonial Context and its Potential Impact on Africa', *African Journal of International and Comparative Law*, 4 (1992), pp. 592, 596–601.

32. See Article 20 (2). The text of the Charter is reprinted in International Legal Materials, XXI No.1, 59, 1982.

33. See Zarate, 'The Emergence of a New Dog of War', p. 143, citing Manuel R. Garcia-Mora, *International Responsibility for Hostile Acts of Private Persons Against Foreign States*, p. 121, Note 282.

34. CEMA, paragraphs 2 and 3. The issue of the role of mercenaries in the political destabilisation of Africa is also amplified in UN Conventions and Resolutions. For example, in Resolution 44/81, the General Assembly noted in the preamble that the activities of mercenaries are contrary to fundamental principles of international law, such as the non-interference in the internal affairs of states, territorial integrity and independence, and seriously impede the process of the self-determination of peoples struggling against colonialism, racism and apartheid and all forms of foreign domination. The preamble also makes reference to the relevant resolutions

of the OAU and the convention adopted by the Assembly of Heads of State and Government of the OAU at its 14th ordinary session condemning and outlawing mercenarism and its adverse effects on the independence and territorial integrity of African States.

Furthermore, paragraph 1 condemns the increased recruitment, financing, training, assembly, transit and use of mercenaries, as well as all other forms of support to mercenaries for the purpose of destabilizing and overthrowing the governments of Southern Africa and Central America and of other developing states and fighting against the national liberation movements of peoples struggling for the exercise of their right of self-determination. Also in Resolution 43/107, the preamble again underscores the General Assembly's view that the use of mercenaries constitutes a threat to international peace and security; their activities are contrary to fundamental principles of international law, such as non-interference in the internal affairs of states, territorial integrity and independence, and mercenaries seriously impede the process of self-determination of peoples struggling against colonialism, racism and apartheid and all forms of foreign domination. See ibid., paragraphs 4 and 5.

35. See Article III (2).

36. See Zarate, 'The Emergence of a New Dog of War', p. 143.

37. Shearer, 'Dial an Army', p. 204. EO had actually been given four military objectives: to make Freetown safe from the RUF threat; generate revenue for the government by securing the Sierra Rutile area and diamond fields, guarantee payment; eliminate the RUF headquarters; and clear the remaining areas of RUF occupation.

38. On this fourth proposal, see F.J. Hampson, 'Mercenaries: Diagnosis Before Prescription', *Netherlands Yearbook of International Law*, XII (1991) pp. 3–38.

8

Understanding the African Security Crisis

Eboe Hutchful

African politics since 1990 has undergone a wave of formal democratisation, coinciding with and often spurred on by the end of the Cold War. This dual process has facilitated the signing of peace accords which have brought an end to some of the continent's most durable armed conflicts and to the dumping of weapons on the continent by great powers. It also encouraged fleeting thought about a paradigm shift in the security discourse away from the military-based concept. However, the winding down of the Cold War has also empowered internal constituencies within individual African states to challenge former client regimes in ways unimaginable in the past. The vulnerability of the state, coupled with liberalisation, has made possible novel ways of articulating political, commercial and military agendas. The result has been a descent into a new cycle of violence in which private military companies have become a recurring decimal. Over the last decade, Africans have been subjected to an extraordinary variety of sources and forms of violence: civil wars, ethnic pogroms, religious conflict, political repression, forced migrations, and the upheaval associated with structural adjustment. The end of the Cold War has thus sharply undermined existing security paradigms and practices, intensifying the security dilemma of some African states, and giving rise in the process to complex new forms and permutations of force, within and beyond the state. In response to these developments, African states and civil society, as well as foreign donors and powers, are seeking (in collaboration as well as competition) to reconfigure security concepts and architectures.

This chapter seeks to unravel the African security dilemma, including the rise to prominence of non-state military formations and private foreign armies, and to discuss options for addressing the security challenges facing the continent.

The Security Paradox

There is a certain paradox in the way that debates on security in Africa have developed since the last decade. By the end of the 1980s, critics were arguing that 'security' needed to extend beyond its political and military meaning to encompass the wider understanding of the term, such as the satisfaction of basic needs, the right to a sustainable environment, protection of cultural and religious identity, and so on. These critics have also sought to understand not only the systemic sources of violence embedded in the domestic and global orders – including those associated with capitalist economic restructuring – but also how the practices of the state itself have constituted a fundamental source of insecurity, not solely in the political but also in the social and ideological realms. However, the genocidal conflicts in Somalia, Rwanda, Algeria, Liberia and Sierra Leone have led to the discussion being refocused once again on the most primordial meaning of security, which is the right to protection from physical abuse and attack. Physical safety became the pre-eminent concern for many Africans as states collapsed and were unable to provide fundamental conditions for protection of life. Thus both the 'primordial' and 'extended' concepts of security are being problematised in tandem.

The erosion or collapse of security infrastructures in Africa is linked in turn to broader and more longstanding problems of governance and development, as well as to developments at the core of the international system of power and its arms economy. However, while 'governance' and 'development' received considerable emphasis from donors in the 1980s, issues of 'security' (like those of the arms economy) received very little attention – although by the mid-1990s this would change as concerns about the stability of African states grew.

Depending on how it is viewed, 'security' is both a dependent and independent variable, a cause as well as effect of both governance and development outcomes. The relationship between governance and security is at once intimate and obvious. First, governance is both about creating and assuring conditions of security, and at the same time is necessarily underpinned by the management of the instruments of violence. The capacity for legitimate use of coercion is a cornerstone of governance. Secondly, governance involves the effective administration, regulation and control of the instruments of violence. However, we need to speak not only of governance and security but also of governance *of* or *over* security: of governance not only as the projection of security but also as the effective management of the institutions of violence, i.e. the security apparatuses. This is due in part to a paradox of security.

Rulers may discover that the attempt to strengthen internal security also increases the power of 'strongmen' within their security apparatus, and hence the threat to the ruler himself. Extending the monopoly of the state over security is thus liable to increase the insecurity of the ruler, as 'strongmen' arise within the very organs of security. This is the source of what may be termed the 'ruler's security dilemma'. Deliberately weakening the power of the direct controllers of violence is an option, attractive in the short term but potentially fatal in the long term. However, this 'security dilemma' is not confined to the ruler. Managers of the security apparatus, as repositories of power, also face their own governance challenges: first, in terms of the technical efficiency of the production of force, and second, in terms of similar threats of strongmen from within the ranks of the institution. In many African armies and security forces, informal links and structures of power, based on such factors as ethnic, family and political connections, count for much more than formal hierarchy and lines of command. Hence, a key issue in the outcome of governance is how holders of political power and managers of coercion relate to one another, as well as how they respond to institutional and political challenges from 'strongmen' within constituencies under their formal control.

Security as a Racket

The relationship between the three variables of governance, security and development may be seen in the triple basis of the modern state, namely: social and territorial control; monopoly over the legitimate use of force; and accumulation and the drive for revenue. These facets of the foundation of the state are driven by different forces and entail dealing with different constituencies, and are brought together by the political entrepreneur, ideally in a formula that enhances political accountability and responsiveness, security, economic growth and state revenues. Such an outcome is not assured, given the fundamental conflict between popular sovereignty (political legitimacy), market sovereignty (accumulation), and the bureaucratically organised interests of the state itself (power, in other words).

However, *security* is the crucial historical variable in the rise of the state. According to Tilly, states arose as a 'security racket', trading protection to merchants and others in return for revenues and other services, and in the process providing a framework for the organisation of production, exchange and accumulation.[1] A similar thesis has been advanced by Hymer and others to explain the rise of African empires, which were able to extract monopoly rent in return for extending military protection to long-distance traders.[2] War (as well as other forms of external competition) did for the state what competition in the marketplace did for the entrepreneur: it disciplined the state, forcing it to hone its capacities and extend its control over internal populations and resources, the price of failure being collapse or takeover by a rival state.

In administering this 'racket' the ruler has to respond to several challenges. One is what we have already termed the 'ruler's security dilemma', namely, those who *organise* the racket are liable to be displaced by those who actually *execute* the racket. The spate of coups d'état and the use of the early dogs of war in Africa in the 1970s are indicative of this dilemma. Thus the reliability and efficiency of the ruler's security implements are in potential conflict with each other. A second problem is: how to fund the racket? The avenues have included predatory practice, trading monopolies, taxing and borrowing from merchants, seeking protection under the wings of a foreign patron, and growing taxation of citizens (getting citizens to pay for their own protection – or, as is at least as likely to be the case, repression). Tilly refers to taxation to suggest a counterintuitive link between military and coercive power and the development of political accountability: warfare creates expanding demand for taxes and revenue, which in turn allows the lower strata to demand a voice in decision making. Whatever the merits of this argument, we do know that historically, the consolidation of the state's coercive power and territorial and social control has proceeded in tandem with the activation of civil society and its ability to press demands on the state and defend its rights against arbitrary power, together forming the contradictory basis to modern statehood and citizenship.

'Security rackets' also benefited from economies of scale. Herein lay the superiority of the modern territorial state as opposed to, say, princely municipalities and tribal kingdoms, particularly as the growing sophistication and cost of armaments dictated increasing centralisation of the instruments of violence. After World War II, we see an unprecedented further growth in the scale of the security racket, both in territorial terms (supra-nationality), and in the technical sophistication of production. In other words, the security racket went global, with its own hierarchies, centres and peripheries, and patrons and clients. It no longer encompassed merely states and their national populations, but the community of states as such, formalised in global alliances. Pushing wares designed to enhance the security of states became one of the most lucrative political, diplomatic and commercial rackets available. 'Security' was articulated to hegemonic objectives that made it possible for patrons to subsidise at least part of the cost of reproducing the security of the client state. Hence, at the global level, 'security' was placed beyond the logic of the marketplace, in the specific sense that states did not always have to pay the 'going price' for their security. 'National security' became a pre-eminent component of the ideological repertoire of states; at the same time the concern for security was no longer seen as coterminous with the boundaries of the nation-state, as before, but rather with the existence of a particular global order.

'Security' is a racket in another, much more sophisticated sense. Like 'order', 'security' is a highly ideological construct, a normative term which suggests that something good, some utility, results from the exercise of coercion. The ideological force of the discourse is broadened when 'security' is linked to other normative states, such as 'democracy', or posed as a state of common property (such as in the notion of 'national security'), embodied in a supposedly neutral professional force. This illusion, of course, quickly falls apart as soon as we deconstruct the concept. Who are the protected? What is being protected? Against whom? Who defines when a state of security has been achieved, or when a state of insecurity is in force? The attempt to respond to these questions has generated discourses – gendered, classist, ethnicised, racialised, ideological – fundamental to the management of power and order in society. Discourses of security have in turn furnished the basis for altruistic domination, the power exercised by protectors over the protected. In fact, as we well know, 'security' has always been dispensed very selectively within both the national and international community, reflecting socioeconomic status, residency, and state of citizenship (who does not remember French and Belgian paratroopers airlifting white European nationals from Stanleyville in 1964 in the face of the rebel advance, while leaving the natives to stew in a war that was fuelled by mercenaries from Belgium and other European states?); in other words, differential perceptions of human worth.

The ability to project 'security', then, as a common good ('equal protection under the law'), is one of the crucial achievements of the modern state, requiring in turn the ability to engineer perceptions and (to some degree) illusions, of which, surely, the notion of 'private security *forces*' has to be one of the best contemporary examples. For most Africans, however, the reality has been that of force (or worse, violence); Africans have too often seen the 'force', and hardly ever the 'security', in 'security forces'. Popular security has been sought through ethnicity, vigilantism and other primordial and non-formal institutions, away from (and often against) the state.

A subtle form of organised demonology – the ability to create national and global 'others' who are then held as 'threats' to security – is another component in the discursive construction of security and of the power to project 'security' on a mass basis. Such demonisation – of particular races, nationalities, ethnic groups, political beliefs and movements, even genders – has occurred at the level of both popular discourse as well as in the technical language of securocrats. Even as the Cold War has receded, new forms of demonisation have been invented, with scholars and the Pentagon speaking of 'civilisational wars', the 'Coming Anarchy', and 'Rogue States'.[3] Such demonisation persists not simply because of the need for the management of mass psychology, but also because it constitutes the ideological carapace for what has undoubtedly emerged, since the end of the World War II, as the most powerful and successful extractive machinery in history, the military-intelligence-industrial complex. Private security, the central subject of current discourse, is a manifestation of this complex.

'Back to the Future': The OAU and African State-Making

If the state is indeed a security racket, then the African state is manifestly a failed racket. To some extent this is a global phenomenon: all over the world, even within the most 'secure' states, populations are today facing levels of social and criminal violence unprecedented in modern times. But the African state is qualitatively different in its security predicament. The reasons for this are complex in nature, involving historical, structural, as well as political and policy factors,[4] but undoubtedly one reason can be located in the moment of creation of the continental system of states in the 1960s. At the time of the formation of the OAU in 1963, three fundamental issues of statehood confronted Pan-Africanism; the decisions of the nascent organisation in responding to these issues would have far-reaching effects on the ability of its member states to develop the attributes of statehood. The first was with regard to the issue of governance. One of the least known (but most disgraceful) episodes in the founding of the OAU was the rejection by the African Heads of State of a clause – the *only* one rejected – in the preamble of the draft Charter that would have bound them to ensure 'good government' in their respective countries, requiring them to affirm that 'the aim of government is the well-being of the governed'.[5]

The second issue was that of (regional) security. At the first OAU summit conference in Cairo, President Kwame Nkrumah of Ghana reportedly 'spoke forcefully of the need for [an] African High Command. He referred to the breakdown of law and order in Zaire, which led the Zaire government to ask for a UN peacekeeping force, and to the mutiny in the Tanzanian army, which led the Tanzanian government to call in British troops, and said that if there had been an African High Command there would have been no need to go outside Africa to seek help. He referred to the armed conflict between Ethiopia and Somalia and added that if there had been an African High Command, it would have gone in to separate the warring parties or at least supervise the cease-fire between them.'[6]

These were prescient words, but they would be rejected repeatedly by the OAU, even though the casualties would mount inexorably from the 1960s, both in terms of collapsing states and African lives lost. Many African governments preferred to seek protection under the military umbrella of a foreign power, or to ally themselves with such powers to undertake the interventions that Nkrumah had considered the legitimate province of Pax Africana. As the 34th OAU Summit opened in Ouagadougou in June 1998 against the background of the recent conflicts in Liberia and Sierra Leone, the resumption of war between Ethiopia and Eritrea, a military revolt to unseat the government of Guinea-Bissau, and so on, African leaders would undoubtedly have reason to revisit Nkrumah's words.

Thirdly, this rejection of the ideas of Nkrumah was connected with a broader, and barely concealed, normative dispute about the nature of 'development'. Should development take the form of a socialistic society, as favoured by Nkrumah and a minority of African leaders, or of capitalism? Since in the African context the policy instrumentalities of 'socialist' and 'capitalist' development were frequently indistinguishable, it can be assumed that the real issue behind the rejection of Nkrumah's ideas of continental economics – which would have greatly expanded the possibilities of African development -was the extent to which 'development' should be sensitive to issues of distribution and basic needs.

Through the decisions wilfully taken at the dawn of African statehood, the OAU frustrated any possibility of a clear choice between two competing options: of genuine regional collaboration (including more localised and self-sufficient forms of collective security) on the one hand, and competitive state-building on the other, in the first case by sanctioning an organisation so weak as to be unable to co-ordinate or defend the collective interests of member states, in the second by freezing unviable colonial state boundaries, outlawing intervention by African states in the internal affairs of other states, and releasing rulers from the obligation to be accountable to their citizens, thus pre-empting the very tensional forces that could be expected (at least in the realist paradigm) to shape institutional capacities and bonds of legitimacy in African states.[7] In reality, of course, underneath the pretentious carapace of a new 'normative politics', the realist logics of conflict and force operated. New African states interfered extensively, but covertly, in the affairs of neighbouring states (providing arms, rear bases and refuge for dissidents, funneling third party weapons to insurgents, etc.), and participated in regional interventions orchestrated by foreign principals.

New Post-Cold War Configurations of Force

The farcical (if not cynical) pretence of a 'moral Pax Africana' has of course collapsed, and arguably a new and more 'realist' Pan-Africanism rooted in a better understanding of the historical role of force may be emerging. Ironically, the end of the Cold War, in concert with other factors, has led in some respects to a renewed saliency of force – driven, however, by domestic contradictions rather than Great Power manipulation – on the one hand and far-reaching reconfigurations of force structures and paradigms on the other.

Firstly, the winding-down of the Cold War, and folding of the imperial military umbrella of the French and the Soviets, have led to the collapse of patron-client security arrangements. Secondly, neoliberal economic restructuring has undermined existing patronage networks generated and sustained on the basis of public resources. Thirdly, massive retrenchment of military resources has taken place worldwide, reflecting both the end of the Cold War and the global crisis of public finance. At the same time, much of the trade in new and surplus weapons (particularly small arms and light weapons) and retrenched personnel has been diverted from government to private arms dealers and security organisations. From sales of US$2 billion to US$3 billion per annum during the Cold War, private arms sales are estimated to have exceeded US$25 billion in 1996.[8]

Finally, the activation of civil society and pressures for greater accountability and inclusion have forced regimes to reconfigure their arrangements to make them more inclusive and accountable, or to even abandon power altogether. The resurgence of civil society has prompted (depending on the context) state retreat, backlash, and/or power re-composition, leading in some cases to drastic changes in the complexion of national political power. Both developments are connected to some degree with globalisation and the erosion of certain forms of social control, such as communications and the media, on which the state had relied to enforce its will against civil society. How the political transition has been managed has had decisive consequences for both national political order and regional security: the horrific consequences of the failure of democratisation in Burundi and power-sharing in Rwanda, of the attempt to repress political Islam in Algeria and civil society in countries like Togo and the former Zaire, may be contrasted to the salutary outcome, on the other hand, of the transition in South Africa.

As the hegemony of the state has crumbled, it has precipitated rivalry among warlords, visionaries, adventurers, ethno-nationalists, and big power interests to manipulate the vacuum, appropriating and shaping new force structures as well as the nature of the developing geopolitical space. Combined with weapons proliferation from winding-down wars, the decomposition of the state has prompted vicious new forms of conflict, depicted by the media and scholars as a descent into barbarism and anarchy, 'orgies of primordial savagery',[9] and a 'retreat from modernity'.[10] However, the resulting geopolitical conflicts and alignments are not necessarily 'new', but represent a sometimes intricate fusion of old and new cleavages, struggles and alliances, bringing together veterans from the Katangese and successive Zairean rebellions, and the protracted Angolan war, as well as victims and perpetrators of past and current Rwanda and Burundi 'ethnic' pogroms.

However, these same dynamics have encouraged the development of several positive initiatives. One of these is the new geopolitical alliances, African-centred rather than shaped by foreign powers, with a corresponding potential for emergence of regional hegemonies (South Africa, Nigeria). The second is a new regionalism, demonstrated at several levels: a revived interest in regional approaches to issues of trade and economic co-operation (the emergence of more formal and elaborate SADC structures, the expansion of IGAD and, most remarkably, the revival of a new version of the East African Community); and a much more serious interest in mechanisms of regional security.

This regional approach to security is underpinned by a new normative interventionist ethos, best expressed by Mandela at the 34th OAU summit in Ouagadougou. According to Mandela, 'Africa has a right and a duty to intervene to root out tyranny', notwithstanding the OAU's founding principle of non-intervention in the internal affairs of member states: 'we [African leaders] must all accept that we cannot abuse the concept of national sovereignty to deny the rest of the continent the right and duty to intervene when, behind those sovereign boundaries, people are being slaughtered to protect tyranny.[11]

In uttering these words, however, Mandela is giving voice to the increasingly fashionable notion of 'sovereignty as responsibility', the idea that misrule, or defence of democracy and proper governance, constitutes appropriate grounds for intervention into the 'internal' affairs of other states.

The 'Mandela Doctrine' is already manifested in the new willingness with which African countries are prepared, contrary to the historical and ideological legacy of the OAU, to intervene across national borders to restore order, to resist insurrections, and even to overthrow tyrants, as in the case of the SADC in Lesotho in 1994; ECOWAS/ECOMOG in Liberia and Sierra Leone, East Africa in Burundi; Uganda, Rwanda and Angola in Zaire; Angola in the Congo, and, most recently, the armies of Guinea and Senegal in Guinea-Bissau. As this record suggests, however, this security-oriented regionalism is occurring within a variety of formats – collective, bilateral and unilateral, formal and informal, regional and sub-regional[12] – that suggest it has yet to develop the requisite consensus as to framework. But the basic philosophy is an anti-militarist one: that tyranny and political repression are a fundamental cause of domestic and regional instability.

New Forms of Militarisation in the 1990s

The overall positive developments in security thinking since the end of the Cold War have been more than cancelled out by new conflicts and new vulnerabilities. These are characterised by the following features:

(1) The Poor Man's Wars

Africa has been ravaged by vicious new forms of conflict, depicted by the media and scholars alike as 'orgies of primordial savagery'.[13] These new forms of internal war differ in crucial respects from the old. In the past:

> most internal wars were guerrilla wars fought over political ideologies and the gaining of control over the government, and that most employed some form of Maoist mobile-guerrilla strategy. Those wars ... intensely intermix political and military action in the crucial battle for the hearts and minds of men. The effect is to simultaneously make these conflicts sophisticated and, more important, to moderate the violence that occurs in them.
>
> The new wars are clearly not like that. There is no common center of gravity to which the combatants appeal; in many cases it is not clear that the 'insurgents' have any interest in or intent on gaining political power or responsibility; and there is little sense of boundary on the extent of violence both sides would commit. These conflicts seem, indeed, to be a new breed of internal war.[14]

(i) According to analysts, these new conflicts are typically occurring in poor, marginal states, lacking in appreciable economic or strategic resources (in this sense their most obvious paradox is the contradiction between the modest economic stakes and the almost mindless intensity of the conflict) and often distant from the former staging grounds of superpower conflicts. Nevertheless, control over resource enclaves (drugs, diamonds, petroleum, lumber, etc.) is key in these conflicts, making it possible to fund large arms purchases by directly exploiting or taxing the trade in these resources;

(ii) Unlike the earlier liberation wars, this form of rebellion is typically initiated and led by an obscure and nihilistic leader, sometimes with no political organisation or political agenda. The warlord is essentially a political entrepreneur who links violence, territorial acquisition and business, opportunistically dismembering or reconstituting states. His weapons are the deliberate and systematic use of terror and 'ethnic cleansing' (perpetrated mostly against the civilian population), the use of children and youth (often abducted and drugged) as shock troops and cannon-fodder, and the utilisation of violence to colonise natural resources, exploited in turn to drive the war machine. On the other hand, these tactics succeed because they are directed not against strong states, but states already in retreat, with weak legitimacy and coercive ability (in both Liberia and Sierra Leone the rapid overwhelming of the security apparatus of the state by these ragtag bush armies was one of the most notable aspects of these rebellions);

(iii) Conventional armed forces and intervention strategies (which continue to operate in state-state scenarios) are poorly adapted to dealing with these conflicts. A frequent outcome is the collapse and fragmentation of weak armies, remnants of which gravitate to warlord factions or constitute new warlord factions in their own right (e.g. Armed Forces of Liberia, and the so-called 'sobels' in Sierra Leone). However, in part because of the absence of an ideological bond, the warlord himself faces recurrent challenge from within the ranks of his own force, leading to splintering and proliferation of warlord factions. Such conflicts are also notoriously difficult to mediate because of the lack of clear (or any) political objectives, the proliferation of warlord factions, and the fact that, unlike Cold War conflicts, they have limited dependence on official (or great power) sources of arms or patronage.[15] The conclusion is that external intervention will do little to cure these conflicts;

(iv) In these new 'wars' (unlike the previous ones which were essentially political in nature and objectives), civilians and neighbours rather than the state are defined as the 'enemy'. They thus mark a breakdown in social relations as such, or a deliberate intention to destroy, not just the state, but the existing social fabric, and with it any possibility of cohabitation. These conflicts are driven by 'intimate hatreds',[16] the manipulation of historical memories to demonise the 'other', and new ways of communicating xenophobia. Although articulated in 'ethnic' and/or 'religious' terms, the fundamental causes of these conflicts are often economic, political and environmental. Nevertheless, it is important to remember that ethnic, religious, and cultural identities are powerful ideological forces which cannot simply be reduced to their roots in material or historical conditions, and which, irrespective of how demonstrably irrational and manipulative they may be, exercise a powerful and autonomous influence on conduct. Although cynically harnessed by often unscrupulous politicians, their intense appeal seems to suggest a mass attraction rooted in popular alienation. The stress on exclusionary micro-identities suggests disillusionment, real or imagined, both with the state and with the nation-building project. Thus Cock is entirely correct in arguing that the light weapons proliferation, which characterizes and stimulates these conflicts, is an indicator 'of a level of social disintegration'.[17]

This issue also points to a paradoxical relationship between democratic contestation and militarisation. Ironically, struggles for democracy have sometimes stimulated militarisation. Elections can lead to conflict resolution (as in South Africa), but at least as often can aggravate conflict (as in Burundi), particularly where the conflict poses the prospect of drastic reconfiguring of political power, thus confronting existing power centres – such as the military and/or dominant ethnic groups – with the prospect of power-loss or resistance.

Incumbent political elites have varied in the way that they have sought to respond to this challenge: in the case of Burundi, the democratic transition was initially allowed to go ahead, and then eroded and eventually overturned by an increasingly restive military; in Rwanda, Hutu hardliners overthrew the regime to prevent an accommodation with Tutsi rebels; in Nigeria and Algeria the victorious political forces were forcibly prevented from assuming power after the elections, leading in the latter case to a vicious underground war proverbial for its meaninglessness and anonymity. Political forces that have lost democratic contests, such as Nguesso in Congo-Brazzaville, and the pro-apartheid right wing and Inkatha Freedom Party in South Africa, have resorted to private militarisation to offset their losses, often appealing to ethnic and racial chauvinism in the process. The common element in these scenarios is the refusal of power brokers to accept the verdict of democracy. On the other hand, in some countries the political disintegration and 'ethnicisation' has emerged through a fracturing of civil society (such as the churches in Burundi and Rwanda, and political parties in Kenya), in the process of resistance against authoritarianism, or through the pretext that elections have provided for official and unofficial 'ethnic cleansing'.[18] 'Ethnic' conflict has been used by African leaders as a pretext to limit or avoid political liberalisation, and (once again) power-sharing.[19]

(2) Criminality

These conflicts coincide with, and may even be buttressed by, new forms of violent criminality, facilitated by familiar structural factors (pervasive poverty and unemployment, worsened by neoliberal reform), as well as recent political developments (such as democratisation and the dissolution of apartheid) which have reined in the repressive capacity of the state; the presence in several countries of large numbers of unemployed and increasingly marginalised demobilised soldiers, and ready access to weaponry.[20] This criminality is difficult to control, given its scale and financial resources (at least in the case of the drug traffic), international mobility, the weakness and/or complicity of law enforcement structures,[21] and compressed global communications;

(3) Military Companies and the Privatisation of War

Another development – the focus of this book – is the commercialisation of security and the emergence of the defence corporation and security entrepreneur. Unlike the warlord, the latter is an ex-military professional, often with links to Cold War or (in the case of Executive Outcomes) apartheid structures, with no admitted territorial ambitions, loyalties or political interests, claiming to work exclusively in the employ of sovereign entities to fill the security vacuum emerging at the end of the Cold War. The defence corporations for which they work (such as Executive Outcomes, Sandline, Defence Systems Ltd, and Military Professional Resources Incorporated) are diversified, international concerns combining war and business and offering a range of services and interlocking ownership and directorates.

A less glamorous but probably more significant development is the spread of private security companies. As the coercive power of the state has declined – and with it, its ability to project security – social sectors used to protection from the state (and thus by definition privileged) are turning to private security companies. In the process, 'security' has been converted into a commodity, available at a price to those who can afford it, often tied directly to 'rent' from minerals and other productive assets, and no longer a right conferred by citizenship and membership in a political community. However, the phenomenon of privatisation of force paradoxically also reflects the security needs of the state itself, given its historically problematic (and declining) ability to project force; private armies like EO may replace or supplement official armies, which have proven to be difficult to control politically, as well as unreliable on the field of battle.

(4) Effects of the Privatisation of War

Hence, this 'privatisation of force' – or what Cock calls 'private militarisation'[22] – has complex social and political manifestations and is related in equally complex ways to official force structures, instead of being entirely discrete phenomena. The new force structures have been bent to a variety of purposes and agendas, depending on the context: tapped into respectively by states to underpin a status quo which they are no longer capable of defending by military means, by privileged social interests to substitute for the protection that the state is no longer capable of providing, by criminal gangs as well as popular forces seeking to protect themselves (via vigilantism) from them, and finally by a variety of warlords and adventurers to challenge the state. Hence, whether they enhance or detract from 'security' depends on who is doing the reckoning. Examples of this intertwining of official and unofficial forms of force abound: the relationship between the former Rwandan regime and the interahamwe; between the 'third force' and the South African security forces, and between the Sierra Leone government and the kamajors; or, at a different level, the alliances between Taylor and certain West African states, between ECOMOG and certain Liberian and Sierra Leonean warlords, between UN forces and anti-Aideed warlords in Somalia.

(5) Army Rebellion

These new forms of private militarisation have also overlapped with continued manifestations of older, institutional forms of militarism. African armed forces, stimulated by a variety of old and new grievances, corporate as well as political, have continued to challenge regimes (including new democratic dispensations) through coups, insurrections, and rebellions. The list is long: Guinea-Bissau, June 1998 (military revolt following suspension of chief of staff); Zambia, October 1997 (attempted coup by junior officers); Sierra Leone, May 1997 (overthrow of President Kabbah); Central Africa Republic, April, May and November 1996 (pay revolts); Burundi, September 1996 (coup), October 1993 (assassination of President Ndadaye); Niger, January 1996 (coup against President Ousmane); Guinea, February 1996 (pay revolt); the Gambia, 1994 (overthrow of President David Jawara); Lesotho, January and April 1994 (pay revolts), etc.

(6) Unilateral State Intervention

Yet another new form of militarisation is the extent to which African states are themselves prepared to act unilaterally to mount military actions openly across the borders of other countries (Uganda, Rwanda and Angola in Zaire, Angola in Congo-Brazzaville), in contravention of the OAU policy of non-intervention and respect for the territorial integrity of other states. However, the resulting geopolitical conflicts and alignments are not necessarily 'new', but represent a sometimes intricate fusion and culmination of old and new cleavages, struggles and alliances, bringing together veterans from the Katangese secessionist movement, various Zairean rebellions, and the Angolan war, as well as victims and perpetrators of past and current Rwanda and Burundi 'ethnic' pogroms. These actions have been related to the emergence of the 'new breed' of statesman – the so-called 'African Metternichs' (Museveni, Kagame, Zenawi, Afewerki, etc.), who are reshaping the regional geopolitical system while at the same time transforming national politics, economies, and security structures. The leaders of these 'post-insurgency regimes' are men schooled in the diplomacy and *realpolitik* of violence (though, in contrast to the warlord, in the 'older' liberation tradition), having fought in or come to power through protracted war (and in several cases facing internal insurgencies of their own). The unexpected resumption of the war between Eritrea and Ethiopia suggests, however, that this violence can also turn in traditional geopolitical directions.

(7) Imperial Paradigms

Finally (and these developments notwithstanding), continued foot-dragging by African states on issues of regional security is encouraging new forms of imperial intervention. There are several foreign 'peace-keeping' initiatives in the works (including proposals from France and the UK), but the best-known example is the American African Crisis Response Initiative (ACRI). Another initiative being considered by the Americans is an African Centre for Security Studies to be run by the Pentagon and replicating similar centres located in Germany, Washington, and Hawaii, and covering the European, Latin America and Caribbean, and Asia-Pacific regions. The OAU and many African countries complain about the lack of co-ordination between these initiatives, and (in relation to ACRI) the selective and potentially divisive approach taken by the US in targeting certain African states for partici-pation in ACRI while excluding others, thus creating the appearance of an alliance system with the US rather than a regional security system intended to benefit all African states. There are also concerns relating to the limited resources available under the initiative (which focuses on training rather than the equipment and logistics requested by African armed forces), as well as philosophical differences over the approach to peacekeeping, in particular regarding the 'robust' approach preferred by the Americans.

This Great Power concern with issues of regional security in Africa is not in all senses a positive development, however, since it also mirrors a shift once again in the donor discourse from 'governance' (emphasising democracy) back to political stability and 'development': in other words, not democracy *and* security, but democracy *or* security.[23]

Two factors have been most often cited to account for the declining security situation in Africa – even though, as we will suggest, these are perhaps symptoms rather than fundamental causes of the problem. The first is the decomposition of the state and its security apparatus. The end of the Cold War (like its active phase) had been double-edged and ambiguous in its effects on conflict. While on the one hand it facilitated peace accords and political accountability, on the other it has helped to intensify the 'security predicament'[24] of many African States. And as the hegemony of the state has crumbled, it has precipitated rivalry among warlords, political visionaries, adventurers, ethno-nationalists, and big power interests to manipulate the vacuum, appropriating and shaping new force structures as well as the nature of the developing geopolitical space. This loss in its coercive ability has affected the ability of the state to project security, and the emergence of alternative (private) forms of force and security. The second factor is the proliferation of small arms and light weapons in the contemporary world system. Unlike heavy weapons, which require an organised military formation, light weapons, cheap to buy, easy to use and maintain, facilitate a privatisation – indeed, individuation – of violence, and are thus emblematic of the state of social relations par excellence.

The proliferation of small arms reflects a variety of international and regional factors: the huge stock of surplus weapons that have materialised as a result of the end of the Cold War, the economic considerations that increasingly govern arms transactions, not only on the part of the countries of the former Soviet bloc (where such exports reflect critical balance of payments considerations) but also the United States, and the failure of peace accords to ensure effective disarmament of former combatants, allowing soldiers to walk away with their individual weapons, or large caches of arms to go undetected. These weapons subsequently reappear on the black market or are diverted to other areas of conflict. Hence, while significant retrenchment of official military resources and transactions has occurred on a global scale, this has been compensated by diversion of much of the traffic in light weapons and small arms to private arms dealers and end-users. From sales of US$2 billion to $3 billion per annum during the Cold War, private arms sales are estimated to have exceeded US$25 billion in 1996.[25]

Possible Solutions: Reforming Paradigms and Structures

The decomposition of security arrangements of African states and the struggles to democratise the public space have intertwined to produce some sharp questioning of security concepts and architectures, as well as of their underlying relations of power, both on the part of citizens and (less radically) states. African leaders have not only failed to link the security of their regimes with the security of their citizens, but to guarantee the security of even their regimes. The result is that both state and non-state actors are being forced to rethink the meaning and implements of security. On their part, African states have sought to bolster their security arrangements in three principal ways. One, as we have seen, is by 'contracting-out' security functions (including to mercenaries), while continuing to rely on military aid and co-training networks. A second approach has been to revamp security architectures at several levels. A third approach is an increasing shift in the direction of regional security and 'Pax Africana', an autonomous African capacity to deal with regional conflict issues, based on a clearer recognition of the link between domestic and regional stability, and, possibly, emergence of somewhat more flexible attitudes on the question of national sovereignty. However, these new forms of conflict and violence clearly suggest that the actions taken so far to address the problem of militarisation and security are not adequate. What is further required to be done?

First, of course, the regional and national specificities in the processes of militarisation mean that we need to avoid excessive generalisation. Militarisation has, to some degree, divergent roots, manifestations and trajectories in the different regions in Africa. We need also to make dis-

tinctions on a country basis: between states which have avoided militarisation (such as Senegal or Cote d'Ivoire); states which have gone
some way to overcome a previous episode or history of militarisation
(Tanzania or Zimbabwe); those in which there is a reasonable prospect
and purposive set of policies and actions (South Africa, Uganda, Mali);
those in which the issue remains problematic or contested (Uganda); and
those in which there is no short-term prospect of demilitarisation
(Somalia, Angola, Sudan). And in terms of recent transitions, while some
African countries are consolidating the processes of demilitarisation and
democratisation alluded to earlier, others have suffered reversals or been
unable to sustain these processes. These, in turn, reflect different kinds of
political transition and possibilities for a transformative politics.

Nevertheless, several guidelines on how to approach the issue of demilitarisation have some universal applicability. Demilitarisation requires
action at multiple levels, social, political, cultural, and institutional.
Specifically, there is a need to:

(1) Control the Spread of Weapons in the Region

This issue has received wide recognition,[26] and African states are playing
a growing role in attempts to impose international restrictions of the
spread of small arms, such as the Oslo meeting in April 1998 designed to
produce a moratorium on such arms in West Africa (*New York Times*, 5
April 1998). We mention it here, however, in order to support the
argument that we need to move from a focus on the supply of weapons
to the underlying social factors that create a demand for weapons, and
from seeing the problem of weapons proliferation and violence in
technical and legal terms to seeing it in social terms.[27]

(2) Address Fundamental Issues of Governance and Economic Justice

These new conflicts may be traced to structural problems such as high
levels of unemployment, poverty and economic inequality, sharpened
by structural adjustment and the malintegration of former combatants.
Particularly relevant is the generational crisis and criminalisation of
African youth, which is feeding directly into rebellion and warlordism.[28]
Whether among the Somalian 'technicals' or in Taylor's army, the youth
are turning to weapons ('the Kalashnikov advantage') as a subsistence
tool, because of diminishing opportunities for education, employment or
self-realisation. In addition, these bored and alienated youth are particularly susceptible to forms of global cultural communication (such as the
Rambo movies) that glorify violence and demean women. These problems
are sharpened by the failure to address fundamental governance issues:
corruption, human rights abuses, lack of protection for ethnic and
religious identity, autonomy and voice. Many African post-electoral

regimes would qualify as 'illiberal democracies', in many respects barely different from the authoritarianisms that they displaced. Elections have been extensively manipulated.[29] The contradictory linkage between democratic contestation and militarisation in Africa discussed earlier points to the need to transform relations of power – thus to the need for more, not less, democracy – simultaneously with reconstruction and strengthening of 'collapsed states'[30] and African state systems in general, in terms of their legitimacy, institutional capacity, and fiscal and economic vitality;

(3) Attack the Cultural Basis of Militarism

Several sociologists have pointed to the connection (rooted in both African and Western culture) between militarism and masculinity (or constructions of the masculine identity). For instance, 'in some areas [of the Horn] a man without a gun is not considered a "real man"'.[31] In Angola, this relationship between masculinism and militarism was captured in UNITA's symbolism of the 'Cockerel'.[32] Cock has also observed that '[m]any young South Africans understand weaponry as emblematic of manliness'. She suggests, further, that

> [t]his linkage between militarism and masculinity is frequently harnessed to an ethno-nationalism. Many of the men possessing light weaponry have deep-seated fears and insecurities that are grounded in racial and ethnic identities which are antagonistically defined. For many South Africans ethnic identities are the strongest source of social cohesion.[33]

What requires reaffirmation is not the so-called 'warrior traditions' so frequently celebrated and manipulated by militarist politicians, but rather the African cultural traditions of tolerance, reconciliation, and harmony.

(4) Reconceptualise 'Security' to Place the Emphasis on Human Security rather than the Security of the State

Democratisation poses the need to challenge reigning security concepts, redefining the concept away from the current militaristic, state-centric and gendered meanings, to one which sees the human person, rather than the state, as the beneficiary of security, emphasising certain broad entitlements such as access to basic needs and services (education, health, shelter), a clean and sustainable environment, and protection for cultural identity and autonomy. The South African White Paper on Defence comes close to such a formulation when it defines security as:

an all-encompassing condition in which individual citizens live in freedom, peace and safety; participate fully in the process of governance, enjoy the protection of fundamental rights, have access to resources and the basic necessities of life; and inhabit an environment which is not detrimental to their health and wellbeing.[34]

The objectives of security policy are thus seen to include

The consolidation of democracy, the achievement of social justice, economic development and a safe environment, and a substantial reduction in the level of crime, violence and political instability. Stability and development are regarded as inextricably linked and mutually reinforcing. (p. 5)

(5) Reconceptualise, Correspondingly, the Institutions Involved in Security

The logic of this definition of 'security' is that responsibility for territorial defence is no longer confused with overall responsibility for security, or with the military as the sole or primary security institution. According to the White Paper, while the SANDF remains an important security instrument of last resort, it is 'no longer the dominant security institution. The responsibility for ensuring the security of South African people is now shared by many government departments and ultimately vests in Parliament' (p. 6). The state, too, is no longer seen as exercising exclusive responsibility for determining security policy or managing conflict. NGOs, with their proximity to the people, are increasingly seen as essential aspect of an early warning system, particularly with regard to ethnic conflict, and useful as well for monitoring human rights abuses and the conduct of security forces.

(6) Institutionalise Dialogue, Transparency, Public Oversight and Participation and Access to Information in Defence and Security Matters

Reconceptualising defence and security should involve an open and wide-ranging dialogue between the government (executive as well as legislature), political parties, civil associations and NGOs, and defence and security forces. In this area the South African Defence Review process was exemplary. However, it is necessary to go beyond this to actually entrench oversight, participation, and access. The provisions of the Defence Review are once again worth quoting:

Parliament and public scrutiny and debate will be meaningful only if there is sufficient transparency on military matters. A measure of secrecy will undoubtedly be necessary in order to safeguard national security interests, the lives of military personnel and the integrity of military operations. However, the governing constitutional principle is freedom of information. Exceptions to this principle will be limited, specific and justifiable in a democratic society .... (p. 12)

The Minister and the DOD shall consult with interest groups and stakeholders in civil society in the formulation of defence policy, and shall provide the public with adequate information on defence matters (p. 19).

The powers of parliament should be real, substantive, explicit, as far as possible independent of those of the executive, and dominant, and should be combined with the demilitarisation of Ministries of Defence and other security management organs, putting in place civilians with the requisite training and skills in defence analysis, planning and accounting to enforce and monitor defence policies made by government.

(7) Enhance Democratic Principles within the Military to the Maximum Extent Consistent with Military Discipline

As Tsadkan Gebre Tensae, the Chief of Staff of the Armed Forces in the Transitional Government of Ethiopia, observed:

A national army must have sufficient internal discipline and a hierarchic chain of command if it is to function as a fighting machine when the situation demands. Without violating these requirements, however, the internal management of the army must be made democratic. The relationships between members of the armed forces, the rules and governing punishment and promotion, and the non-military activities of the army can all be shaped to reflect democratic principles. [35]

If the army is not run, internally, as a democratic institution, it cannot sustain democratic relations with the rest of the civil society. Members of the armed forces will lose their commitment to defend the democratic rights of the population because they, themselves, are being denied those rights within the army. An army that is inspired by the ideals of the constitution and is internally managed in a democratic way will not only prove to be an asset to democracy in general, but will also be fully satisfied with the goals, objectives and internal life of the institution, and will thus be a highly motivated and effective fighting machine. [36]

For this to be possible, security forces should reflect the complexion of the overall population. [37]

(8) Empower Local Civil Society Organisations to Spur on Demilitarisation

Finally, these measures contributing to security and demilitarisation require the support of a social movement if they are to become a reality. And by 'social movement', we refer not just to urban workers, academics, NGOs, etc. but the community as such and women in particular, whose contribution has received little recognition. The role of the community in the reintegration of demobilised soldiers in Mozambique, and in peacemaking and reconciliation in Mali (in relation to both the Tuareg rebels and the armed forces), has been dramatically highlighted,[38] as has that of women in the peace movement in Angola.[39]

In turn, this social movement needs to be energized by new cultural norms and a broad vision of transformation. Cock could not have put it better when she argued in her discussion of the Southern African region that demilitarisation involves

> more than institutional defence restructuring. It also involves new forms of social integration, solidarity, identity and citizenship. It means creating alternative cultural values and meanings ... demilitarization goes beyond the restructuring of state institutions to involve the recasting of social relations in a much broader project of social transformation in the region ... .[40]

Notes Chapter 8

1. See C. Tilly, 'War Making and State Making as Organized Crime', Peter Evans, Dietrich Rueschmeyer, and Theda Skocpol (eds), *Bringing the State Back In* (Cambridge University Press, 1985).
2. S. Hymer, 'Economic Forms in Pre-Colonial Ghana', *Journal of Economic History*, Vol.1, p. 20, 1970. Also: E. Terray, 'Long Distance Trade and the Formation of the State: the Case of the Abron Kingdom of Gyaman', *Economy and Society*, Vol. 3, 1974, p. 3.
3. See, for example, Robert Kaplan, 'The Coming Anarchy', *Atlantic Monthly*, February 1994, p. 273.
4. M. Ayoob, *The Third World Security Predicament: State Making, Regional Conflict, and the International System* (Lynne Rienner, 1995). In this excellent and comprehensive book, Ayoob argues that unlike the older (European) states the principal security threats to Third World states are internal (the result of poorly integrated multi-ethnic societies, irrational borders, weak and disarticulated economies, and external dependency), though shaped by broader historical and structural factors both external and domestic in origin. Third World state formation occurs within a radically different historical and international context than had confronted earlier states: an environment dominated by mature (and stronger) states, with the further challenge of developing the attributes of statehood in a much more compressed time-frame than older states had been required to do.

5. C. O. C. Mate, *Inside the OAU: Pan-Africanism in Practice* (Macmillan, 1986), p. 60.
6. Ibid., p. 172.
7. The realist paradigm might conceivably argue that in defining their security in terms of relationships with external power centres, African states broke the organic link between war-making and state-making: between security, accountability to citizens, and accumulation processes. For more on the realist paradigm and security in Africa, see Chapter 1.
8. Raymond Bonner, 1998. 'For U.S., Gun Sales are Good Business'. *New York Times*, 6 June 1998.
9. Barbara Crossette in the *New York Times*, 14 June 1998.
10. Ali Mazrui, 'Conflict as a Retreat from Modernity: A Comparative Overview', O. Foley (ed.), *Conflict in Africa* (I.B. Tauris, 1995).
11. *Cape Times*, 10 June 1998, p. 4.
12. None of these classifications are watertight. In the case of ECOWAS, intervention in theory took a collective form, but in fact there were strong elements of unilateralism (on the side of Nigeria). In the case of Central Africa, the intervention into the former Zaire did not assume such a formal structure (and thus could technically be considered as 'unilateral'); nevertheless there was strong collaboration between the governments of Uganda, Rwanda, and Angola in the exercise. The subsequent intervention into Congo-Brazzaville demonstrated further differences in modality.
13. Barbara Crossette in the *New York Times*, 14 June 1998.
14. Donald Snow, *Uncivil Wars: International Security and the New Internal Conflicts*, Boulder (Lynne Rienner, 1996), p. ix. However, in the African case, RENAMO may well constitute the opening wedge in this new form of warfare and the critical linkage between the old and new forms of internal war.
15. Patrons may find these warlords difficult to control, as suggested by Houphouet-Boigny's experience with Taylor, and Museveni and Kagame in relation to Kabila.
16. Diane Jean Schemo, 'In Colombia, Accord comes without Peace', *New York Times*, 28 June 1998, p. 5.
17. Jacklyn Cock, 'The Cultural and Social Challenge of Demilitarisation', G. Gawthra and Bjorn Moller (eds), *Defence Restructuring of the Armed Forces in Southern Africa* (Ashgate, 1996).
18. *African Journal of Political Science*, Special issue on Elections in Africa, New Series, Vol.2, No.1, June 1997; R. Kadende-Kaiser and P. Kaiser, 'Modern Folklore, Identity and Political Change in Burundi', *African Studies Review*, Vol. 3, No. 40, December 1997.
19. The demonisation of ethnicity by politicians has come under increasing attack by some African and Africanist scholars. At a 1995 symposium on 'Elections and Conflict Resolution in Africa' in Washington, it was acknowledged that 'ethnicity is often a major source of conflict' in Africa, but it was also argued by one participant that 'parties based on ethnicity, race, region, religion and gender need not be divisive or destructive of national unity, but can instead be part of a system of power-sharing', and by another that to 'suppress such parties undermines democracy'. *PeaceWatch* (United States Institute of Peace, Washington) Vol. 1, No. 5, p. 6. This conference may want to take this crucial issue of ethnicity on board in its deliberations.

20. See Virginia Gamba (ed.), *Society Under Siege: Crime, Violence and Illegal Weapons* (Halfway House: Institute for Strategic Studies, 1997).

21. In some cases the law enforcement officials have formed a target of this criminality, which is to some extent politically inspired. In the Kwa-Zulu province, 39 police officers have so far been shot dead this year, while in Gauteng province another 43 have been killed. Overlapping political motives are also apparent in frequent robberies of weaponry from military and police armouries.

22. Cock, 'The Cultural and Social Challenge of Demilitarisation'.

23. James C. McKinley Jr, A New Model for Africa: Good Leaders Above All', *New York Times*, 25 March, 1998.

24. M. Ayoob, *The Third World Security Predicament*.

25. However, it is important to qualify this stress on the spread of weaponry. The genocide in Rwanda – considered the worst in contemporary times, was notable for the low-tech instruments employed: machetes, clubs, and other 'traditional' implements!

26. *The New Field of Micro-Disarmament: Addressing the Proliferation and Build-Up of SmallArms and Light Weapons*, Brief 7, BICC [Bonn International Centre for Conversion] Bonn, September 1996.

27. Cock, 'The Cultural and Social Challenge of Demilitarisation', pp. 121 and 141. The following discussion has benefited immensely from the thought-provoking analysis in Cock's article.

28. See Paul Richards, 'Rebellion in Sierra Leone: A Crisis of Youth?' in O. Foley (ed.), *Conflict in Africa*, London, 1995 (I.B. Tauris, 1995).

29. *African Journal of Political Science*, Vol. 2, No. 1, June 1997.

30. William Zartman (ed.), *Collapsed States: The Disintegration and Restoration of Legitimate Authority* (Lynne Rienner, 1995).

31. Kees Kingma and Vanessa Sayers (eds), *Demobilization in the Horn of Africa: Proceedings of the IRG Workshop, Addis Ababa, 4–7 December 1994*, Bonn: BICC, June 1995.

32. Horace Campbell, 'Angolan Women in Search of Peace', *African Journal of PoliticalScience* (forthcoming, 1999).

33. Cock, 'The Cultural and Social Challenge of Demilitarisation', p. 119.

34. 'Reallocating Defence Expenditures for Development: the South African Experience', *African Security Review*, Vol. 7, No. 2, 1998, p. 5.

35. General Tsadkan Gebre Tensae (n.d.). *A Vision of a New Army for Ethiopia* (text of speech. No further details available).

36. Ibid., p. 2.

37. Ibid., p. 4.

38. Robin-Edward Poulton and Ibrahim ag Youssouf, *A Peace of Timbuktu: Democratic Governance, Development and African Peacemaking* (United Nations, Geneva, 1998).

39. Horace Campbell, 'Angolan Women in Search of Peace'.

40. Cock, 'The Cultural and Social Challenge of Demilitarisation', p. 117.

Arresting the Tide of Mercenaries: Prospects for Regional Control

'Funmi Olonisakin

Private armies[1] exist because there is a demand for them. Their supporters argue that they are effective and that their decisive use of force can gain victory for their clients over their adversaries, thereby serving to end armed conflict and the accompanying human suffering.[2] This has been the case in some conflict areas like Sierra Leone (albeit temporarily). If mercenaries are so effective, why should they not continue to operate? What makes them attractive in some conflict areas and not in others? The issues at the root of the resurgence of these private armies in Africa have not been subjected to close examination by many advocates of this activity. Opponents, whilst clearly articulating its ills, have yet to engage in rigorous analysis on how best to control the activity. Some questions require proper attention. Is it enough to regulate private armies, or should they be abolished and, if so, how? If they are abolished, what will replace them? Who will handle the present crisis areas and how can the necessity for private armies be removed? This chapter seeks to address some of these questions. In particular, it examines the present regional security arrangements within Africa, and how these arrangements can bring about the establishment of a structure that could eliminate the need for mercenaries in the region.

Regulation or Abolition?

Advocates of mercenarism argue that the need for the activity, along with the military effectiveness of private armies, makes nonsense of any argument for its elimination. All that is required, they maintain, is for private security firms to be regulated. The firms themselves, whilst agreeing that they should be subjected to some form of regulation, believe that existing companies hardly need external regulation as they are already self-regulating.[3] These armies are seen to already comply adequately with international humanitarian law, and to abide by codes of conduct commensurate with those of professional armies. Nonetheless, they see external regulation as being in their own interest – it would serve to curb the excesses of a few aberrant private military companies or, indeed, wayward newcomers into the business, thus keeping the perceived 'credibility' of the business intact. There is an implicit assumption in this argument that regulation of mercenaries will nullify the criticisms levelled against the activity. The real issues, however, are whether regulation will go a long way to address the real concerns of the opponents of the activity; and whether its continuation, albeit with external regulation, addresses the issues at the core of the debate.

What type of regulation is being advocated? Private Military Companies are more concerned with regulation which seeks to measure their performance in terms of technical competence; adherence to the law of armed conflict; and respect for human rights in the area of operation. It has to be said that in some of the countries where they have operated, allegations and reports of gross human rights violations by some of the warring parties have far exceeded allegations against the private armies – a factor which is often capitalised upon by the private security firms in their bid to gain the support of the sceptics.[4] However, the nature and terms of their contracts with their prospective clients are not included as one of the areas to be subjected to regulation. Yet, any regulation that does not remove the justification for those who see the activity as morally indefensible would be grossly inadequate.

Why is the Activity not Morally Defensible?

Regulating the conduct of private armies in this way does little to address the issues that make them unappealing – in particular, the perceived immorality of the trade – for two crucial reasons. The first is the huge financial cost to the target nation and the exploitation of natural resources which might otherwise have been beneficial to the citizens of these countries, many of who suffer due to scarcity of resources. The second is that the ending of armed conflict creates only superficial peace and invariably serves to delay conflict resolution and regional stability.

Mercenaries come at a tremendous material cost to citizens of the target states and this is at the core of the concerns of the abolitionists. Sierra Leone, for example, paid Executive Outcomes US$35 million (for 20 months from April 1996), while the country's foreign exchange earnings in 1995 were at an all time low of US$39 million.[5] In addition, the vast majority of Sierra Leoneans languished while their natural resources were being plundered by a small group of people. Thus, when supporters of private security forces argue that it is cheaper to employ their services than to deploy UN peacekeepers, they fail to consider a fundamental question – for whom is it cheaper? The argument that private armies are cheap in comparison to the cost of UN operations overlooks the paralysing effect on the target countries, which are confronted by huge bills that they can ill afford. It can only be considered cheap for the UN and for the countries that would normally bear the cost of peace operations. Yet it is arguable that for these countries the cost is negligible, not only because it is shared, but because the contributors are not impoverished like Sierra Leone and those that have to pay dearly for the services of the mercenaries.

This problem might be alleviated were it possible to regulate the income of the security companies, so that restrictions are placed on the amount that they can levy a country or client for their services. This is not a viable option, first, because it removes the core reason for the existence of the trade. While this result may be the desired intention of the abolitionists, it is unlikely that it will be effective. Secondly, all that is likely to result from such a process is that trading partners would keep details of payment secret or difficult to trace in the same way as is done in the illicit transfer of arms. Alternatively, given the failure to address the problem of long-term stability, a problem that makes UN-type peace operations expensive, the question has not been asked as to whether the UN should pay the costs of private armies, if they are more effective and are likely to come at less financial cost to the organisation. And in this case, should not these private armies be absorbed into the UN system so that they can be properly regulated? This may also not augur well for either side, as the UN will not be able to justify huge payments to the mercenaries, which is their main motivating factor in enforcement operations. Even then, other questions remain and some of them are discussed below.

Beyond the immediate financial and economic impact on impoverished countries, however, accusations of immorality also stem from fears about the long-term impact of mercenarism on security and stability on the African continent. This potential must be understood in order to appreciate the argument of the abolitionists. Mercenarism must fundamentally be seen as a trade. The activity exists purely for profit and not for any altruistic reasons. Success in the trade can only generate further economic interest in the activity. It is not improbable that a successful private security firm can expand, and attract significant international economic interest and investment as has already been demonstrated (see Chapters 2 and 3). It is conceivable that where economic interests are threatened, the beneficiaries of such a system will not wish to see an end to the trade. Thus, the trade has enormous expansion potential, if allowed to grow. Private armies have a real potential to become a serious political problem on the international agenda. Those who benefit from the trade economically and financially, both within Africa and particularly outside will go to any lengths to protect it and before long the campaign against it will become an expensive process.

This growth of private armies can (and will) only occur at the expense of African countries, as it will effectively mean that they are maintaining a status quo without any lasting solution to conflicts. The cost of abolition will become higher the longer the trade is left to thrive. After all, the slave trade started on a relatively small scale and by the time the campaign to abolish it began, the world faced a difficult challenge, given the huge scale of the trade and the enormity of the interests in it. Those who see these scenarios as unrealistic should only stop to consider that slavery began with only a few exchanges of goods for people; before long it ravaged an entire continent for hundreds of years while the rest of the world lived in relative prosperity.

Mercenaries, given that they are motivated by profit, can ultimately employ their services on the side of an immoral cause. It is not sufficient to say that they will only work for recognised governments. Such governments, though recognised, may not serve the interests of their people and may not be legitimate in the eyes of the people. Moreover, there is a limited extent to which legitimate governments will want to employ the services of private armies. This was the case in Sierra Leone, where Kabbah's elected government could no longer justify Executive Outcomes' presence in the country. Additionally, even if there were legislation to control them, as seen in South Africa, it may be difficult for governments to enforce such legislation. Furthermore, private armies are considered morally indefensible for a number of other reasons. Firstly, they can be used to subvert a sovereign, legitimate state, because an illegitimate government suppressing its people may employ them. Secondly, mercenaries serve as an extension of the international trade in arms, and are a useful tool for the indirect transfer of arms to conflict areas for sellers who do not wish to attract great attention. The mere employment of the services of mercenaries in a conflict area could lead to the importation of their preferred arms and ammunition to that area. This ultimately creates a ready market for arms-producing nations (often the home states of the security firms or of their partners) and along with this, a reluctance to challenge their activities in these environments.

Another crucial factor, which is often ignored, is that it is difficult to reconcile the use of private armies with the enforcement of international humanitarian law. If, for example, the use of international criminal tribunals as seen in Rwanda and former Yugoslavia is to be developed, it will be virtually impossible to use private armies for the purposes of apprehending those suspected of crimes against humanity – a role played by the peacekeeping force in former Yugoslavia. It will be difficult, if not impossible, to trust private armies to be impartial in their apprehension of suspects because they could not possibly be expected to hand over their paymasters or their allies for trial if they were named amongst the suspects.

Advocating the use of mercenaries, even with a certain degree of regulation, would still amount to a lack of understanding of the real problems and threats that this trade poses. At the root of the argument for regulation is the assumption that the activity is useful and will continue to be needed in conflict environments, particularly in Africa. The question of what needs to be done to address the circumstances that have created the need for mercenaries in Africa, and how best to deal with the root causes of the security problems on the continent, is not often asked. If a comprehensive approach were to be adopted in dealing with African conflicts, would the space be created in which mercenaries could operate? Can the structures be created for the implementation of a comprehensive approach? Those who argue for the abolition of mercenarism look at the problem from the perspective of seeking lasting solutions to the conflicts on the continent. The continuation of mercenarism will not create long-term stability in the conflict areas, and there are no lasting benefits to be derived from the use of private armies.

The presence of mercenaries in Africa will serve only to prolong insecurity, as it is virtually impossible to deal with the root causes of conflict in an environment where one party enjoys the protection of private armies and is therefore reluctant to consider the long-term security needs of the community. This will often result in the peaceful concentration of power in the hands of a particular party and in less accountability. The employment of mercenaries in conflict-ridden regions is an indication that there are serious security problems which only an effective regional mechanism can deal with. Regardless of their pronouncements, accountability cannot be assured simply because mercenaries say so. If security is left in private hands, instead of with a group of sovereign states, an uncontrollable trend towards letting unaccountable forces determine internal power relations may be let loose on the world. In addition, concepts of sovereignty and statehood, which have long guided the international system and created some form of order (no matter how flawed), will become severely threatened.

It is easy to dismiss the above arguments and argue that this phenomenon is restricted to specific regions and has no effect on the larger international environment or on continents such as Europe and America. Whilst it is arguable that the African region is presently the most attractive to mercenaries, and indeed the most prone to the involvement of private armies, the impact of these developments on other regions should not be underestimated. Failure to put a stop to this phenomenon, which has slowly evolved since the 1960s when many African states emerged, can have serious repercussions for countries of the West, some of which are beneficiaries of this trade. The countries of the West cannot ignore the destabilisation caused to Africa by the activities of mercenaries. They will have to bear the burden of responding to the resulting humanitarian crises and the economic and social cost of dealing with those migrating from the insecurity of the African continent.

Thus, in thinking about mercenarism, particularly in its new form, it is arguable that states not just within Africa, but also outside it, have a responsibility to stop this tide. Seeking to render mercenaries accountable by regulating them in terms of their conduct on the field will not bring an end to the problem. It will simply legitimise an act driven by the desire of, initially, a handful of people, and in time, possibly a group of nations, to profit financially from the plight of poor communities. The desire for profit will only prolong this suffering.

However, passionate pleas for the ending of a profitable trade and potential landmark historical developments will not alter the realities on the ground. Armed conflicts remain, and the mercenary business is alive and well. The crucial question that the rest of this chapter seeks to address is what must be done? Can new regional security structures eliminate the need for private armies?

COMBATING THE NEED FOR PRIVATE ARMIES – SHORT- TO MEDIUM-TERM STRATEGIES

A number of developments created the space for mercenaries to operate on the scale recently witnessed in Africa. One is the failure of the international community to deal effectively with the crises which followed the exit of the superpowers. This is manifested most glaringly in the execution of peace support operations that adhere rigidly to old (and at times new) doctrines that serve no useful purpose in the conflict environment in which they are deployed. This resulted in long drawn-out and expensive operations, which have provided the supporters of private armies with strong arguments in favour of the trade. In some conflict environments, the UN stuck to traditional peacekeeping methods, even when they bore little relevance to the situation on the ground, as was initially the case, for example, in Somalia and Rwanda.

Traditional peacekeeping, which entailed the deployment of lightly armed, neutral and impartial military personnel, with the consent of the host nation and co-operation of conflicting parties, only proved useful in a number of cases in the early post-Cold War years. These included Namibia and Mozambique, where UN forces played a monitoring role. In the cases where there was no agreed peace, and the conflicts had escalated, a different type of response was needed, particularly one where peace forces would be more heavily armed and equipped to use force to end the violence where necessary, especially when there was immense human suffering.

However, the UN continued with its traditional peacekeeping approach. Its troops were severely punished in Somalia and Cambodia, for example, for carrying light arms and for the small sizes of the forces. Even when a new approach was considered, such as second generation peacekeeping, it was still based on the fundamental assumption that the consent of the host state would be sought. This was bound to be ineffective in conflict environments where there was no recognisable central authority; it added to untold human suffering, and several warring factions emerged which had no desire for negotiated peace and no regard for the rules of war. This was demonstrated in Liberia from 1989 to 1997. Such conflict situations dictate a response that entails the use of force to end the atrocities and to create an atmosphere where peaceful negotiations can be resumed. Of course, it is to be expected that any engagement that entails peace enforcement would result in higher casualty levels for the peace force, but the goal must be that such enforcement action would do more good than harm.

The result of such conflict situations, which private armies have taken advantage of, is that many of the usual troop-contributing countries are reluctant to participate in this type of operation. Understandably, it is difficult for states that do not have a strategic interest in the conflict area to embark on operations that would lead to the loss of men and material. The collective interest of the majority, or of the most powerful members of the UN, cannot always be guaranteed – particularly when they are asked to contribute men and materiel to conflicts that do not have a direct impact on their national security and interest. This was a problem even with traditional peacekeeping operations, where the risk of losing men in the operations was low, let alone when these men are required to operate in dangerous environments with a real possibility of high casualties.

The slow response to African conflicts and the reluctance of potential troop-contributing nations to participate in these complex operations have opened the way for mercenaries, who do not shy away from the use of force. However, their own doctrine is also not without flaws. Fighting to achieve victory for one party over other(s) in a conflict will not necessarily bring about peace and long-term stability. Unless the use of force is accompanied with sincere and effective peacemaking strategies among the victor(s) and the vanquished, armed conflict will re-emerge at the earliest opportunity. Thus, governments which rely on the use of force by mercenaries (or any other organisations, for that matter) to achieve victory in armed conflict without genuine attempts at reconciliation, are likely to suffer a reversal once this military support is withdrawn. Beyond the use of force, mercenaries have no interest in peacebuilding, whereas international organisations are more likely to pursue an all-encompassing strategy. The latter course is likely to be more time-consuming and costly, but it provides the best chance for lasting peace and stability.

Relying on Regional Hegemonies

It is likely that only states with a vested interest in the outcome of a conflict, or which are affected by that conflict, would be willing to find effective ways of responding to it. This suggests that the most effective responses to the crisis in Africa will in all probability come from within, and not from outside the region. Africans are not completely incapable of developing effective responses to crises, although the capacity levels vary from one sub-region to another. The capacity to respond is usually determined by the ability of the sub-regional hegemony. In sub-Saharan Africa, Nigeria and South Africa stand out as the clear regional hegemonies. Apart from the recent events in the Democratic Republic of the Congo (DRC) and Lesotho, South Africa has had little experience of responding to crises. But Nigeria, which has had more experience in UN peacekeeping, or of responding to crisis in the West African sub-region, has been prevented by domestic problems from executing these operations in a more efficient manner. Even so, the Economic Community of West African States (ECOWAS), under Nigeria's leadership, managed to restore a measure of order to Liberia, and later to Sierra Leone, although the war in Sierra Leone continues.

Was ECOMOG effective in Liberia? Arguably, the involvement of the Nigerian-led ECOMOG force in the Liberian conflict prevented the involvement of mercenaries on the scale later seen in Sierra Leone and the Democratic Republic of the Congo.[6] ECOMOG's first enforcement operation in September 1990 achieved some positive results. Perhaps the most notable was that it halted the atrocities (although not permanently), particularly the killing of innocent civilians, which was widespread in the Liberian conflict. This created a situation in which humanitarian assistance could be delivered to thousands of Liberians and foreigners who were caught in the crossfire. In addition to this, the warring factions, recognising that they could not achieve a military victory on the battlefield, agreed to a cease-fire. From this point, negotiations ensued.

However, success did not accompany the multinational efforts to achieve a settlement in this conflict until 1997, when elections were successfully conducted in Liberia. In the years between 1990 and 1997, the warring parties in the Liberian war multiplied and over a dozen peace agreements were broken, while several more eruptions of violence had to be put down by ECOMOG. The failure to achieve an early settlement can be attributed to numerous factors. Amongst these were the lack of peacemaking and peacekeeping precedents on the part of the sub-regional organisation, ECOWAS; the lack of commitment by the warring factions; operational problems suffered by the peace force, ECOMOG; and the inadequacy of the main contributing state, Nigeria, which played controversial roles at different times and was plagued by domestic crisis. Indeed, the state of the Nigerian contingent in ECOMOG, which was the largest, was seen to have impacted negatively on the ECOMOG operation overall. As a result of a political army at home, which paid greater attention to domestic politics than to military professionalism, the Nigerian contingent in ECOMOG was unable to realise its true potential in Liberia. Many commentators believe that a responsible and accountable government in Nigeria would have produced a more professional response to the Liberian operation and the mission would have been concluded much earlier.

Nonetheless, the ECOMOG operation in Liberia showed that when a regional hegemony is willing and able to employ the required strategy (including the use of force) in response to regional crisis, it is likely to meet with some success and to eliminate the need for mercenaries. Nigeria was crucial to the success of this operation. The ECOMOG operation was not designed to achieve victory for one party over others and was not intended to be an end in itself as interventions by private armies often are. Rather, it was part of a more comprehensive framework, designed to achieve long-term stability in Liberia, despite the obvious flaws in the process. A regional arrangement seeking to avoid the mistakes of the ECOMOG operation would have to move towards eliminating the need for private armies.

However, two main factors make it difficult to rely on the notion of a willing and able regional hegemony throughout Africa. One is that the same proximity argument which prevents states outside of Africa from intervening in areas that do not serve their national interests, also applies within Africa. Only states within a particular sub-region, who are affected by a conflict, are likely to be willing to participate in efforts to address such conflicts, particularly when it means that their soldiers will operate in dangerous environments. Regional hegemonies are likely to be more interested in intervening in their immediate sphere of influence. Thus, it is more probable that South Africa will consider intervention in Lesotho than in Guinea-Bissau. The implication is that in regions where there is no clear sub-regional power, or a state able to bear the financial burden of intervention, response to a conflict may be slow or not forthcoming at all. The slow and ineffective international response to the Rwanda crisis can be attributed in part to this problem. Under such circumstances, a space is created for mercenaries to operate in.

The second factor is that in Africa, states have to contend with an additional problem: even when some states are willing, they may not be able to respond effectively. Many lack the resources to contribute significantly to simple traditional peacekeeping operations, let alone the complex ones demanded by the difficult conflict situations of recent years. Although countries like Mali and the Gambia contributed troops to ECOMOG for almost the entire duration of the Liberian operation, their combined troops were less than a platoon – the number they could contribute without compromising national security. Indeed, their contribution (which was vital) represented a political gesture rather than a significant military presence. Some countries are willing to contribute men but unable to provide logistical support for them. For example, Senegal contributed two battalions in the third year of the ECOMOG operation, largely with provision of logistics by the United States.

In addition, as Nigeria's experience has demonstrated, even subregional hegemonies do not have unlimited resources to spend on these regional operations. Nigeria spent 30 billion Naira (US$3 billion) on the ECOMOG operation in Liberia.[7] It was already fully committed in Liberia when the Sierra Leone war started in 1991. Thus, Nigeria was only able to send a limited number of men to Sierra Leone and was in dire need of additional contributions from other member states. A small ECOMOG force was unable to respond effectively to the civil war before it escalated. The regime therefore had to depend on Executive Outcomes to repel the impending rebel onslaught – at a horrendous cost to the country, one of the poorest in the world. Nigerian troops in Sierra Leone, who numbered about two battalions on the ground at the time of the 1997 coup, could not reverse events. The forces of the junta in Sierra Leone were eventually overwhelmed after additional reinforcements in February 1998. By the time the crisis in Guinea-Bissau began, Nigeria and other ECOWAS states were already stretched thin.

Apart from the problems of logistics and other resources required by African countries, the ECOMOG operations also reflected problems of training and doctrine. The lack of harmony in training and doctrine and non-standardisation in communications equipment affected command and control, in addition to creating other operational difficulties at different stages of the ECOMOG operation. Thus, some observers argue that under responsible regional leadership, and with adequate training and logistic support, African military personnel can respond effectively to the crises in the region. With the appropriate responses and capability, it is possible that mercenaries will have little role to play in these environments.

The African Crisis Response Initiative

One way in which the regional alternative to mercenaries can be properly developed is to enable African states to respond to crises in their own region. The US made the first attempt in 1996 to improve the capacity of African peacekeeping forces (with the anticipated co-operation of other donor countries). Its proposal for an African Crisis Response Force initially sought to train and equip a Crisis Response Force that would perhaps serve as a standing army. Countries targeted in this scheme were those with a demonstrable willingness and capability to participate in peacekeeping operations, with an initial size of about ten battalions. It was envisaged that such a force would increase the capacity of Africans to respond swiftly to future crises in Africa. However, following criticism from some African and European countries, the concept was modified.

The revised concept, which is referred to as the African Crisis Response Initiative (ACRI), appears on the surface to have undergone only minor modifications. However, they removed what would have been the most effective aspects of the ACRI. One is that ACRI will not create a standing force. The second is that it will not conduct enforcement operations. Rather, it is aimed at increasing the capability of military units from the target states, which would continue to operate as part of the command of the home military establishment. In addition, the focus is on training for chapter VI operations, which would entail traditional and in some cases second generation peacekeeping operations. This new approach is not particularly relevant to many of the current conflict situations in Africa. The events in Sierra Leone, the DRC and Guinea-Bissau, among others, suggest that a peace force would probably need to use force. Thus, ACRI has focused on a doctrine that would prove inadequate in these conflict environments. In addition, a standing force made up of contingents from different countries, with harmonised training and doctrine, would be far more effective in responding to crisis. The ACRF would have been more relevant to the needs of the region.

What is perhaps the greatest flaw of the ACRI concept is that this capacity-building exercise has focused on states which are unlikely to contribute troops in the event of crisis. The states targeted thus far are Ethiopia, Ghana, Malawi, Mali, Senegal, Tunisia and Uganda. Ghana is perhaps the most likely to provide men in reasonable numbers for peace-keeping operations. Mali is willing, but as its experience in ECOMOG has shown, it cannot contribute up to a platoon for a long period in a peace operation. The fact that these countries are spread over several sub-regions also makes it unlikely that there can be any credible response by any of these countries to crises in their sub-region. Uganda is unlikely to intervene in crises in the Maghreb, nor is Tunisia likely to deploy men in East and Central Africa as indicated above. Uganda (alongside Tanzania) contributed a battalion to ECOMOG in 1994 and withdrew this contingent a year later. It is arguable that Uganda and Nigeria did not have the same level of interest in Liberia. ACRI might have been more promising if countries were targeted in sub-regional clusters. Indeed, any effort that excludes the powerful countries able to provide human and material resources for peace operations (such as Nigeria and South

Africa) will have little chance of success, unless patron-states such as the US provide adequate logistic and manpower support. The latter is highly unlikely. The ACRI project would have been a most useful tool for eliminating the need for private armies on the continent if it had not been for the conceptual flaws outlined above.

However, apart from these short- to medium-term strategies for responding to African crises in place of private armies, the complete eradication of mercenarism on the continent will depend on a long-term strategy. Much will depend on whether and how the region is able to make an effective collective effort. Such a response must come first from the region, if the wider international community is to emerge with any meaningful complementary response.

MAKING MERCENARIES PERMANENTLY UNNECESSARY – LONG-TERM STRATEGIES

In developing a long term approach that will lead to the permanent elimination of mercenarism on the African continent, one of the key questions that may guide this process is: why do mercenaries not operate in Western Europe? A simple answer may be that violent conflicts are not rampant in this region. In Northern Ireland and Spain, where nationalist aspirations by some groups led to years of terrorism and violent conflict, it is unthinkable that mercenaries could have been deployed in these countries. Even where mercenaries are attracted by the presence of mineral resources in the conflict area, this is not a sufficient cause for private security firms to get involved in a conflict region. Rather, it is because the existing structure of that area creates the opportunity for intervention. This allows some commentators to declare blatantly that African countries are not like Western European ones and as such, they must not be compared. They are not worthy to be called sovereign states in the way that their Western European counterparts are. Thus, advocates of mercenaries see many African countries as states that will never do well, and must rely on the permanent support of security firms. They do not address the obvious questions, however, of why violent or armed conflict is absent in many Western European countries. How can Africa get to the stage of managing conflict without violent manifestations? The desire for personal security is universal to human beings and not a prerogative of any group of people or nations.

One of the issues at the core of the African crisis is the absence of a structure, both at the national and regional level, for managing conflict in ways that will maintain the integrity of the state. As a result, conflicts are allowed to escalate and human suffering and massive destruction become the order of the day. Thus, even if the short-term measures discussed above were employed in response to conflicts in Africa, violence would persist in the absence of a framework that seeks comprehensive solutions to the conflicts in the region. Such a framework would be a definite way to make mercenaries permanently unnecessary.

The most ideal way to abolish the use of mercenaries is to eliminate the need for them – in the same way that it has become inconceivable that private armies will be needed in Western Europe or North America. States would need to alter their approaches to security both at the individual state level and collectively at the regional level. Unless security is seen in a wider context, allowing regimes to cater to the overall security of the people through inclusive strategies, thereby eliminating the need for violent conflict, mercenary firms will always find work on the African continent. The problem of mercenarism must therefore be addressed within a comprehensive framework.

Revisiting the Roots

Any attempts to develop effective regional security structures which can remove the need for mercenaries must re-examine the root causes of Africa's security crisis, as well as conduct a critical analysis of existing security structures in Africa. When the majority of African states emerged in the 1960s, they were confronted with security problems that were in large part the legacy of colonial domination. Many states faced the threat of instability either as a result of the need of others to redraw the artificial boundaries or in cases where some groups sought self-determination. Other states remained under colonial domination. Although Africans sought to find African solutions to these problems, the attempts did not result in an effective security structure in the region.

The OAU, founded in 1963, had an opportunity to evolve such a structure but could not reach a consensus. Kwame Nkrumah of Ghana proposed the formation of an African High Command for the purpose of strengthening and consolidating the political independence of the new African states, protecting them against all forms of imperialist aggression, and assisting in the liberation of African states still under colonial rule. Nkrumah sought to address potential threats created by the possibility of external interference in African affairs as seen in the Congo, and the danger posed by then apartheid South Africa.[8] He proposed a political union of Africa under an 'African Union Government', within which a common defence plan could be evolved, as well as a common foreign policy and a fully integrated economic programme for the whole continent. It was envisaged that such a structure, if established, would serve as a deterrent against any foreign intervention.[9] The OAU failed to enshrine such a structure in its Charter at the founding meeting of the organisation. Instead, the founding members opted for a Charter with loose arrangements. Before long, military intervention, which also included the use of mercenaries, became commonplace in Africa.[10]

The type of Charter that the OAU planners drafted determined the organisation's success in dealing with the region's security problems. Two fundamental assumptions guided the formation of the OAU: pan-Africanism and sovereignty. These assumptions were reflected in the objectives of the OAU specified in Article II of the Charter.[11] At the political level, the principle of the inviolability of inherited colonial boundaries was intended to uphold the sovereignty of these states, and it served to prevent potential border wars and secession attempts by groups seeking self-determination. However, there was no effective response to those conflicts which could not be prevented by this political stance, as was the case in Western Sahara and in the Ogaden War between Ethiopia and Somalia and within Ethiopia itself. By according pride of place to sovereignty, however, the principle of Pan-Africanism became an unattainable ideal and it was difficult to collectively address Africa's security problems in any effective manner. The lack of a stronger organisation with enforcement mechanisms also affected the OAU Convention for the Elimination of Mercenaries.

In particular, the OAU Charter was not concerned with internal security threats in member states. Many African leaders initiated and implemented unpopular policies and carried out gross violations of human rights without any repercussions. The misconduct of many African leaders and governments was overlooked on the grounds that the OAU could not interfere in the internal affairs of member states. This position was also encouraged by a similar clause in the UN Charter, and by a Cold War system which paid little attention to the internal conduct of governments.[12] The collective failure to address these problems led to a gradual decay of many of these societies, characterised by corruption of elites, mismanagement, group divisions and unresolved conflict.

The OAU's adherence to the principles of sovereignty and territorial integrity seriously limited the role of the organisation when internal conflicts turned into civil wars as seen, for example, in Chad and Sudan. In Chad, it was not until the Libyan threat to carve up the country that the OAU began to play a more prominent role. The reluctance of the OAU to become involved in internal conflicts was also obvious in Angola and Mozambique. Indeed, its chances of participating in the conflict resolution process in these cases were hindered by the OAU support for the recognised governments in those countries, in its bid to uphold its principles.[13] Like the UN, the Charter of the OAU made no provisions for peacekeeping, despite the prominence of the activity since the previous decade. The Congo experience discouraged many African leaders from engaging in peacekeeping activities at the time. In Chad, the use of force by the OAU Force to stop hostilities was unthinkable. It would have been against everything that the organisation stood for. However, the huge scale of human suffering witnessed in the years since the end of the Cold War was not apparent in that conflict.

Post-Cold War Opportunities

The end of the Cold War provided an opportunity for Africans to rethink their security arrangements. The autocratic regimes shored up by the Cold War powers no longer enjoyed the backing of their powerful allies. In Africa, there were hopes that conflicts previously exacerbated by the superpower rivalry could now be resolved. This happened to some extent in Namibia, South Africa, Angola (which suffered a reversal) and Western Sahara. Africa has, however, been confronted with new realities. Its relatively limited strategic importance to the great powers during the Cold War has further diminished drastically. What is perhaps the greatest difficulty is that the unresolved conflicts that were suppressed in the previous era have now found expression in the post-Cold War environment. Added to this is the recognition that Africa will now be responsible for its own security problems. The security problems (military and economic) of the former Soviet Union assumed priority over those of the African continent. Thus, Africa could not rely on a sustained effort by the West to deal with its new security problems.

The changes in the international environment are recognised among the regional organisations in Africa. The OAU has seen the need for a relaxation of its strict rule of non-interference. The establishment of a mechanism for conflict prevention, management and resolution is an indication of its determination to develop a more meaningful response to the current security problems.[14] However, the OAU continues to be plagued with the problems that prevented it from being effective in the Cold War era. It is still unlikely that the collective interests of states coincide on security issues and, therefore, it is almost impossible to obtain a consensus or a common political will to deal effectively with these problems. With a membership of 53 states with diverse interests, it is difficult to implement a common security agenda. It is easier, for example, to convince South Africa of the need for military intervention in neighbouring Angola, rather than in Guinea-Bissau. Invariably, only a group of states in a region with common interests – common borders and similar security problems – are likely to show interest in security problems virulent in their sub-region. Thus, structures with the potential to deal effectively with the security challenges on the continent are to be found in the sub-regions. A regional security system that is to deal effectively with the problem of mercenarism is therefore likely to come from the sub-regional level. Two prominent sub-regional organisations form the focus of the rest of this chapter.[15]

The Economic Community of West African States (ECOWAS)

A number of obstacles prevented the realisation of the central objective of regional integration, which lay at the heart of the founding of ECOWAS in 1975.[16] Amongst these were, firstly, the divisions caused by colonialism, which place ECOWAS member states within two often opposing camps – Anglophone and Francophone. Secondly, there was a duplication of effort due to the creation of rival groupings, which sought to pursue some of the same objectives outlined by ECOWAS.[17] Attempts are being made to harmonise the functions of the different groupings in the sub-region.

However, many of the obstacles to progress in ECOWAS come from intra-state sources. When the member states took steps to address external security issues through the establishment of the protocols on mutual defence and non-aggression, the internal security threats were not considered. Such threats varied from regimes' suppression of minority groups and failure to address group concerns about corruption of elites, all of which led to the breakdown of the agreement. The poverty and mismanagement that occurred at the internal level would invariably affect the collective effort on the external level. Like the OAU, the Charter of ECOWAS contained the non-interference in internal affairs clause. It was convenient for the leaders of member states, which were undemocratic and had a poor human rights record. The clause served to entrench them in power. This was compounded by the poverty level in these states and the lack of an industrial base, which ensured that they continued to trade with and depend largely on products from regions outside of West Africa.

The failure to address the internal problems, which had the potential to grow into a serious security problem and destabilise the sub-region, meant that ECOWAS could not achieve much progress, and its members were indeed heading for a security crisis. If the concept of sovereignty had been slightly relaxed, in a way that allowed states to address these internal issues, some progress might have been made. Moreover, no state appeared to have the moral authority to persuade other states to follow its example. It seemed such progress could only be made when each individual state had achieved stability and could boast of an accountable government. A collection of accountable and stable governments could only lead in the direction of a stable sub-region, where countries can focus their attention on development. A supra-national arrangement, where sovereignty was relaxed, with a holistic approach to the security problems of the sub-region, both internal and external, seemed to be the way forward. However, since the states were unwilling, as in the case of the OAU, to surrender a measure of sovereignty, not much progress was made. In particular, the authority of Heads of State and Government needed to lose its supremacy, for example to an elected parliament as seen in the European Union.

The events in Liberia in 1989 confirmed the need for a comprehensive approach to security in West Africa. A war that had been under the surface for more than a century (but in part shored up by the Cold War system) exploded in Liberia. A civil war erupted in Sierra Leone (triggered by the war in Liberia), from 1991. The Casamance insurrection in Senegal has slowly escalated, and recently Guinea-Bissau was thrown in turmoil. Together, the crises in the sub-region have resulted in millions of refugees and displaced peoples.

The security consequences of the Liberian war for the West African sub-region were grave. Within months of the war, there were more than half a million refugees in neighbouring states. This war, which resulted in the death of more than 200,000 civilians, including foreign nationals, was the bloodiest the sub-region had witnessed so far. In the absence of intervention from the US, Liberia's traditional ally, it was left to Liberia's neighbours to intervene in the conflict. Following unsuccessful mediation efforts and continuing civilian casualties, ECOWAS (through its Standing Mediation Committee – SMC), deployed a peace force to Liberia, with the purpose of restoring order, rescuing foreign nationals trapped in the crossfire and eventually monitoring a peace agreement.[18] Due to the proliferation of conflict parties and their intransigence, the war would take seven years to resolve. ECOWAS would later send an intervention force (consisting mainly of Nigerians) to Sierra Leone, which is still fighting rebel soldiers in parts of the country. The organisation is presently dealing with a fragile truce in Guinea-Bissau. With these developments, ECOWAS has found it almost impossible to focus on the economic issues, which were initially its main focus.

The experiences in Liberia pushed ECOWAS to confront a number of issues, both political and military. Firstly, internal security threats in member states cannot be disregarded. In particular, the conduct of states, such as their human rights record, accountability and participation of the people in governance, can only enhance overall security in the region. Secondly, a proper structure must be established for the planning and execution of peace operations, whether as preventive deployments, second generation peacekeeping or peace enforcement. The ECOMOG experience indicated the flaws in ECOWAS. It is basically an economic organisation without the capacity for conducting a traditional peacekeeping operation, let alone the difficult combination of enforcement and peacekeeping it carried out in Liberia. The Secretariat does not have a section comparable to the Department of Peacekeeping Operations in the UN (DPKO), charged with the task of co-ordinating peace operations. These structural weaknesses made it easier for the military regime in Nigeria, the most powerful nation in West Africa, to hijack control of the operation from ECOWAS. This resulted in a number of operational problems, some of which were a reflection of Nigeria's domestic problems. The organisation will need to address these structural problems in further operations. Finally, Liberia, Sierra Leone, Senegal and recently, Guinea-Bissau, show that ECOWAS policy makers must confront the need for a standing regional force or earmarked units, with standardised training and doctrine and logistics, to deal speedily and effectively with crises and potential security threats in the sub-region.

Although the extent to which ECOWAS leaders have realised these challenges is difficult to assess, the revised ECOWAS Treaty of 1993 reflects some changes along these lines. Among them, the new treaty provides for the establishment of a commission on co-operation in political, judicial and legal affairs, regional security and immigration,[19] a parliament, and a clause on the need for democracy and human rights in member states. However, a number of issues remain unaddressed. Firstly, there is no proper structure in place and no indication as to how the parliament would work, or how the commission on regional security would work effectively. Secondly, the willingness of states to move toward supra-nationality, or indeed to implement good governance and the rule of law is cast in doubt. Thirdly, there is the need for the strong states to guarantee both the political and military arrangements. Nigeria's domestic problems and continuing mutual distrust among states make this a difficult task.

Interestingly, the strength of ECOWAS lies not in its new structure, given the fact that the same problems that trouble the OAU (albeit on a smaller scale) are also likely to prevent progress here. However, ECOWAS has the advantage that its most powerful nation can push the other states towards developing an effective regional security system. An accountable and responsible hegemony would set a good example. If the hegemony is not accountable, the opposite effects could be achieved. So far, Nigeria, though capable, has failed to instil positive standards in other member states, given the deplorable state of its own domestic environment.

The major obstacle which has impeded progress in the security realm in West Africa, and which will determine the success of any ECOWAS effort to establish a new security regime, is the role of Nigeria. It is conceivable that an accountable Nigeria would be more likely to promote the cause of regional security. Firstly, it could keep obstinate states in check. Secondly, it could contain security threats before they escalate to levels where affected states would solicit the assistance of mercenaries. Thirdly, by acting as a positive influence on leaders of ECOWAS member states towards good governance, it could prevent the escalation of conflicts, hence eliminating the need for mercenaries.

The Southern African Development Community

Unlike the West African sub-region, Southern Africa began its experiment in regional cooperation on the political front, with the creation of the Front Line States (FLS) in 1974. The policies of the Heads of State of the FLS, initially comprising Tanzania and Zambia and gradually expanding to include other states, were implemented by the Inter-State Defence and Ministerial Committee (ISDMSC).[20] The idea of economic co-operation came about half a decade later with the Southern African Development Co-ordination Conference (SADCC). The states sought to achieve greater economic cooperation in order to reduce their dependence on South Africa. In 1992, the SADCC was transformed into the Southern African Development Community (SADC). The treaty, signed in 1992, signifies a determination to achieve regional integration and to harmonise policy in the member-states.

Regional Security under SADC

The SADC made a move towards a comprehensive approach to regional security in 1994 (as provided for in the treaty) when the FLS took the decision to 'dissolve and become the political and security wing of the SADC'.[21] The FLS member states then created the Association of Southern African States (ASAS).[22] Over the course of several meetings between July 1994 and January 1996, this idea went through different stages. In March 1995, SADC Foreign Ministers proposed that ASAS should become the political arm of SADC.[23] It would operate independently of the secretariat of SADC and report directly to the SADC Summit, and would also operate informally and flexibly with its chairmanship rotating every two years.[24] What was not absolutely clear at this stage was whether this arm would be operated as a sector by an individual state.

In June 1996, the SADC Summit approved the establishment of a political and security wing of the SADC, the Organ on Politics, Defence and Security, which would operate flexibly and informally. The Heads of State outlined objectives for this security wing which signified that a more than flexible approach was needed if these objectives were to be pursued in earnest. They also indicated that a more complex structure had to be put in place. Amongst the objectives set out were:

- preventive deployment (as part of conflict prevention)
- collective security capacity – which may entail the use of force against a dissenting member state

- regional peacekeeping – which may differ from traditional peace-keeping and entail the use of enforcement and second generation peacekeeping
- international peacekeeping as part of the UN
- restoring law and order in conflicts on the territory of a member state – which implies the use of force when necessary, to ease human suffering
- co-operating with the civil sector to collectively combat other security problems such as terrorism, drug trafficking, and cross-border crime.
- combating aggression from outside the region against a member-state – part of a mutual defence arrangement. This implies the use of force if not outright war.

If these objectives are properly implemented, it is conceivable that the sub-regional organisation can successfully eliminate the need for mercenaries in Southern Africa. Added to this is the fact that South Africa has taken the first step in controlling the activities of private armies within its borders, through the Foreign Military Assistance Bill. However, SADC faces fundamental problems in terms of projecting what seems to be the positive influence of regional hegemonies in the sub-region. This sub-region is significantly different in character from West Africa.

Stumbling Blocks

The organ on politics, defence and security has remained inoperative since its establishment. A dispute relating to its structure and mode of operation has resulted in a stalemate. There are differing views within the SADC, particularly between South Africa and Zimbabwe, on how the organ should function. One position (advanced by South Africa) is that the organ should remain under the jurisdiction of the SADC Summit, according to the provisions of the treaty. The other view (Zimbabwe's) is that the organ, which will operate flexibly and informally, should be placed under the control of a separate summit, with a separate chairperson. This stalemate has been attributed to a number of factors. One relates to the ego of some of the personalities involved. The president of Zimbabwe, Robert Mugabe, is thought to desire a leading role in the SADC, expressing preference for the flexible system of the FLS, whereby the longest-serving head of state would be chairperson.[25] South African diplomats, on the other hand, see such a move as preventing their country from making any significant contribution to the organ.[26] The opposing points of view of South Africa and Zimbabwe are seen to be the result of a struggle for Southern African leadership.[27]

In addition, the stalemate has resulted in part from the structure of the SADC. The system whereby one state is in charge of co-ordinating each sector makes it difficult to find a place for the organ in the sectoral system. Member states would be reluctant to entrust one state with the management of military security issues, which are often considered sensitive. In a system such as that in ECOWAS, where all member states collectively manage each sector, it would perhaps have been difficult for any one state to claim ownership of the organ. However, this may not resolve one issue – the need to bypass the secretariat.

As a result of the deadlock, the ISDSC has continued to function without a political head. Military cooperation within the SADC has continued in a number of areas, which gives an indication as to the objectives that an organ on politics and security will pursue in the short to medium term. Joint military observer courses have been held,[28] in addition to an agreement to co-operate on fighting cross-border crime. Moves have been made towards a regional peacekeeping force. A delegation of defence, foreign affairs and home affairs ministers from ten SADC countries visited Bosnia and Denmark to observe Danish participation in multinational peacekeeping operations in Bosnia.[29]

Even if the current dispute over the structure of the SADC organ were to be resolved, a number of obstacles would have to be addressed. Firstly, although the organisation has adopted a holistic approach to security, its objectives are ambitious ones which members cannot hope to achieve in the short to medium term. Although SADC states are faring much better than their ECOWAS counterparts in terms of adoption of democratic practices, many of the states are not at the same level of political development. Even though South Africa is clearly the most powerful state in the sub-region and has put in place a system that emulates many of the ideals set out in the SADC treaty, its influence on the other member states is not as strong as Nigeria's in West Africa. This may be related both to population and the way the country uses its resources. For example, Nigeria's population is larger than the combination of all other 15 member states of ECOWAS. In addition to this, despite 'silent resistance' from some member states, Nigeria is able to cajole them into compliance by making side payments – a reflection of its use of oil wealth as a tool of foreign policy. Thus, analysts are often quick to conclude that a stable and secure Nigeria, with an accountable government will be a positive influence on other states in the sub-region, in this same way that it is currently a negative influence and seen to export a 'coup culture' to other countries. It remains to be seen whether South Africa has this level of influence on other SADC states. Secondly, like ECOWAS, SADC will require a level of supra-nationality in order to achieve many of its objectives. Many of these states, democratic or not, may be unwilling to relinquish part of their sovereignty.

Conclusion

All the avenues through which mercenaries can be controlled in Africa have serious loopholes that make any such attempt a daunting task. At the heart of the short- to long-term strategies is the need to have clearly identifiable, responsible and accountable regional hegemonies. Even where they are identifiable, responsible and accountable, they may be incapable of coping with the sheer magnitude of crisis. The assistance of the international community will be crucial in this regard. The nature and structure of their assistance must be seriously modified to reflect the needs on the African continent if they are to have any impact and go some way towards eliminating the need for private armies. To achieve this objective through the creation of strong regional structures is a daunting but not insurmountable task. Much will depend on what happens at the sub-regional level.

ECOWAS and SADC will have to overcome serious political problems, particularly the fear and distrust of smaller states about regional dominance by the most powerful states. Yet the more powerful states will have to be relied upon to act as responsible guarantors for many of the important economic and security issues. Equally important, member states of these organisations must confront the issue of supra-nationality if greater progress is to be achieved in the area of regional integration. The SADC is, however, thought to enjoy some advantages over ECOWAS. The first is that, with few exceptions, the SADC states have one language, English, in common. This is seen to be a unifying factor, unlike the ECOWAS, which has to contend with Anglophone-Francophone divisions.[30] Second is the fact that many of them were unified against apartheid, which provided them with many years of experience of working together.[31] The third is that SADC can boast of two regional (but not equal) powers – South Africa and Zimbabwe.[32] This will serve to prevent complete domination by one state, as is the case in West Africa, where Nigeria enjoys near-complete dominance.

While Southern and Western African countries are grappling with ways to improve their own security arrangements, such elaborate security structures are non-existent in East and Central Africa. It is also difficult to identify a clear hegemony in this region. Thus, maintaining security in this part of the continent may depend on the extent to which the other two regional organisations are willing to contribute. In addition, the capacity-building exercise provided through arrangements like ACRI may be given collectively to states in these regions, so as to increase their ability to respond effectively to conflict. Furthermore, a country like the DRC, which is strategically important in that region, is already a member of SADC. The problem, however, is that in the absence of a clear regional hegemony, it will be difficult to find a country to provide leadership in terms of good governance.

Notes Chapter 9

1. The term 'private armies' is used here interchangeably with 'mercenaries'.
2. See, for example, Herbert M. Howe, 'Private Security Forces and African Stability: the Case of Executive Outcomes', *Journal of Modern African Studies*, 1998; and David Shearer, *Private Armies and Military Intervention*, Adelphi Paper 316.
3. See David Shearer, *Private Armies*; and Sandline International, *Private Military Companies: Independent or Regulated?*, March 1998.
4. Sandline International, *Private Military Companies*, pp. 2–3.
5. See *West Africa*, 24–30 June 1996, p. 985.
6. Even then, there were reports of Israeli and Korean involvement at the early stages of the conflict. Interview with NPFL representatives in November 1993.
7. Figure quoted by the late General Abacha, in a speech at the inauguration of Charles Taylor as President of Liberia, after the 1997 elections.
8. See K. Nkrumah, *Africa Must Unite* (Panaf, 1963), pp. 133–49 and pp. 173–93.
9. Nkrumah's proposals for an African High Command and the rationale behind them are contained in many of his speeches and writings. See for example, *Speech at The Hall of Trade Unions*, 9 July 1960; *An Address to the National Assembly On Afffican Affairs*, 8 August 1960; *Africa Must Be Free*, speech on Africa Freedom Day, 15 April 1961; *The Fight on Two Fronts*, 1 May 1961; *Africa's Glorious Past*, speech at the opening of the Congress of Africanists, Legon, 12 December 1962; and *True Freedom For All*, speech at 4th Afro-Asian Solidarity Conference, 10 May 1965.
10. See Chapter 1 for details of some of these instances.
11. They are: to promote the unity and solidarity of African states; to coordinate and intensify their cooperation, and efforts to achieve a better life for the peoples of Africa; to defend their sovereignty, integrity and independence; to eradicate all forms of colonialism from Africa; and to promote international cooperation having due regard to the Charter of the UN and the Universal Declaration of Human Rights.
12. The OAU Charter provided for strict non-interference in the internal affairs of member states. Six of the seven fundamental principles enumerated in Article III are designed to serve mainly the autonomy of member states from interference or coercion by other member states individually or collectively. These principles are: the sovereign equality of all member states; non-interference in the internal affairs of states; respect for the sovereignty and territorial integrity of each member state and for its inalienable right to independent existence; peaceful settlement of disputes by negotiations, mediation, conciliation and arbitration; unreserved condemnation in all its forms of political assassination as well as of subversive activities on the part of neighbouring states or any other states; dedication to the total emancipation of dependent African territories; and affirmation of a policy of non-alignment with regard to all blocs.
13. See A. Berhanykun, 'OAU-UN Relations in a Changing World', in Yassin El-Ayouty (ed.), *The OAU after Thirty Years* (Praeger, 1994), p. 123.
14. See the *'Report of the Cairo Consultation on the OAU Mechanism for Conflict Prevention, Management and Resolution'* (May 1994, p. 19.

15. Much of the analysis of ECOWAS and SADC conducted in this chapter relies on the findings of an earlier research work. See 'Funmi Olonisakin, 'Rethinking Regional Security in Africa: An Analysis of ECOWAS and SADC', in *Strategic Review for Southern Africa*, November 1998.
16. For a summary of the achievements and failures of ECOWAS in its first 20 years, see K. Bash-Duodu, *The Economic Community of West African States*, Occasional Paper, Africa Group, No. 2, August 1995.
17. Ibid., pp. 12–13.
18. For details of the ECOWAS mediation and the initial activities of the ECOWAS Monitoring Group (ECOMOG), see M. A. Vogt (ed.), *The Liberian Crisis and ECOMOG: A Bold Affempt at Regional Peacekeeping* (Gabumo Publishing, 1992).
19. See Chapter X, Articles 56–59 of the *Revised Treaty of ECOWAS*, signed 24 July 1993.
20. M. Hough and A. Du Plessis (eds), *Africa: Selected Documents on Political, Security, Humanitarian and Economic Issues* (Institute for Strategic Studies, University of Pretoria, 1996), p. 32.
21. J. Cilliers, 'The Evolving Security Architecture in Southern Africa', *African Security Review*, 4(5), 1995, p. 40, as cited in M. Malan, *SADC and Sub-Regional Security: Unde Venis et Quo Vadis?*, ISS Monograph Series, No.19, February 1998, p. 12.
22. Communiqué of the Meeting of Foreign Ministers of the Frontline States, held in Windhoek, Namibia, 30 July 1994.
23. ISSUP/SASA Conference Paper, *Security 95 – Security Co-operation in Southern Africa: What are the Prospects?* (8 June 1995).
24. Ibid.
25. RSA, *The Star*, 23 August 1995, p. 19.
26. Ibid.
27. RSA, *Mail & Guardian*, 25 August 1995.
28. Angola, *SADC News*, June 1998.
29. Ibid.
30. Interview with Major-General Hamman.
31. Ibid.
32. Ibid.

Conclusion

On the threshold of the twenty-first century, Africa continues to intrigue and shock the world with its ability to engineer ever-new security dilemmas even as it recycles old ones. With the formal end of the Cold War, Africa is experiencing progressive value depreciation in relation to the rest of the world. Economic liberalisation, touted as the only viable option to regenerate the weak political economies of the continent, has set forth conflict-laden methods of resource appropriation. Shorn of big power protection and too weak to exercise their coercive prerogatives, African states are increasingly relying on private military companies to guarantee their sovereignty. However, these PMCs – repackaged and cor-poratised mercenaries of old – provide military security shields at the expense of the national wealth, the unfair appropriation and distribu-tion of which lie at the base of conflicts.

The closing years of the Cold War saw two security-related develop-ments on the continent. On the one hand, some of the most intractable inter-state wars, fuelled largely by antagonistic global ideologies, saw a dramatic decrease. On the other, within individual state borders, entrenched dictatorships were crumbling under the dual impact of civil society action and international pressure. There was optimism that Africa was at last turning the corner. To consolidate the 'peace dividend', political economists, security and development experts and global powers sought to reconfigure Africa's security paradigm towards non-military variables. By combining good economic housekeeping (read neo-liberalism) with good governance (Western-style electoral democracy), it was believed that Africa would be able to fight poverty, sustain the environment, promote human rights and ensure ethnic, religious, cultural and gender equity, thus eliminating the underlying causes of physical insecurity.

A decade after the demise of direct dictatorship in much of Africa, the most evident replacement is illiberal democracies to which it has given prominence. More worryingly, the winding down of inter-state and proxy wars has drawn attention to, and intensified, internal struggles that had hitherto been overshadowed by Cold War priorities. Thus, Africa is being wracked by internal wars of genocidal proportions – Algeria, Angola, Liberia, Burundi, Rwanda, Somalia and Sudan – and regionalised wars within states – the Democratic Republic of Congo, Sierra Leone and Guinea-Bissau. Imperceptibly therefore, the 'derivative' notion of security (political, social, economic and spiritual well-being) is yielding ground once more to the 'generic' (physical) even before the critics have had time to rid scholarship of this neorealist paradigm of security. Confronted with this conundrum, policy-makers and critics alike have identified two broad frameworks for ensuring security on the continent – multilateral and unilateral.

Multilateralism derives its justification from collective responsibility, given the weakness of individual states of the continent. It requires concerted practical efforts to promote, strengthen and defend institutional and policy frameworks that ensure peace while combating instruments and actions that spread instability. Practical manifestations of this approach include the new lease of life the Organisation of African Unity seems to be enjoying in the post-Cold War era and the creation or restructuring of subregional economic and security umbrellas such as the Economic Community of West African States. Increasingly, Africans are jettisoning traditional notions of sovereignty in favour of pragmatic approaches to solving internal problems. The boundaries of 'territorial integrity' and the limits of 'internal affairs' are being stretched to encompass the entire continent. In this context, the goals of multilateralism include the following:

- Promoting, democratising and empowering multilateral institutions as key vehicles of continental stability.
- Encouraging good governance in member states.
- Defence and promotion of human/peoples' rights, strengthening and applying international norms of behaviour.
- Promoting and defending African rights in the global economy.
- Dismantling the physical instruments of violence and destruction – limiting arms transfer through legislation and moratoria and combating the illegal trade in weapons.

Multilateralism, as some of the chapters in this book have demonstrated, has its problems. For example, rivalries inherent in sub-regional organisations as well as the emergence of regional hegemons have the potential to undermine the cohesion of these structures. Also, the military may be emphasised at the expense of the political where irresponsible hegemonies dominate sub-regional arrangements.

Unilateralism, in opposition to multilateralism, sees the state as the master of its own sovereign destiny. Where the state is relatively strong militarily and economically, it frowns on any attempts to interfere in its internal affairs. Where it is weak but resource-laden, it will seek to ensure its independence and military security through bilateral arrangements with a neighbouring hegemon or a pretender-state. That failing, or in combination with that, the state may seek to bolster its internal security by contracting the services of external corporate armies – mercenaries. Usually, the recourse to mercenaries is underscored by the existence of violent civil strife, the unreliability of national security forces and intense competition for internal resources.

This book has sought to caution against the apparent international complacency towards the proliferation of mercenary activity in Africa by pointing out the ramifications of this trade on internal and regional security within the global context.

Interrogating Mercenarism

As demonstrated in the chapters of this book, mercenaries (or private military companies as they refer to themselves) need conflicts to exist and rich natural resources to induce intervention. They have been active in the post-Cold War era due to big power disinterest in out-of-area conflict-policing and the hype about strategic advantages they bestow to their employers in internal conflict. Ironically both corporate mercenary organisations and some mercenary-watchers in the North proffer the same philosophical and rational arguments to carve out a role for mercenaries in 'policing' these very conflicts, predicting that the role of mercenaries in conflict intervention will be on the increase in the post-Cold War era. The pro-mercenary lobby, with which this book takes issue, sees the use of private military companies by governments as little different from (their) purchase of weapons from private sources and so mercenarism should only be subjected to regulation similar to that governing the arms trade.

It is true that weapons exert an exacerbating influence on conflict, make peaceful management of conflicts more difficult and encourage banditry. In this lies the legitimacy of the campaign against weapons pro-liferation. However, equating the significance of a gun to the hand that pulls the trigger is hypocritical. Weapons, without bullets and without the mercenary misusing them, are only political symbols. While weapons are thus only potential instruments of violence and abuse, private military companies (corporate mercenaries, if stripped bare of the euphemism) are both instruments and perpetrators of violence. And they engage in violence for a share of the afflicted state's wealth. Mercenary-hiring companies and their intellectual supporters have often called on the international community to engage with mercenary companies so as to jointly create structures to monitor these companies' adherence to the laws of war in their areas of operation. However, these advocates for engagement hardly ever add the financial aspects of mercenary activities – the *raison d'être* of mercenary activity – to the list of areas to be scrutinised.

Mercenaries cannot be an alternative or a supplement to multilateral conflict management. This is the simple message most of the contributors to this book have tried to convey to the reader. It is our belief that mercenary intervention in conflicts, in contradistinction to multilateral efforts, is based almost exclusively on cold financial calculations. While there may be other considerations, the pecuniary motive is central to every mercenary involvement examined in this book. Multilateral peace enforcement is not funded by the afflicted state, neither is it hired by any particular party to an internal conflict. Multilateral peace enforcers take sides in extreme circumstances and only when neutrality has become impossible. Mercenaries, on the other hand, will always take sides, or even swap sides, in conflict based purely on the calculations of the mercenaries as to which party has the potential to control and disburse national resources. Usually, this party happens to be the governing party. Flowing from this, it is not difficult to see the basis of EO and Sandline propaganda that they only intervene on the side of legitimate governments. In conflict-stricken states of Africa, there exists only a very thin line between 'legitimate' governments and insurgents or rebels as the question of legitimacy depends on who defines it. More often than not, the recognised government is the rebel or coup plotter who contrives to 'legitimise' seizure of power through fraudulent ballot-politics. The facts, as detailed in this book, suggest that the 'legitimate government' argument is nothing more than a propaganda stunt. Private military companies, or segments within them, have switched sides in conflicts. EO in particular, started as a service provider to the apartheid war machine in the area of clandestine warfare operations and guerrilla warfare behind enemy lines. At various times, it has served as the backbone of UNITA insurgency or the MPLA counter-insurgency in Angola. The announcement by EO that it has formally dissolved is yet another ploy to evade the recent South African legislation on mercenaries. EO, as this book shows, is still active under different names and forms under the convoluted Heritage-Branch network.

Besides the underpinning pecuniary motive of mercenary involvement in conflicts, the variety of existing mercenary structures and the ad-hoc nature of armies mercenary-hiring companies assemble for specific operations, make regulation impossible. The chapters of this book exposed a wide variety of mercenary outfits. The main category comprises the established Western groups – Defence Systems Ltd, Executive Outcomes, International Chartered Incorporated, International Defence and Security, Military Professional Resources Incorporated and Sandline International. Insecurity in Africa has also thrown up loose mercenary groups put together by the Mafia in the former Soviet Union and Eastern Europe. Lone mercenaries sell their services on the market, hopping from one company to the rival, from one side in a conflict to the other. Non-combat security companies also exist, several of which are actually brigades of the combat companies; in fact, not a single mercenary-hiring company boasts a standing army. The legitimate question that flows is 'how does a regulatory mechanism streamline the activities of these motley mercenary groups?'

The experiences documented in this book point to the fact that mercenaries are becoming at best the manifestation of the failure of common security and at worst the avant-garde of attempts at corporate recolonisation. Controlled by mining multinationals, the major private companies ensure that beleaguered African leaders have guns pointed at their heads while signing away cheap mineral concessions to mercenary-backed mining concerns. This shuts out fair competition and constitutes crude blackmail. Within this context, private military security shields become more expensive to the weak state if considered against the fact that the cost of multilateral peace enforcement is not borne by the state in conflict.

At the base of the argument that other states must pay to rescue a sister state from crisis irrespective of the state's strategic importance or physical contiguousness, lies the morality of the international community. First, in this era of globalisation gross abuse of human rights, such as ethnic cleansing and genocide, is no longer confined to internal, or even regional, borders. The ripples of internal crisis are global. Secondly, the *raison d'être* of the United Nations is the security and well being of the global community. That is why African states have consistently contributed troops to multilateral peacekeeping efforts in regions and countries as far flung as the Middle East and Cambodia. Thirdly and more importantly, African sub-regions have initiated moves to put in place structures to deal with local crisis. What they do not require, above all, is for the powerful states of the international community to undermine the efforts of ECOMOG, SADC and others by promoting alternative structures based on mercenary forces. Rather, they should help financially and technically to upgrade the capacity of these indigenous structures.

Eventually, private security is proving a costly luxury for the hiring states in Africa in another way. Mercenary outfits, far from being security-enhancing technical tools in weak states, are becoming important internal political players. As analyses in some of the chapters in the book indicate, we are seeing the beginning of instances of private security involvement in palace coups in Africa.

At another level, the proliferation of mercenary groups in Africa is inextricably linked with the fierce battles between rival financiers and mining companies for concessions on the continent. The prospect of inter-mercenary wars on African soil will exacerbate, rather than ameliorate, the precarious security landscape. Besides, considering the fact that instability constitutes the primary demand factor for the mercenary business, why should mercenaries be interested in peace? This book has exposed human rights abuses associated with mercenary intervention in conflicts in Africa. It has also established the links between the mercenary business, crime, the narco-business and illegal arms transfers. Finally, the availability of private security shields increases the degree of impunity among African leaders and leads to the entrenchment of irresponsible governance. Together, these facts constitute an indictment of complacency towards private security and by extension, the regimes that hire them. Together, they build a strong case against any form of recognition of this trade, including through international regulation.

In arguing against regulation though, we also take cognisance of the impotence and failure that have characterised national and international legislation aimed at suppressing mercenary activity. The general observation of the authors is that it is virtually impossible to outlaw the trade in the absence of stability and efficacious alternative security structures. The appreciation of the complexity of the mercenary trade led to the conclusion that, ultimately, the objective should be to make mercenarism redundant by rekindling the debate on and faith in multilateral common security structures. In the immediate and medium term, the book seeks to place the responsibility for the security and protection of weak and small states squarely on sub-regional structures. We note the inherent and operational shortcomings of African sub-regional security structures. Instead of throwing out the baby with the bathwater as many canvassing for private military do, we have sought to analyse the difficulties confronted by regional peacekeeping and enforcement initiatives and suggest measures to enhance their efficacy in ensuring stability in Africa.

The responsibility of the international community towards conflict management and peace-building lies in initiating and overseeing the implementation of integrated measures that are capable of ensuring sustainable peace-ending wars, building confidence, cementing cracks of ethno-religious animosities and eliminating the underlying cause of conflict – poverty. Mercenaries have no conceivable role in this chain. Shortcomings in past multilateral peacekeeping operations should not constitute an excuse for abdicating this duty to corporate armies driven by the lure of hyper-profits and exotic adventure. At the very least, researchers in security have a duty to advance concrete suggestions about security sector reform borne out of the analyses of past failures to make multilateral peacekeeping more result-oriented and cost-effective. They should not choose the easiest way out by consigning multilateral intervention to the dustbin of history and championing the cause of private security firms. To succumb to these simplistic options would constitute a monumental failure on the part of the United Nations and world policy makers, a missed opportunity to set up an accountable, firm, fair and effective international framework to manage conflicts in the next millennium.

If, on the other hand, powerful nations of the North lose interest in Africa's internal conflicts out of fear of negative domestic opinion or for lack of economic and strategic returns, then they would be driving nails into the coffin of international morality. Such a scenario would inevitably sound the death knell of the United Nations as a collective security agency and diminish the chances of making the twenty-first century a more peaceful era than the outgoing millennium. The reverberations of refugee flows, ecological catastrophes, drugs and terrorism are not limited to areas of conflict alone.

Despite the consistent reservations expressed in this volume, we do not entirely rule out some role in conflict intervention for private military groups outside our preferred multilateral framework. If countries, out of sensitivity to public opposition to foreign deployment of their troops, can create and finance a special crisis force bound by the laws governing such a country's regular forces, they could deploy such a force in conflict zones under the supervision of a pre-existing multilateral force or a United Nations Observer Team with UN regulations for such intervention. At the moment, no private military company qualifies for such a role, not even MPRI and DSL which are often seen as very close to their respective governments.

The first years of the post-Cold War period are still fraught with dangers and the proliferation and intensification of internal conflicts bear testimony to this. The period also offers many challenges and opportunities for a more stable global village based on consensus and arbitration. Mercenaries promote the former and frustrate the realisation of the latter. It requires political will and courage from the international community to make mercenaries a relic of the past.

Ultimately, it is the role of civil society and its organisations at the local, regional and international levels, to give voice to the menace of mercenaries and push the issue onto the centre-stage of international discourse. This book is meant as potent ammunition to all forces advocating the empowerment of sub-regional bodies as guarantors of local stability within the context of long-term peacebuilding strategies. The Centre for Democracy and Development has been monitoring the triggers of instability on the African continent for a number of years. The Centre appreciates the fact that while mercenaries are not the root cause of instability in Africa, they nonetheless exert huge exacerbating influences on conflicts. Our encounter with mercenarism on the continent convinces us that the trade is a menace and no effort should be spared in exposing it as such. The Campaign against Mercenaries in

Mercenaries: Africa's Experience 1950s–1990

Country	Dates	Mercenaries Involved	Nationality	Recruited by	Objective	Outcome
Sierra Leone	1950s			Harry Oppenheimer of De Beers	Sir Percy Stilltoe (UK) to conduct anti-smuggling activities	
Kenya, Malawi, Tanzania	1960s–70s	Kulinda Security Ltd (WatchGuard)	Various	Governments	Military training	
Zambia	1967–69	WatchGuard	Various	Zambian Government	Provision of forces for counter-coup; guarding national border	
Belgian Congo (now DRC)	1960–1961	Compagnie Internationale	200 mercenaries, mainly Belgian and South African	Secessionist leader, Moise Tsombe and Belgian company, Union Minière (UM)	To promote secession of Katanga (now Shaba) from the Congo and protect UM concessions	Mercenaries defeated by combined UN and Congo forces

Country	Dates	Mercenaries Involved	Nationality	Recruited by	Objective	Outcome
Belgian Congo	1964–1965	'Five Commando', led by Col. 'Mad Mike' Hoare	1,000-strong force, mainly South African, Belgian, French and Rhodesian.	The CIA and then President Moise Tshombe	To fight patriotic nationalist forces, led by PM Patrice Lumumba	Nationalists defeated. Lumumba assassinated, neo-colonialism established
Belgian Congo	1967	Remnants of 'Five Commando' led by Bob Denard and Jacques Schramme	Same as above	Unknown	To overthrow President Sese Seko Mobutu	Seized border town Bukavu but failed in plot
Rhodesia (now Zimbabwe)	1965–80	Mostly former UK soldiers, recruited into Rhodesia Light Infantry (RLI) and special forces (SAS)	Mainly British	Rhodesian government	Support white minority rule against Mugabe's ZANU and Nkomo's ZIPRA	High casualties; minority government defeated, elections held in 1980, power transferred to Mugabe
Biafra, Nigeria	1967	Group of French mercenaries led by Robert Faulques, veteran of Katanga, and later Rolf Steiner	53-strong; mainly French and German	French Secret Service and secessionist leader, Col. Ojukwu	To help Biafra secede from Nigeria	Defeated by federal troops, 5 killed
Angola, Zaire	1975–	Security Advisory Services Ltd	UK, former para-troopers	Donald Telford, UK and recruiting agency run by John Banks	Force recruiting; support CIA-backed FNLA in Angola against Moscow-backed MPLA and South Africa-backed UNITA	Defeated by MPLA, some tried and executed as war criminals (1976)
Benin	1977	'Force Omega', led by Bob Denard	'60 white and 30 black mercenaries'	Opponents to the Government	Overthrow the president	Coup failed, some killed; Denard later stages successful coup on Comoros Islands

Mozambique	1980s	Lonrho	Gurkhas	International organisations (World Bank, UN, humanitarian NGOs)	Rural security	
Mozambique, Sudan, Kenya and others	1980s–early 1990s	Defence Systems Ltd (DSL)	British		Installation security and force training	Ongoing
Seychelles	1986–?	Longreach Pty Ltd			Military intelligence support	
Southern Africa	1986–90	KAS Enterprises (WatchGuard)			Anti-poaching operations	Terminated
South Africa	1989–	Executive Outcomes (EO)			Special Forces training	
1990–						
Algeria	1992	Eric SA			Oil pipeline security	
Angola	1992	EO	Southern African and European		Covert reconnaissance for De Beers and other South African mining houses; discussions begun with Heritage Oil & Gas (Soyo)	
Angola	1993	EO	Southern African and European	Angolan MPLA-government	Defeat UNITA, using air force and high-tech communications	UNITA substantially weakened
Angola	1993–	EO	Southern African and European	Angolan MPLA Government	Military training	TERMINATED January 1997
Angola (Soyo)	1993–94	EO	Around 500 South African	Oil Companies	Oil facilities seizure from UNITA rebels and security	Successful: recaptured by UNITA after EO left
Angola	1994	EO	Southern African and European	MPLA Government	Air and logistical support for Angolan army (FAA)	Diamond fields and rebel stronghold seized from UNITA; EO withdrew January 1996

Country	Dates	Mercenaries Involved	Nationality	Recruited by	Objective	Outcome
Angola	1994	Capricorn Systems International	RSA, Others		Air logistical support; combat air support	
Angola	1994–95	Saracen International	RSA, Others	FINA	Mining facilities security	
Angola	1994	Teleservices International	RSA, Others		Mining and oil facilities security	
Angola	1995–96	Alpha-5	RSA, Others		Mining facilities security	
Angola	1995–98	Teleservices International			Mining facilities security	
Angola	1995–96	Ibis Air International (formerly Capricorn Systems International)	Southern Africa and Others		Air logistical support; combat air support Demining; mining	
Angola	1996	Saracen International	as above		facilities security; establishing special rapid reaction police unit	
Angola	1996	Stabilco	as above		Mining and other commercial facilities security	
Angola	1996	Omega Support Ltd	as above		Mining and other commercial facilities security	
Angola	1996–?	Panasec Corporate Dynamics; Bridge Resources; COIN Security; Corporate Trading International			Mining and other commercial facilities security	
Angola	1996	Shibata Security			Demining; mining facilities security	

Country	Year	Company	Personnel/Backing	Activities	Status
Angola	1996	Longreach Pty Ltd		Rural security, mining and other commercial facilities security	
Angola	1996	Defence Systems Ltd (DSL)	1,000 personnel by October 1997, mostly Angolan	Oil facilities security (Soyo); mining facilities and transportation security; humanitarian assistance security	
Angola	1996	Military Professional Resources Incorporated (MPRI)	US, others	Negotiations about military training	FAILED
Angola	1997–98	Stabilco (1998: SafeNet)	Southern African. Others	Military support to UNITA	
Angola	1997–98	Omega Support Ltd (Strategic Concepts Pty Ltd)		Military support to UNITA	
Angola	1997–98	Panasec Corporate Dynamics		Military support	
Angola	1997–98	International Defence And Security (IDAS)	Various	Military support; mining facilities seizure and security	
Angola	1997–98	AirScan	Angolan government	Military support and training; air combat support	
Angola	1998	IRIS	Recruited 300 South African and British personnel; Backed by British, South African and US businesses	Military support to UNITA?	
Botswana	1991	EO	De Beers	Covert reconnaissance for De Beers	

Country	Dates	Mercenaries Involved	Nationality	Recruited by	Objective	Outcome
Botswana	1992	EO		De Beers, Others	Covert reconnaissance for De Beers and other South African mining houses	
Cameroon	1992	Secrets			Military training of Presidential Guard	
Congo-Brazzaville	1994–96	Levdan	65 former Israeli military personnel	Brazzaville government	Military training to Presidential Guard and armed forces	
Liberia	1995	MPRI		Government	Military training	
Malawi	1995	EO		Government	Military training	
Mozambique	1995	EO		Government	Military training, demining	
Mozambique	1996–97	Ronco Consulting Corporation			Demining	
Namibia	1991	EO		De Beers	Covert reconnaissance for De Beers	
Sierra Leone	1991	Specialist Services International		Sierra Leonean Government	Port security	
Sierra Leone	1992	Marine Protection Services		Sierra Leonean Government	Policing of fisheries; tax collection	
Sierra Leone	1994	Special Protection Services		Mining Companies	Mining facilities security	FAILED
Sierra Leone	1994	Frontline Security Services		Sierra Rutile and SIEROMCO	Mining facilities security	FAILED

Country	Year	Company	Personnel	Client	Activity	Notes
Sierra Leone	1995	EO, assistance from South African Army soldiers and Russian aircraft pilots	500 South African EO-personnel	Sierra Leonean Government	Against rebel movement Revolutionary United Front; military training and support; siege relief, humanitarian assistance and repatriation, mining facilities seizure and security	Rebels split up and driven back to Liberian border
Sierra Leone	1995–96	Ibis Air International (formerly Capricorn Systems International)			Air logistical support; combat air support	
Sierra Leone	1995	J&S Frankllin (Gurkha Security Guards Ltd)		Sierra Leonean Government	Military training and support for Republic of Sierra Leone Defence Forces (RSLDF) and mine protection	Terminated a few weeks later following rebels killing the leader
Sierra Leone	1995	Control Risks Group; Group 4			Mining facilities security	TERMINATED
Sierra Leone	1995	DSL			Military training, mining facilities security	TERMINATED
Sierra Leone	1996	EO			Military training and support; rebel HQ seized	
Sierra Leone	1996	Sandline International (EO UK)		Government	Military training	
Sierra Leone	1996	LifeGuard Management (from EO)			Mining facilities security	
Sierra Leone	1996–97	Teleservices International			Mining facilities security	

Country	Dates	Mercenaries Involved	Nationality	Recruited by	Objective	Outcome
Sierra Leone	1997	EO			Operations transferred to LifeGuard and Sandline International, January 1997	
Sierra Leone	1997	LifeGuard Management	285 employees (Zimbabwean, Namibian, Angolan, Mozambican and South African mercenaries		Mining facilities security, military training and support; intelligence gathering	
Sierra Leone	1997–98	Sandline International	At least 30 personnel		Logistics, intelligence and air support to Nigerian ECOMOG force based in Freetown; intelligence and reconnaissance; military training and logistics for Kamajor fighters	
Sierra Leone	1997–98	Cape International Corporation		Sam Norman, head of Kamajor fighters	Military training, mining facilities security	
Sierra Leone	1997–98	ICI (Pacific Architects Engineers)			Logistics and air transport security; personnel protection	
Sierra Leone	1998	EO			Management services to LifeGuard and Sandline	

Country	Year	Company	Client	Activities	Outcome
Sierra Leone	1998	LifeGuard Management		Mining facilities security, military training and support; intelligence gathering; counter-insurgency operations	
Sierra Leone	1998	DSL	UNDP	Security for UN humanitarian relief convoys	
Uganda	1995	EO		Military training, mining facilities security	
Uganda	1996	Saracen Uganda		Mining facilities security	
Uganda	1997–98	AirScan		Military support to rebels	
Rwanda	1994–?	Ronco Consulting Corporation	Rwandan Patriotic Front	Demining; limited military training for Rwandan forces	
South Africa	1990–91	EO		Special Forces training	
Togo	1995	Service and Security		Police paramilitary training	
Zaire	1996	2000 mercenaries with tanks, helicopters and jets involved. EO denied involvement allegedly French security officials recruited around 300 mercenaries for Mobutu's so-called White Legion	French, British, South African, Angolan, Serbian, Moroccan, Belgian, Mozambican	Military support against rebels	Mobuto was defeated White Legion disappeared into Congo-Brazzaville

Country	Dates	Mercenaries Involved	Nationality	Recruited by	Objective	Outcome
Zaire	1996–97	Omega Support Ltd		Kabila	Military training and support to rebels	
Zaire	1997	EO/Sandline			Military support and training	FAILED
Zaire	1997	International Defence and Security (IDAS)		Kabila	Military training and support to rebels	
Zaire	1997	Stabilco			Military support	FAILED
Zaire	1997	GeoLink			Military support	TERMINATED
Zaire/DRC	1997	Intercon Consulting Services			Military training, mining facilities security	
DRC	1998	DSL		DRC-Government	Oil facilities security; protection of US embassy	
DRC	1998	SafeNet (formerly Stabilco)			Military support and training against Kabila government	FAILED
DRC	1998	IRIS			Military support and training	
DRC	1998	Silver Shadow		DRC-Government	Military training	
DRC	1998	Angolan EO spin-offs			Aerial reconnaissance, intelligence-gathering, operations planning	CANCELLED
DRC	August 1998 ?		More than 100 South African and French troops	DRC-government?	Defence of mining installations in Katanga province	
Zambia	1995	EO			Military training	

Convention for the Elimination of Mercenarism in Africa

Preamble

We, the Heads of State and Government of the Member States of the Organization of African Unity,

Considering the grave threat which the activities of mercenaries present to the independence, sovereignty, security, territorial integrity and harmonious development of Member States of the Organization of African Unity;

Concerned with the threat which the activities of mercenaries pose to the legitimate exercise of the right of African People under colonial and racist domination to their independence and freedom;

Convinced that total solidarity and co-operation between Member States are indispensable for putting an end to the subversive activities of mercenaries in Africa;

Considering that resolutions of the UN and OAU, the statements of attitude and the practice of a great number of States are indicative of the development of new rules of international law making mercenarism an international crime;

Determined to take all necessary measures to eliminate from the African continent the scourge that mercenaries represents;

Have agreed as follows:

Article 1

Definition

1. A mercenary is any person who:
a) is specifically recruited locally or abroad in order to fight in an armed conflict;
b) does in fact take a direct part in the hostilities;
c) is motivated to take part in the hostilities essentially by the desire for private gain and in fact is promised by or on behalf of a party to the conflict material compensation;
d) is neither a national of a party to the conflict nor a resident of territory controlled by a party to the conflict;
e) is not a member of the armed forces of a party to the conflict; and
f) is not sent by a state other than a party to the conflict on official mission as a member of the armed forces of the said state.
2. The crime of mercenarism is committed by the individual, group or association, representative of a State or the State itself who with the aim of opposing by armed violence a process of self-determination, stability or the territorial integrity of another State, practices any of the following acts:
a) Shelters, organizes, finances, assists. equips, trains, promotes, supports or in any manner employs bands of mercenaries;
b) Enlists, enrols or tries to enrol in the said bands;
c) Allows the activities mentioned in paragraph (a) to be carried out in any territory under its jurisdiction or in any place under its control or affords facilities for transit, transport or other operations of the above mentioned forces.

Article 2

Aggravating Circumstances

The fact of assuming command over or giving orders to mercenaries shall be considered as an aggravating circumstance.

Article 3

Status of Mercenaries

Mercenaries shall not enjoy the status of combatants and shall not be entitled to the prisoner of war status.

Article 4

Scope of Criminal Responsibility

A mercenary is responsible both for the crime of mercenarism and all related offenses, without prejudice to any other offence for which he may be prosecuted.

Article 5

General Responsibility of States and their Representatives

1. When the representative of a State is accused by virtue of the provisions of Article 1 of this Convention for acts or omissions declared by the aforesaid article to be criminal, he shall be punished for such an act or omission.
2. When a State is accused by virtue of the provisions of Article 1 of this Convention for acts or omissions declared by the aforesaid article to be criminal, any other party to the present Convention may invoke the provisions of this Convention in its relations with the offending State and before any competent OAU or International Organization, tribunal or body.

Article 6

Obligation of States

The contracting parties shall take all necessary measures to eradicate all mercenary activities in Africa.

To this end, each contracting State shall undertake to:
a) Prevent its nationals or foreigners on its territory from engaging in any of the acts mentioned in Article 1 of the Convention;
b) Prevent entry into or passage through its territory of any mercenary or any equipment destined for mercenary use;
c) Prohibit on its territory any activities by persons or organizations who use mercenaries against any African State member of the Organization of African Unity or the people of Africa in their struggle for liberation;
d) Communicate to the other Member States of the Organization of African Unity either directly or through the Secretariat of the OAU any information related to the activities of mercenaries as soon as it comes to its knowledge;

e) Forbid on its territory the recruitment, training, financing and equipment of mercenaries and any other form of activities likely to promote mercenarism;

f) Take all the necessary legislative and other measures to ensure the immediate entry into force of this Convention.

Article 7

Penalties

Each contacting State shall undertake to make the offence defined in Article 1 of this Convention punishable by severest penalties under its laws, including capital punishment.

Article 8

Jurisdiction

Each contracting State shall undertake to take such measures as may be necessary to punish, in accordance with the provisions of the Article 1 of this Convention and who is found on its territory if it does not extradite him to the State against which the State has been committed.

Article 9

Extradition

1. The crimes defined in Article 1 of this Convention are not covered by national legislation excluding extradition for political offenses.

2. A request for extradition shall not be refused unless the requested State undertakes to exercise jurisdiction over the offender in accordance with the provisions of Article 8.

3. Where a national is involved in the request for extradition, the requested State shall take proceedings against him for the offence committed if extradition is refused.

4. Where proceedings have been initiated in accordance with paragraph 2 and 3 of this Article, the requested State shall inform the requesting State or any other State member of the OAU interested in the proceedings, of the result thereof.

5. A State shall be deemed interested in the proceedings within the meaning of paragraph 4 of this Article if the offence is linked in anyway with its territory or is directed against its interests.

Article 10

Mutual Assistance

The contracting States shall afford one another the greatest measure of assistance in connection with the investigation and criminal proceedings brought in respect of the offence and other acts connected with the activities of the offender.

Article 11

Judicial Guarantee

Any person or group of persons on trial for the crime defined in Article 1 of this Convention shall be entitled to all the guarantees normally granted to any ordinary person by the State on whose territory he is being tried.

Article 12

Settlement of disputes

Any dispute regarding the interpretation and application of the provisions of this Convention shall be settled by the interested parties in accordance with the principles of the Charter of the Organization of African Unity and the Charter of the United Nations.

Article 13

Signature, Ratification and Entry into Force

1. This Convention shall be open for signature by the members of the Organization of African Unity. It shall be ratified. The instruments of ratification shall be deposited with the Administrative Secretary-General of the Organization.
2. This Convention shall come into force 30 days after the date of the deposit of the seventeenth instrument of ratification.
3. As regards any signatory subsequently ratifying the Convention, it shall come into force 30 days after the date of the deposit of its instrument of ratification.

Article 14

Accession

1. Any Member State of the Organization of African Unity may accede to this Convention.
2. Accession shall be by deposit with the Administrative Secretary-General of the Organization of an instrument of accession, which shall take effect 30 days after the date of its deposit.

Article 15

Notification and Registration

1. The Administrative Secretary-General of the Organization of African Unity shall notify the Member States of the Organization of:
a) The deposit of any instrument of ratification or accession;
b) The date of entry into force of this Convention.
2. The Administrative Secretary-General of the Organization of African Unity shall transmit certified copies of the Convention to all Member States of the Organization.
3. The Administrative Secretary-General of the Organization of African Unity shall, as soon as this Convention comes into force, register it pursuant to Article 102 of the Charter of the United Nations.

IN WITNESS WHEREOF, We the Heads of State and Government of the member States of the Organization of African Unity have appended our signatures to this Convention.

Done in the City of Libreville, Gabon, on the 3rd of July 1977.

Appendix III

OAU Resolution on the Activities of Mercenaries

AHG/Res. 49 (IV) September (1967) (Kinshasa)

This Resolution stated that:

The Assembly of Heads of State and Government meeting in its Fourth Ordinary Session in Kinshasa, Congo from 11 to 14 September, 1967;

Determined to safeguard and ensure respect for the integrity and sovereignty of Member States;

Considering that the existence of mercenaries constitutes a serious threat to the security of Member States;

Recognizing their sacred and solemn responsibilities to spare present and future generations the scourge of racial hatred and conflict;

Conscious of the danger that the presence of mercenaries would inevitably arouse strong and destructive feelings and put in jeopardy the lives of foreigners in the continent;

1. STRONGLY CONDEMNS the aggression of the mercenaries against the Democratic Republic of the Congo;

2. DEMANDS that the mercenaries who are now in Eastern Congo (Bukavu) leave immediately the territory of the Congo, if necessary with the help of competent international bodies;

3. CALLS UPON all Member States that in case this generous offer is not accepted, to lend their wholehearted support and every assistance in their

power to the Government of the Democratic Republic of the Congo in its efforts to put an end to the criminal acts perpetrated by these mercenaries;

4. CALLS UPON the UN to deplore and take immediate action to eradicate such illegal and immoral practices;

5. APPEALS urgently to all States of the world to enact laws declaring the recruitment and training of mercenaries in their territories a punishable crime and deterring their citizens from enlisting as mercenaries.

Appendix IV

OAU Declaration on the Activities of Mercenaries in Africa

The Declaration stated that: We Heads of State and Government of Member States of the Organization of African Unity, meeting in Addis Ababa, Ethiopia from 21 to 23 June 1971;

Considering the great threat which the activities of mercenaries represents to the independence, sovereignty, territorial integrity and the harmonious development of Member States of the OAU;

Recalling resolutions CM/Res.49 (IX) and ECM/Res. 17 (VII) on Mercenaries;

Considering that, to perpetrate their crimes against Member States of the OAU, the mercenaries often use African territories still under foreign domination;

Considering that activities of mercenaries and the forces behind them constitute an element of serious tension and conflict between Member States;

Considering that total solidarity and co-operation between Member States are indispensable for putting and end, once and for all, to the subversive activities of mercenaries in Africa;

Considering the undertaking made by various non-African States to take the appropriate steps to prevent their nationals from returning to Africa as mercenaries and to ensure that their territories should no longer be used for the recruitment, training and equipping of mercenaries;

1. REAFFIRM the determination of African peoples and States to take all the necessary measures to eradicate from the African continent the scourge that the mercenary system represents;

2. REITERATE our irrevocable condemnation of the use of mercenaries by certain countries and forces to further jeopardize the independence, sovereignty and territorial integrity of Member States of the OAU;

3. FURTHER EXPRESS our total solidarity with States which have been victims of the activities of mercenaries;

4. PROCLAIM our resolve to prepare a legal instrument for co-ordinating, harmonizing and promoting the struggle of the African peoples and States against mercenaries;

5. PLEDGE OURSELVES to co-operate closely to ensure immediate implementation of the previous decisions and directives of the policy-making bodies of the OAU before the proposed Convention on the subject enters into force;

6. DRAW the attention of world opinion to the serious threat that the subversive activities of mercenaries in Africa represent to the OAU Member States;

7. REITERATE the appeal made to Member States to apply both in spirit and letter, Resolution ECM/Res.17 (VII) of the Seventh Extraordinary Session of the Council of Ministers held in Lagos in December 1970, and consequently invite them:
i) to take appropriate steps to ensure that their territories are not used for the recruitment, drilling and training of mercenaries, or for the passage of equipment intended for mercenaries and that,
ii) to hand over mercenaries present in their countries to the States against which they carry out their subversive activities.

8. INVITE all States which had pledged not to tolerate the recruitment, training and equipping of mercenaries on their territory and to forbid their nationals to serve in the ranks of the mercenaries, to fulfil their undertakings. Also invite other non-African States not to allow mercenaries, be they their nationals or not, to pursue their activities on their territory;

9. REQUEST the Chairman of the Assembly of Heads of State and Government to do everything possible to mobilize world opinion so as to ensure the adoption of appropriate measures for the eradication of mercenaries from Africa, once and for all;

10. APPEAL to all Member States to increase their assistance in all fields to freedom fighters in order to accelerate the liberation of African territories still under foreign domination, as this is an essential factor in the final eradication of mercenaries from the African continent.

Appendix V

OAU Convention for the Elimination of Mercenaries in Africa
OAU Doc. CM/433/Rev.L., Annex 1 (1972)

Preamble

We the Heads of State and Government of Member States of the Organization of African Unity

Considering the grave threat which the activities of mercenaries represent to the independence, sovereignty, territorial integrity and harmonious development of Member States of OAU,

Considering that total solidarity and co-operation between Member States are indispensable for putting an end, once and for all, to the subversive activities of mercenaries in Africa,

Decided to take all necessary measures to eradicate from the African continent the scourge that the mercenary system represents.

We agree on the following:

Article One

Under the present Convention a 'mercenary' is classified as anyone who, not a national of the state against which his actions are directed, is employed, enrols or links himself willingly to a person, group or organization whose aim is:

(a) to overthrow by force of arms or by any other means, the government of that Member State of the Organization of African Unity;
(b) to undermine the independence, territorial integrity or normal working institutions of the said State;
(c) to block by any means the activities of any liberation movement recognized by the Organization of African Unity.

Article Two

Offence

1. The actions of a mercenary, in the meaning of Article One of the present Convention, constitute offenses considered as crimes against the peace and security of Africa and punishable as such.
2. Anyone who recruits or takes part in the recruitment of a mercenary, or in training, or in financing his activities or who gives him protection, commits a crime in the meaning of paragraph I of this article

Article Three

Duties of State

The Member States of the Organization of African Unity, signatories to the present Convention, undertake to take all necessary measures to eradicate from the African continent the activities of mercenaries

To this end, each State undertakes particularly:
(a) to prevent their nationals or foreigners living in their territory from committing any of the offenses defined in Article Two of the present Convention;
(b) to prevent the entry to or the passage through their territory of any mercenary or equipment intended for their use;
(c) to forbid in their territory any activity by organizations or individuals who employ mercenaries against the African States Members of the Organization of African Unity;
(d) to communicate to other Member States of the Organization of African Unity any information relating to the activities of mercenaries in Africa;
(e) to forbid on their territory the recruitment, training or equipment of mercenaries or the financing of their activities;
(f) to take as soon as possible all necessary legislative measures for the implementation of the present Convention

Article Four

Sanctions

Every contracting State undertakes to impose severe penalties for offenses defined in Article Two of the present Convention.

Article Five

Competence

Every contracting State undertakes to take the measures necessary to punish any individual found in its territory who has committed one of the offenses defined in Article Two of the present Convention, if he does not hand him over to the State against which the offence has been committed or would have been committed.

Article Six

Offenses Calling for Extradition

In accordance with the provisions of Article Seven of the present Convention, the offenses defined in Article Two above should be considered as offenses calling for extradition.

Article Seven

Extradition

1. A request for extradition cannot be rejected, unless the State from which it is sought undertakes to prosecute the offender in accordance with the provisions of Article Five of the present Convention
2. When a national is the subject of the request for extradition, the State from which it is sought must, if it refuses, undertake prosecution of the offence committed.
3. If, in accordance with sections 1 and 2 of this Article, prosecution is undertaken, the State from which extradition is sought will notify the outcome of such prosecution to the State seeking extradition and to any other interested Member State of the Organization of African Unity.
4. A state will be regarded as an interested party for the outcome of a prosecution as defined in section 3 of this Article if the offence has some connection with its territory or militates against its interests.

[Articles 8 to 11 are formal and are therefore omitted]

Appendix VI

The Report by the UN Special Rapporteur on the Use of Mercenaries, 1998

Distr.
GENERAL
E/CN.4/1999/11
13 January 1999
ENGLISH
Original: ENGLISH/SPANISH

COMMISSION ON HUMAN RIGHTS
Fifty-fifth session
Item 5 of the provisional agenda

THE RIGHT OF PEOPLES TO SELF-DETERMINATION AND ITS
APPLICATION TO PEOPLES UNDER COLONIAL OR ALIEN
DOMINATION OR FOREIGN OCCUPATION

Report on the question of the use of mercenaries as a means of violating
human rights and impeding the exercise of the right of peoples to self-
determination, submitted by Mr Enrique Bernales Ballesteros (Peru),
Special Rapporteur pursuant to Commission Resolution 1998/6

CONTENTS (PARAGRAPHS)

Introduction 1–5

I. ACTIVITIES OF THE SPECIAL RAPPORTEUR 6–23

A. Implementation of the programme of activities 6–8

B. Correspondence 9–20

C. Correspondence regarding mercenary activities against Cuba 21–23
II. MERCENARY ACTIVITIES IN SIERRA LEONE 24–31
III. PERSISTENCE AND EVOLUTION OF MERCENARY ACTIVITIES 32–62
A. The present situation 35–44
B. Current international legislation and its limitations 45–55
C. Terrorism and mercenary activities 56–62
IV. PRIVATE SECURITY AND MILITARY ASSISTANCE COMPANIES AND MERCENARY ACTIVITIES 63–76
V. CURRENT STATUS OF THE INTERNATIONAL CONVENTION AGAINST THE RECRUITMENT, USE, FINANCING AND TRAINING OF MERCENARIES 77–78
VI. CONCLUSIONS 79–92
VII. RECOMMENDATIONS 93–101

Introduction

1. At its fifty-fourth session, the Commission on Human Rights adopted resolution 1998/6 of 27 March 1998, in which, *inter alia*, it reaffirmed that the recruitment, use, financing and training of mercenaries were causes for grave concern to all States and violated the purposes and principles of the Charter of the United Nations. The Commission called upon all States that had not yet done so to consider taking the necessary action to ratify the International Convention against the Recruitment, Use, Financing and Training of Mercenaries and urged them to cooperate fully with the Special Rapporteur in the fulfillment of his mandate.

2. Taking note of the report of the Special Rapporteur (E/CN.4/1998/31 and Add.1), the Commission also urged all States to take the necessary steps and to exercise the utmost vigilance against the menace posed by the activities of mercenaries and to take necessary legislative measures to ensure that their territories and other territories under their control, as well as their nationals, were not used for the recruitment, assembly, financing, training and transit of mercenaries for the planning of activities designed to destabilize or overthrow the Government of any State or threaten the territorial integrity and political unity of sovereign States or to promote secession. The Commission welcomed the adoption by some States of national legislation that restricted the use of mercenaries and the cooperation of those countries that had issued invitations to the Special Rapporteur to visit their countries. It also

requested the Office of the United Nations High Commissioner for Human Rights, as a matter of priority, to publicize the adverse effects of mercenary activities on the right of peoples to self-determination and, when requested and where necessary, to provide advisory services to States that were affected by the activities of mercenaries.

3. The Commission on Human Rights also decided to extend the mandate of the Special Rapporteur for three years and requested him to submit a report, with specific recommendations, to the Commission at its fifty-fifth session.

4. On 5 November 1998, the Third Committee of the General Assembly adopted a draft resolution on the question of the use of mercenaries as a means of violating human rights and impeding the exercise of the right of peoples to self-determination.

5. Pursuant to the provisions of resolution 1998/6, the Special Rapporteur has the honour to submit this report for the consideration of the Commission on Human Rights.

I. ACTIVITIES OF THE SPECIAL RAPPORTEUR

A. Implementation of the programme of activities

6. The Special Rapporteur submitted his reports (E/CN.4/1998/31 and Add.1) to the Commission on Human Rights on 18 March 1998. While in Geneva, the Special Rapporteur had consultations with representatives of various States and held meetings with members of non-governmental organizations. He also held meetings with officials of the Office of the United Nations High Commissioner for Human Rights, notably the Activities and Programmes Branch.

7. The Special Rapporteur returned to Geneva on three occasions, from 26 to 29 May, 17 to 21 August and 16 to 20 November 1998, to hold various consultations, to draft his report to the General Assembly and the present report and to participate in the fifth meeting of special rapporteurs and special representatives, independent experts and chairmen of working groups of the Commission on Human Rights. He also went to United Nations Headquarters in New York to submit his report to the Third Committee of the General Assembly. He made his presentation on 23 October 1998.

8. During the reporting period, the Special Rapporteur received invitations to undertake official missions to the United Kingdom of Great

Britain and Northern Ireland and the Republic of Cuba. He would like to express his appreciation for the invitations and hopes to carry out the missions during 1999.

B. Correspondence

9. In pursuance of General Assembly resolution 52/112 of 12 December 1997 and Commission on Human Rights resolution 1998/6 of 27 March 1998, the Special Rapporteur sent a communication to all Member States of the United Nations on 6 July 1998 requesting them to send information on the existence of mercenary activities, the possible participation by nationals of their country in such activities and domestic legislation currently in force outlawing such activities. He also requested suggestions to help enhance the international approach to the topic, arrive at a better definition of mercenary and regulate the private companies which offer security services and assistance and military advice and which sometimes recruit mercenaries.

10. The Special Rapporteur received valuable assistance from the Governments of Ecuador, the Eastern Republic of Uruguay, the Federal Republic of Yugoslavia (Serbia and Montenegro), Honduras, Ireland, Portugal, Sweden and the Syrian Arab Republic, which provided useful information and comments (see the report of the Special Rapporteur to the fifty-third session of the General Assembly, paras. 8–16).

11. After having drafted his report to the General Assembly, the Special Rapporteur received the following communications. By note verbale dated 18 August 1998, the Permanent Mission of Finland to the United Nations Office at Geneva replied as follows: 'The Government of Finland is committed to the fight against the use of mercenaries. In this regard, the Government of Finland is currently considering whether it is possible and desirable in the light of the present Finnish legislation to accede to the International Convention against the Recruitment, Use, Financing and Training of Mercenaries, adopted by the General Assembly on 4 December 1989.'

12. Mr Miroslav Milosevic, then Chargé d'Affaires ad interim of the Permanent Mission of the Federal Republic of Yugoslavia (Serbia and Montenegro) to the United Nations Office at Geneva, provided additional information on mercenary activities and terrorist actions in Kosovo and Metohija in a letter dated 27 August 1998. The letter reads verbatim as follows: 'With the terrorist activities in Kosovo and Metohija further escalating, data have been officially registered on the active participation of foreign mercenaries in the terrorist activities of the so-called

'Kosovo Liberation Army' (KLA). Evidence is available on the support and assistance that the Republic of Albania has been providing for terrorists active within the so-called 'Liberation Army'. Thus, for example, northern Albania has been transformed into a number of recruitment centres for terrorists' training and into depots for the so-called KLA. The Yugoslav authorities possess documentation on the nationality, training expertise, tasks, experience and payment of foreign mercenaries who have been taken into Serbia to carry out activities in its territory (Kosovo and Metohija). In June and July of 1998, criminal charges were filed against seven Albanian nationals (mostly from the municipality of Tropoja, taken prisoners in our country on the occasion of smuggling large quantities of arms for committing terrorist activities in Kosovo and Metohija). 'Apart from Albanian nationals, most other terrorists are the so-called 'mujaheddins', nationals from some Arab countries – Afghanistan, Sudan, Russian Federation (Chechnya) and others. Many of them have participated in the war in Bosnia and Herzegovina on the Muslim side. These are highly professional and well-trained fanatics who take a direct part in capturing and cleansing the region, in kidnapping citizens and members of police, in committing horrendous tortures and liquidations. They have served as special training instructors for the so-called KLA. 'Apart from Albania, the Republic of Macedonia has also been used as a canal for the illegal entry of mercenaries into Yugoslav territory. 'In Germany, Switzerland, Austria, the Netherlands, Bosnia and Herzegovina, Croatia and some other countries, highly trained professionals with substantial experience acquired in war operations and sabotage-terrorist activities in various parts of the world have been recruited. Mercenary recruitment and training centres and weapons pooling centres in Bosnia and Herzegovina have been in the vicinity of the cities of Zenica, Tuzla and Travnik. The Yugoslav authorities possess also data on the training of members of separatist-terrorist gangs in Kosovo and Metohija in the village of Mehurici, municipality of Travnik, Bosnia and Herzegovina. 'Members of Muslim military formations in the war in Bosnia and Herzegovina have organized the transfer of Muslim mercenaries from Bosnia and Herzegovina into the territory of Serbia (Kosovo and Metohija).'

13. By letters dated 16, 17 and 29 September 1998, Mr Branko Brankovic, Chargé d'affaires ad interim of the Federal Republic of Yugoslavia (Serbia and Montenegro) to the United Nations Office at Geneva, enlarged upon the information provided by his Government regarding what it considered to be terrorist activities carried out by foreign mercenaries and Islamic fundamentalists in the Autonomous Province of Kosovo and Metohija and submitted two documents entitled 'Facts regarding the situation in Kosovo and Metohija' and 'Conclusions of the National Assembly of the Republic of Serbia on the situation in

Kosovo and Metohija'. The Special Rapporteur regrets that he is unable to reproduce these communications in their entirety owing to the strict maximum number of pages allowed for Commission on Human Rights reports. Nevertheless, he considers it important to point out the following paragraphs contained in the letter dated 16 September 1998: '(...) The Islamic influence upon terrorist activities perpetrated by the so-called Kosovo Liberation Army (KLA) has been manifested in a number of ways, first of all through aid for purchasing armaments and military equipment provided by radical Islamic countries, but also through Mujahedin fighting within the KLA units. There is irrefutable evidence on the linkage between terrorism in Kosovo and Metohija and the Mujahedin in the Middle East, Africa and Asia, aimed at forced secession of Kosovo and Metohija from Serbia and the Federal Republic of Yugoslavia. '(...) In 1998 the leaders of the so-called Kosovo Liberation Army (KLA), with the idea of providing professional, experienced commanders and perpetrators of terrorist acts, have directed their activities to recruiting mercenaries and volunteers from Islamic countries and from Bosnia and Herzegovina. At the beginning of 1998, in the region of Zenica and Kalesija (Bosnia and Herzegovina), posts of the so-called KLA were set up, charged with recruiting mercenaries and volunteers, mostly among the demobilized members of the Muslim army of Bosnia and Herzegovina, particularly ex-members of its special units. According to the information obtained, their monthly pay ranges from DM3,000 to DM5,000. Mujahedin from Arab and other Muslim countries are in charge of the training. '(...) Yugoslav authorities have obtained evidence whereby, from May to July 1998, in the region of Drenica in Kosovo and Metohija, a joint Mujahedin unit, 'Abu Bekir Sidik' was operating. This unit was first set up in mid-1997 in Bosnia and Herzegovina, under the instructions coming from Saudi Arabia and Turkey. Some DM 300,000, earmarked for arms purchasing and illegal transfer into Kosovo and Metohija have been provided by the 'World Office for Islamic Appeal'. The first contingent was illegally brought into the Federal Republic of Yugoslavia (Serbia and Montenegro) in July 1997, and the first Mujahedin unit was infiltrated in May 1998 to the village of Donji Prekaz. The unit, under the command of Ekrem Avija, consisted of 120 Mujahedin, divided into seven groups. A separated group consisted of Mujahedin from Saudi Arabia, the former Yugoslav Republic of Macedonia, Albania, Bosnia and Herzegovina and other countries, with an Egyptian national, Abu Ismail, heading it (he had been in charge of the former Mujahedin unit in the Bosnian war). '(...) I should like to point to the increasing presence of the Islamic factor in Kosovo and Metohija, which indicates to broader strategic interests to exert influence upon Albania, and the whole of the Balkans.'

14. By a letter dated 12 August 1998, Ms Sima Eivazova, Permanent Representative of the Azerbaijani Republic to the United Nations Office at Geneva, submitted a list of persons she described as mercenaries who had fought on the side of Armenia in the armed conflict against Azerbaijan.

The list includes the following: Aleksandr Yuryevich Karavaev, born in Kazan; Sergei Konstantinovich Turchenko, born in Pskov; Nikolai Semenovich Chimpoev, resident of Tiraspol; Sergei Grigoryev, resident of Moscow; Stanislav Stefanovich Semenchuk; Nikolai Ivanovich Bukhnarev, resident of Moscow; Vladimir Vikentyevich Semenov, born in Kamyshin, Volgograd; Konstantin Immanuilovich Voevodsky, resident of St Petersburg; Vladimir Maiorov; and Sergei Kuznetsov. The list includes the names of the following combatants killed in action in 1992: Nikolai Anatolyevich Shamkov; A.F. Voronin; S.M. Mukhaev; R.G. Chechnikov; S.M. Gladilin; E.S. Skrizhalik; and A.M. Kurzyukov. It also includes the names of the following nationals of the Russian Federation: Igor Evgenyevich Babanov; Aleksandr Viktorovich Shitko; Vladimir Zoltan; Sergei Ivanovich Kidalov; Akhmed Zhumagaliev; Daud Khamrasovich Lusenov; Bashir Akhmedovich Nalgiev; and Abukhar Akhmedovich Nalgiev.

15. The Permanent Representative of Azerbaijan also gave the Special Rapporteur a list containing the names of 13 persons considered to be mercenaries, captured while fighting on the side of the Armenian forces: Vladimir Nikolaevich Semion; Nikolai Vitalyevich Goncharov; Vladimir Aleksandrovich Polyakov; Aleksandr Yuryevich Korenko; Igor Chernenko; Sergei Veniaminovich Suchkov and Oleg Fedorovich Serdik. The list includes the names of the following persons, sentenced to death by the Military Division of the Supreme Court of Azerbaijan: Konstantin Vladimirovich Tukish; Yaroslav Leonidovich Evstigneev; Andrei Anatolyevich Filippov; Mikhail Stepanovich Lisovoi; and Vladislav Petrovich Kudinov. She also reported that Vasili Vladimirovich Lugovoi, who had been sentenced to 14 years' imprisonment, had been handed over to the authorities of the Russian Federation on 8 May 1996.

16. By note verbale dated 3 September 1998, the Permanent Mission of Mauritius to the United Nations Office at Geneva replied to the Special Rapporteur's request for information as follows: '(i) There is not the slightest evidence of the existence of any mercenary activities in Mauritius; '(ii) The Government of Mauritius has not enacted any domestic legislation nor is it a party to any international treaties outlawing mercenary activities; and '(iii) The Government of Mauritius is in favour of the International Convention against the Recruitment, Use, Financing and Training of Mercenaries. However, Mauritius has a

well-organized police and a paramilitary force to ensure the security of the country.'

17. In reply to a letter dated 8 July 1998, from Mr Tony Lloyd, Minister with responsibility for human rights and United Nations affairs at the United Kingdom Ministry of Foreign Affairs, the Special Rapporteur sent the following letter on 20 August 1998: 'I have the honour to acknowledge and thank you for your letter of 8 July 1998, in which you accept my request to visit your country on an official mission for the purpose of continuing my current investigations into private companies providing security services and military assistance, pursuant to the mandate conferred by the Commission on Human Rights. I am particularly interested in meeting you and officials from the relevant governmental sectors and departments, specially Foreign Affairs, the Interior, Justice and Defence. I should be grateful if your Government would cooperate in drawing up a schedule of interviews with governmental authorities and officials, preferably in the morning, in order that I may have the afternoons for meetings with non-governmental organizations and academic research centres. As regards the dates of my visit, I would propose the week of 12 to 16 October 1998, arriving on 11 October and leaving on 17 October. I would be accompanied by Mr. Miguel de la Lama, who assists me in carrying out my mandate, and two interpreters. The organizational details will be worked out with your country's Permanent Mission to the United Nations Office at Geneva.'

18. Mr Roderic M.J. Lyne, CMG, Permanent Representative of the United Kingdom of Great Britain and Northern Ireland to the United Nations Office at Geneva, sent the following letter, dated 24 September 1998, to the Special Rapporteur: 'Thank you for your letter of 20 August to the Minister of State at the Foreign and Commonwealth Office, Mr Tony Lloyd MP. I have been asked to reply on Mr Lloyd's behalf. 'My Government looks forward to your visit, and will do all that it can to help to facilitate the meetings you have requested. We have explored the possibility of setting up these meetings during the period suggested by you in mid-October. Unfortunately, our enquiries have established that many of the key interlocutors in the government departments you have specified will not be easily available during the week in question, because they are already committed to conflicting engagements. We think it important that your visit should take place at a time when the right people are available, and I fear that the notice is simply too short for us to organise the sort of programme you would like by 12 October. We would therefore like to propose that your visit should take place at a slightly later date. I understand that my authorities in London will soon be letting me have some dates we could put forward with confidence in our ability fully to meet your requests. I shall write to you again as soon

as I have a new timing to propose. In the meantime, if there is any particular period later in the year in which, for your part, you would not be free to visit the United Kingdom, perhaps you could be so good as to let me know.'

19. The Special Rapporteur and the United Kingdom authorities and Permanent Mission are discussing the possibility of a visit in early 1999. The Special Rapporteur is particularly anxious for a visit to take place then and is grateful for the cooperation of the United Kingdom Government and the Permanent Mission in preparations for the visit and, in general, the fulfilment of his mandate.

20. During 1998 the Special Rapporteur continued to receive the cooperation of various non-governmental organization (NGOs), including Amnesty International, Human Rights Watch and International Alert. He also received communications from, inter alia, the Bahrain Human Rights Organization, in Copenhagen; the Humanitarian Law Centre, in Belgrade; the Muttahida Quami Movement (MQM), in Edgware, United Kingdom; and the Organization for Defending Victims of Violence, in Tehran. He was also assisted by members of various institutions, such as Mr David Shearer, of the International Institute of Strategic Studies. The Special Rapporteur once again underlines the importance of being able to rely on help from NGOs in carrying out his mandate, particularly at a time when the activity and growth of private companies offering security services and military assistance and making use of mercenaries, are a threat to traditional human rights protection systems around the world.

C. Correspondence Regarding Mercenary Activities Against Cuba

21. In his last report to the Commission on Human Rights (E/CN.4/1998/31, para. 20), the Special Rapporteur reproduced a letter from the Ministry of Foreign Affairs of Cuba dated 1 October 1997 concerning attacks on hotel and tourist facilities in Havana, particularly those carried out by the Salvadoran citizen Raúl Ernesto Cruz León, which resulted in the death of an Italian tourist. He also reproduced a letter replying to his request for information on this matter from the Permanent Representative of the United States of America to the United Nations Office at Geneva, dated 13 January 1998 (E/CN.4/1998/31/Add.1). In his recent report to the General Assembly, the Special Rapporteur reproduced a letter dated 3 August 1998, in which the Permanent Representative of Cuba to the United Nations Office at Geneva drew attention to an interview given to the North American

newspaper the *New York Times* by Luis Posada Carriles on 12 and 13 June 1998.

22. In a further communication dated 15 October 1998, the Permanent Representative of Cuba to the United Nations Office at Geneva, Mr Carlos Amat Forés, made the following statement: '(...) I would like to take this opportunity to inform the Special Rapporteur of the latest developments and facts relating to the use of mercenaries that have recently affected Cuba. As we have repeatedly said in various international forums, the use of mercenaries and their recruitment, financing and training are offences that deeply concern the Government of Cuba, since they violate the fundamental principles of international law, while mercenary activities constitute gross violations of fundamental human rights. On 4 March 1998, two Guatemalan citizens, Nader Kamal Musalam Bacarat and María Elena González Meza, were arrested in Havana for attempting to bring into the country the means to make explosive devices, such as detonators, plastic explosive, timers, interface circuits and batteries. Their mission was to explode these devices in public places. On 20 March, María Elena's husband, Mr Jazid Iván Fernández Mendoza, was also arrested in Havana for helping to disguise the components and plan the actions. In the course of the investigation, they admitted that their motives for carrying out such terrorist actions were financial, since, on their return to Guatemala, the activities were to have been paid for by a Salvadoran citizen, Francisco Chávez Abarca, whom they identified as the person who had recruited and trained them and supplied the necessary components. We would like to take this opportunity to draw the Special Rapporteur's attention to Mr Chávez Abarca, since it was he who organized the terrorist acts perpetrated by Raúl Ernesto Cruz León in Havana in 1997. Moreover, on 10 June 1998, another Salvadoran citizen, Otto René Rodríguez Llerena, was arrested in Cuba while attempting to bring in components for two explosive devices for the purposes of terrorist actions. Rodriguez Llerena admitted that he had caused the explosion that took place on 4 July 1997 in the lobby of the Cohiba Hotel in Havana and that he had been recruited, trained and supplied by Luis Posada Carriles, a terrorist of Cuban origin, who had paid him around US$ 1,000 for that action. He stated that, on this occasion, too, his motive was financial and that it was once again Posada Carriles who had supplied the components and financed the trip. Given that these operations are linked with Miami-based extremist organizations, as Posada Carriles himself admitted in his *New York Times* interview, the Government of Cuba once again urges the Special Rapporteur to use his good offices to request the United States authorities to take firm and decisive action to put an end to these objectionable activities. The Special Rapporteur should also continue the work he has been doing in his latest reports, of analysing the causes and conse-

quences of mercenary practices and their increasingly close links with terrorist activities.'

23. On the same date, the Permanent Representative of Cuba to the United Nations Office at Geneva, Mr Carlos Amat Forés, sent the Special Rapporteur an official invitation from his Government to visit Cuba. He noted that 'this visit is a part of my Government's traditional cooperation with the Office of the Special Rapporteur which you direct as part of the global mechanisms for the protection of human rights throughout the world'. The Special Rapporteur appreciates the cooperation offered by the Government of Cuba in the discharge of his mandate and hopes to visit Cuba on an official mission during 1999.

II. MERCENARY ACTIVITIES IN SIERRA LEONE

24. In previous reports, the Special Rapporteur has discussed the armed conflict in Sierra Leone and the presence there of private companies offering security services and military assistance and advice, and of mercenaries recruited by such companies. As is well known, Sierra Leone has for seven years been in the grip of an armed conflict which, having begun as a spillover from the civil war in Liberia, has made the country one of the poorest in Africa and the world and caused the displacement of 420,000 of its nationals. The security firm Executive Outcomes, which is registered in South Africa and made up of former members of the 32nd battalion of the South African army, was contracted by Valentin Strasser's Government, but left Sierra Leone at the time the peace accord was signed in November 1996, after having provided military assistance services for several months.

25. In a coup d'état on 25 May 1997, led by Commander Johnny Paul Koroma, the constitutional President Alhaji Ahmed Tejan Kabbah was overthrown and a revolutionary council was formed. Lower ranks of the armed forces took part in the coup. A number of foreign companies involved in diamond, titanium, gold and bauxite mining left the country or suspended their operations. The Governments of the region not only condemned the coup, but also isolated the new de facto Government and demanded the return of the deposed President. The de facto Government in its turn called on Nigeria to return Foday Sankoh, the leader of the Revolutionary United Front (RUF).

26. The Special Rapporteur has learned that, while in exile in Guinea, the deposed President signed a contract with Sandline International, a company registered in the Bahamas and with offices in Chelsea, London, to provide him with military support, advice and assistance in regaining

power. This company is already known to the Special Rapporteur and has been mentioned in previous reports in connection with its unsuccessful intervention in Papua New Guinea (see, for example, E/CN.4/1998/31, paras. 93–99). In 1997, Sandline International signed a contract for US$36 million with the Government of Papua New Guinea, whose Prime Minister was then Sir Julius Chan, to carry out offensive military operations in Bougainville, engage the rebels of the Bougainville Revolutionary Army and provide mercenaries, sophisticated military equipment and military assistance. Shortly afterwards, Sir Julius Chan's Government was legally ousted and the mercenaries sent in by the company were expelled.

27. According to the information received by the Special Rapporteur, a number of financial and mining companies with debts and interests of various kinds in Sierra Leone gave their support and even partial financial backing to the contract with Sandline International. After signing the contract, the company prepared and dispatched a document giving its view of the operations to be carried out and various strategic and tactical plans. Helicopters and military equipment were subsequently exported to Sierra Leone, allegedly via Bulgaria, Nigeria and Liberia, in breach of the embargo imposed under Security Council resolution 1132 (1997); military experts were also sent and are still providing tactical and operational advice on the ground. The first arms sent by Sandline arrived on 22 February 1998.

28. On 10 March 1998, after bloody fighting in which the forces of the Economic Community of West African States (ECOMOG) took part, the coalition Government formed by the Armed Forces Revolutionary Council (AFRC) and the Revolutionary United Front (RUF) was overthrown, and President Tejan Kabbah was able to return to Freetown. The war continued, however. Soldiers loyal to the ousted military junta have committed atrocities against the civilian population in their flight to the east of the country. The decapitated, dismembered or burnt corpses of more than 100 people were seen in the areas of Bo, Lunsar, Kenema and Makeni, as witness to the vengefulness of those who had been dislodged from power; 1,500 supporters of the junta were arrested and 59 were accused of treason. Of these, 24 were sentenced to death at a trial without right of appeal and were executed on 19 October 1998.

29. While ECOMOG forces devote themselves to keeping order in Freetown, Sandline International employees advise the Government as it plans the creation of a new army with no links to those involved in the coup d'état. Efforts are also being made to organize the population in civil defence or self-defence forces. Hinga Norman, a chief of the Mende Tribe, and educated in the United Kingdom, has set up a 20,000-strong para-

military force called Kamajor with the aim of stamping out the rebellion. This paramilitary force is also reported to be committing gross violations of human rights with the acquiescence of the Government and after training and advice from Sandline International mercenaries. The Special Rapporteur has been informed of appalling acts of cruelty committed by mercenaries on captured rebels and on civilians suspected of collaborating with the insurgents. During the week of 30 November 1998, 70 rebels were killed in battle in Gberay, a rebel base 100 kilometres north of the capital. Many of the bodies were mutilated and incinerated.

30. As the Special Rapporteur has noted in previous reports, the presence in Sierra Leone of the company Executive Outcomes, which had fought the rebels and made possible the November 1996 peace agreement, did not help to avert the coup d'état of 25 May 1997 or the formation of an alliance of former enemies from RUF and AFRC in a coalition Government. Hiring private companies providing security and military assistance and advice is no substitute for maintaining a collective regional security system and genuinely professional national armed forces and security forces loyal to the democratic legal order. It is a false solution. When companies of this type leave the country, they also leave behind the structural problems they found when they arrived – unsolved, if not actually worse.

31. The recruitment, financing and use of mercenaries by such companies are unacceptable under any circumstances, even when the aim is claimed to be the restoration of a constitutional regime overthrown by a coup d'état. This means, however, that the international community must promote the development of effective regional and global collective security mechanisms and give its backing to the work of the United Nations, which has recently opened an office in Sierra Leone (the United Nations Observer Mission in Sierra Leone -UNOMSIL) to work for peace and human rights in that country.

III. PERSISTENCE AND EVOLUTION OF MERCENARY ACTIVITIES

32. One question needs to be asked: why is it that mercenary activities persist when they have so often been condemned by various United Nations bodies, when no State will admit publicly that it uses mercenaries and when anyone working as a mercenary knows that he will be an outcast even among his closest acquaintances? To say that this is due to the attractively high pay is only half the story. Mercenarism is

only in part individual behaviour for which the mercenary himself is solely responsible.

33. The years that the Special Rapporteur has spent looking into the phenomenon of mercenarism and observing how it has metamorphosed outwardly, yet without changing in essence, have led him towards an alternative hypothesis, namely, that it occurs in inverse proportion to peace, political stability, respect for the legal and democratic order, the ability to exploit natural resources in a rational manner, a well-integrated population and a fair distribution of development which prevents extreme poverty. Where all these factors coincide, the risk of mercenary activity is minimal. Conversely, when these factors are not present or occur in haphazard, insufficient, intermittent or contradictory ways, the likelihood of mercenary intervention increases, either because violence, intolerance and the lust for power create conditions that facilitate instrumental links of some kind with mercenaries; or because a third Power, which does not want to be directly involved or to be accused of interventionism, resorts to such action for its own advantage.

34. On the basis of this hypothesis, the Special Rapporteur monitored a number of situations in which mercenaries were involved and found that, while mercenaries might be recruited, trained and financed from within solid, stable countries, they are in fact used chiefly in countries affected by political violence, internal armed conflict, insurrection or insurgency and lacking the necessary financial or technical capacity to prospect for and exploit natural resources on an industrial scale. Their presence and behaviour are not those of heroic saviours, but those of criminals, experienced in the use of arms, those activities generally have an impact on the self-determination of the nation in which they interfere.

A. The Present Situation

35. The Special Rapporteur has established that, unlike mercenaries such as Colonel Bob Denard or Mike Hoare in Katanga during the 1960s, today's mercenaries do not work independently. They are more likely to be recruited by private companies offering security services and military advice and assistance, in order to take part or even fight in internal or international armed conflicts. This is because the parties in a conflict have specific military needs that require the participation and hiring of professional soldiers. Mercenaries are usually, or have been, soldiers, combatants or, more frequently, members of special units and have experience with sophisticated weapons; this applies particularly to those recruited to take part in combat and to train those who are to make up battalions, columns or commando units. Armed conflicts, terrorism,

arms trafficking, covert operations by third Powers, Governments' inability to establish or guarantee security in their own countries and violence connected with positions of extreme intolerance all encourage the demand for mercenaries on the global market.

36. Within the historical structure of the nation State, which is still the basis of international society, it is inadmissible for any State legally to authorize mercenary activities, regardless of the form they take or the objectives they serve. Even where legislation is lacking or deficient mercenarism is an international crime. Mercenary activity arises in the context of situations that violate the right of peoples to self-determination and the sovereignty of States. In practice, mercenaries commit atrocities and impede the exercise of human rights. The mere fact that it is a Government that recruits mercenaries, or contracts companies that recruit mercenaries, in its own defence or to provide reinforcements in armed conflicts does not make such actions any less illegal or illegitimate. Governments are authorized to operate solely under the Constitution and the international treaties to which they are parties. Under no circumstance may they use the power conferred on them to carry out acts that impede the self-determination of peoples, to jeopardize the independence and sovereignty of the State itself or to condone actions that may do severe harm to their citizens' lives and security.

37. The fact that military units made up of mercenaries are supposed to be more efficient, that the use of mercenaries helps to preserve the life of young conscripts or that it is cheaper to recruit mercenaries than to maintain a regular army are weak arguments, and legally and ethically questionable. If such arguments were used as the grounds for praxis, States would reach a stage when they would have to abolish their military forces or cut them back drastically and invite mercenary organizations in to take charge of not only border control, but possibly also the maintenance of law and order.

38. Rather than yield to dangerous arguments such as those that aim to 'rehabilitate' the mercenary at a time of globalization, the undermining of the sovereignty of the nation State and the privatization of conflicts, war and the maintenance of law and order, it must not be forgotten that mercenaries, by definition, act without regard for ideals or legal or moral commitments. They are mercenaries not because order must be restored, armed conflicts must cease or peace must be built, but because they have been paid significant amounts by third parties to involve themselves militarily in conflicts that do not concern them. Their involvement is thus directly motivated by financial gain. The same is true of modern business groups offering security services on an industrial scale around the world

and recruiting, financing and using mercenaries for some of their activities.

39. For mercenary activity to occur, there must be third parties with an interest in resorting to the employment of mercenaries in order to carry out activities that are to their advantage, even though such activities may violate the legislation in force and the international obligation not to interfere in the internal affairs of another State. Clearly, for such a relationship to exist, there must also be recruiting organizations and companies and organizations to act as intermediaries between those who supply and those who demand the service.

40. The Special Rapporteur is of the opinion that investigations into mercenary activities must be objective, encompass all those involved and seek to determine the nature of the act, without accepting any formal legal limitations that may be invoked precisely to conceal the mercenary component. The investigators must also establish the actual identity and nationality of the mercenary, go through the files, rule out altruistic voluntary enlistment, compile information on recruitment and training centres, follow the trail of covert operations, obtain reliable data on aspects relating to the payment and other benefits agreed upon and detect the simultaneous use of other nationalities and passports, etc. In cases where nationality is granted to foreigners taking part in an armed conflict, the investigation must establish the length of time, circumstances and legal grounds for the good faith and legitimacy of the new nationality.

41. The issue of mercenary activity has so many ramifications nowadays that attention must focus on the matter of nationality, which hitherto has been considered as a means of differentiation and a determining factor in the definition of a mercenary. Indeed, a foreign Power can avail itself of nationals of the country it intends to attack in order to do it serious harm. In such a case, the rules of international law as they now stand would not allow the act to be defined as mercenary, even if there was evidence of, for example, recruitment and payment. Even though existing international law may be excessively rigid or full of gaps or does not lend itself to the formal definition of a criminal as a mercenary, it would be wrong to invoke the existing rules either in too restrictive a manner or in such a way as to justify mercenary acts and behaviour.

42. Without obviating the need to clarify, refine, update and expand the rules of customary international and treaty law to combat mercenary activity, it should be established as a principle that, in essence, the aim of such rules is to condemn a mercenary act in the broad sense of the buying and selling of military services that are not subject to the human-

itarian standards that apply to armed conflicts and that are likely to lead to war crimes and human rights violations. It should not be forgotten that, in addition, current international law condemns interference by one State in the internal affairs of another State and the impeding of the self-determination of peoples and that it is, if anything, an aggravating factor if the State interfering employs nationals of the other country for that purpose. Such nationals would not, strictly speaking, be considered mercenaries, but, on the part of those recruiting them, the aim of using them as mercenaries is objectively undeniable, as is the willingness of such nationals to accept a relationship that turns them into mercenaries.

43. The definition does not change if a national group organized abroad for purposes of opposing their country's Government politically and militarily hires and pays mercenaries, based on their military experience or experience in the use of arms and explosives, in order to carry out attacks against the country and its Government. In any case, a distinction must be made between political opposition to a regime, which is the right of any member of a national community, and the employment of methods that are inherently unlawful, such as the use of mercenaries.

44. The Special Rapporteur repeats his view that unlawful activities in which a person's nationality is used by a Power that recruits, prepares and pays that person to mask the mercenary nature of the act should be analysed from the standpoint of a broad, up-to-date interpretation of the international provisions on the subject and the general principles of international law. Since the General Assembly, the Security Council, the Economic and Social Council and the Commission on Human Rights have repeatedly condemned mercenary activities and since, in addition, Member States have condemned such activities and some countries have national laws making the use of mercenaries a crime, where there are no laws or only inadequate laws, a case can be made for the existence of customary international law that condemns and prohibits mercenary activities based on the nature of the acts and not on the fact of having a different nationality. The criterion of being foreign should thus be analysed from the perspective of and in line with higher provisions and the general principles of international law.

B. Current International Legislation and its Limitations

45. The Special Rapporteur deems it necessary to remind the Commission on Human Rights of the need for the international community to examine, study and reflect on the apparent connection between the persistence of and increase in mercenary activities and the obvious gaps in the international legislation currently in force. Furthermore, the

increasing tendency of mercenaries to hide behind modern private companies providing security, advice and military assistance may be due to the fact that international legislation has not taken account of new forms of mercenary activities.

46. In the Special Rapporteur's experience the topic calls for a review as outlined below. Issues on which the relevant United Nations bodies must take a stand include: what is the status of a foreigner who enters a country and acquires its nationality in order to conceal the fact that he is a mercenary in the service of a third State or of another party in an armed conflict? Of a non-resident national who is paid by a third State to carry out criminal activities against his own country of origin? And what about a dual national, one of whose nationalities is that of the State against which he is acting, while he is being paid by the State of his other nationality or by a third State? What are the limits of *jus sanguinis* in an armed conflict when it is invoked by persons who are paid and sent to fight in a domestic or international armed conflict taking place in the country of their forebears? These questions are not simply casuistic. The Special Rapporteur's earlier reports contain specific references to situations such as those described and, even though the evidence pointed to mercenary activities, legal inadequacies and gaps made it difficult accurately to classify an act and the person who committed it.

47. The Commission on Human Rights has already drawn attention to the need to review and update the proposals intended to make the laws proscribing mercenary activities more effective. Furthermore, earlier resolutions of the General Assembly recommended that expert meetings should be convened to study more closely the ambiguities or inadequacies of the international legislation in force and to propose recommendations for a clearer legal definition that would make for more efficient prevention and punishment of mercenary activities. These meetings have not yet been held. The Special Rapporteur recommends to the Commission on Human Rights that it should request the Office of the United Nations High Commissioner for Human Rights to convene, organize and plan such meetings in the context of its activities for the 1999–2000 biennium. The Commission must have clear, unambiguous criteria that would enable it to propose to the Economic and Social Council, and, through it, to submit to the General Assembly, new clear and effective legal proposals for preventing and punishing mercenary activities, particularly in their new forms. Statements formally condemning mercenary activities have not served to prevent an increase in calls on the services of mercenaries and recruiting companies of doubtful lawfulness and legitimacy. What is now needed is an improvement in the normative system to enable it to cope with the development of new criminal methods.

48. An analysis of the factors behind the recurrence and expansion of the phenomenon must consider the problems caused by gaps in existing international legislation and by flexibility with regard to classification as a mercenary. The persistence of mercenary activities, the range and variety of the forms in which they are carried out and the hidden networks of complicity behind these activities suggest that States, particularly the smallest and weakest ones, the least developed, those forming archipelagoes and those faced with armed insurrection and internal conflicts are not adequately protected against mercenarism in its various forms. International legal instruments that characterize mercenary activities negatively do exist, but their configuration and classification leave something to be desired. In other words, they contain gaps, inaccuracies, technical defects and obsolete terms that allow overly broad or ambiguous interpretations to be made. Genuine mercenaries take advantage of these legal imperfections and gaps to avoid being classified as such.

49. Article 47 of the 1977 Protocol I Additional to the Geneva Conventions of 1949 is the only universal international provision in force that contains a definition of mercenaries; paragraph 1 punishes the mercenary by excluding him from the category and rights of combatant or prisoner of war, which amounts to condemning him for his participation, for pay, in armed conflicts; and paragraph 2 then states the definition. The first point to emphasize is that, because of its placement and contents, article 47 of Additional Protocol I does not legislate on mercenary activities, but, rather, limits itself, from the standpoint of international humanitarian law, to providing for the possibility of mercenarism and defining the legal status of the mercenary if he takes part in an armed conflict. As may be seen, the purpose is not to eliminate or proscribe mercenary activities in general, but simply to regulate a specific situation. There is no other existing universal law. Hence the above-mentioned gaps.

50. In addition, the definition of mercenary contained in article 47 of Additional Protocol I lists the cumulative and concurrent requirements that must be met in order to determine who is a mercenary and who is not. Given the variety and complexity of the armed conflicts of the past three decades, however, the wording of this provision has not always been suitable for classifying mercenary activities.

51. According to the information provided directly to the Special Rapporteur by Governments, the laws of most Member States do not classify mercenarism as a crime. In others, although it is classified as a crime, there is no known case in which the law has been enforced against anyone accused of being a mercenary. For instance, the 1870 Foreign

Enlistment Act, a law in force in the United Kingdom of Great Britain and Northern Ireland, prohibits British citizens from becoming mercenaries and from recruiting them. However, the last case in which a person was tried under that law dates back to 1896.

52. The International Convention against the Recruitment, Use, Financing and Training of Mercenaries adopted by the General Assembly on 4 December 1989 has not yet entered into force, although it has been more than nine years since its adoption as barely 16 States have ratified or acceded to it. While its provisions contain measures which are a step forward towards the eradication of this reprehensible activity, it should be noted that article 1, paragraph 1, reproduces almost verbatim the text of article 47 of Additional Protocol I on the definition of a mercenary. Added paragraph 2 relates to mercenary violence against the constitutional order or territorial integrity of a State. No progress has therefore been made with regard to a better and simpler definition of the concept of mercenary, which would allow quicker and more direct action to be taken against mercenary activities. In any event, the Special Rapporteur must point out that it would be easier to improve this important instrument if it were to enter into force in the near future.

53. In this context of the gaps in and the limitations of universal international legislation, the countries of Africa enjoy better legal protection thanks to the Convention for the Elimination of Mercenarism in Africa, which was adopted by the Organization of African Unity (OAU) at its 1977 meeting in Libreville and entered into force in 1985. But 'better legal protection' does not mean full protection against all the current shapes, forms and manifestations which mercenary activities may take on that continent.

54. Clearly then, the gaps in and the inadequacy of existing legislation are manifest and to prolong the situation increases the risks and threats to the self-determination of peoples, the sovereignty of States and the enjoyment of human rights. It is superfluous to point out that it is precisely this context of legal ambiguities which is making mercenary activities more frequent and increasing the hiring and recruitment of mercenaries by private companies providing security services and military advice and assistance. Recourse is being made to mercenaries and companies that recruit, finance and employ them, without any real legal consequences for those who do the hiring or those hired. Most of the mercenaries who fought in wars in the 1990s in the Former Yugoslavia, Angola, Georgia, Nogorny Karabakh or the Democratic Republic of the Congo (then Zaire) now live comfortably in their homes, quite unbothered by the justice system, awaiting new offers to fight.

55. For the above reasons, the Special Rapporteur maintains that the relevant international legal instruments are but imperfect tools for preventing and punishing mercenary activities. There are difficulties in applying article 47 of the 1977 Additional Protocol I to the 1949 Geneva Conventions, particularly in cases where the person acquires the nationality of the country in which the fighting is taking place; mercenarism is not classified as an offence under the internal criminal law of many countries, and the International Convention against the Recruitment, Use, Financing and Training of Mercenaries has yet to enter into force, given the small number of States that have expressed an interest in becoming parties to it. Consequently, the international community is faced with a situation that actually affects it and the time has come for the consideration of the issue by the Commission on Human Rights and also to include the need to review and update international legislation on mercenary activities.

C. Terrorism and Mercenary Activities

56. A mercenary is a criminal; he acts not out of altruistic motives, but to earn money in exchange for his tactical and strategic skills and his handling of weapons and explosives. In this regard, the material connection between a mercenary's activity and the commission of terrorist acts has been established through many terrorist attacks in which the perpetrator was proven to be one or more mercenaries hired to commit the crime.

57. In paragraph 116 of the report submitted to the Commission on Human Rights, document E/CN.4/1997/24, the Special Rapporteur maintained that various forms of terrorist attacks are carried out by highly specialized criminal agents who are hired to blow up aircraft, mine bridges, destroy buildings and industrial complexes, assassinate and kidnap persons, etc. While in many cases the terrorist agent comes from fanatic groups espousing extremist ideologies, it must be remembered that terrorism is also a criminal activity in which mercenaries participate for payment, disregarding the most basic considerations of respect for human life and a country's legal order and security.

58. In parallel with this conclusion, the recommendation contained in paragraph 125 of the report maintained that 'The international community must take into account the connection existing between terrorism and mercenary activities and the participation of mercenaries in criminal acts of a terrorist nature. It is suggested that commissions and working and study groups for the prevention and punishment of

terrorism should be recommended to include mercenary activities in their analyses and conclusions'.

59. The whole world was shaken by last year's terrorist attacks in Kenya and Tanzania. We know that they were by no means the first and that the madness of fanatical sects may cause further mass crimes against humanity. In this context, the Special Rapporteur considers it crucial to investigate the connections between the commission of terrorist attacks and the presence of mercenaries as material agents of these acts. To fail to consider a hypothesis such as that outlined here or to use a different criterion to assess the two phenomena, on the grounds that the motivation is different, would be a serious error that would weaken the line of prevention of terrorism and mercenary activities.

60. A mercenary involves himself in armed conflicts as a function of his military experience, doing so for substantial reward. Many mercenaries who are experts in the use of explosives and technical devices with destructive material effects are hired to commit deadly attacks that cause collective fear and dread or, in other words, indiscriminate terror. Accordingly, although the mercenary is not involved in the formulation of extremist ideologies that allow terror as a means of intimidation for the achievement of its objectives, he assumes the status of terrorist when, for pay, he agrees to become the instrument of terror and commits acts that cause death and destruction with horrible efficiency. Without ceasing to be one, a mercenary can also become a terrorist.

61. Political, racial, religious and other types of extremist organizations whose thirst for vengeance or hatred leads them to call for the destruction of anyone who opposes them do not resort exclusively to their fanatical militants to carry out objective actions aimed at spreading terror. The search for morally reprehensible 'efficiency' usually leads them to attempt to hire experts in explosives or assaults who agree for substantial pay to hire themselves out as mercenaries.

62. These organizations, which have made their criminal terrorist practices explicit, usually invite persons of various nationalities to join. Consideration must be given to the possibility of their recruiting and hiring mercenaries and of some of their affiliates or members hiring themselves out as mercenaries. Consequently, the Special Rapporteur emphasizes in his recommendation to the Commission on Human Rights that the study and analysis of this matter should be conducted with the utmost care and rigour.

IV. PRIVATE SECURITY AND MILITARY ASSISTANCE COMPANIES AND MERCENARY ACTIVITIES

63. In the previous report he submitted to the Commission on Human Rights at its fifty-fourth session (E/CN.4/1998/31), the Special Rapporteur dealt fully with private companies operating in the international market offering security services and military assistance and advice, normally matters reserved for the State and for which the State must assume responsibility. All the arguments and reservations expressed in that report are still valid and the Special Rapporteur therefore reiterates his position on and his concern about a matter that can affect State sovereignty, the self-determination of peoples, the stability of constitutional Governments and, more particularly, the current system of human rights protection and international humanitarian law.

64. These companies are developing their offers more and more aggressively, putting forward arguments for legitimacy based on military efficiency, cheaper operations, their personnel's proven experience and an alleged comparative advantage that would make it feasible or desirable to hire them for peace-building or peacekeeping operations such as those conducted by the United Nations or ECOWAS. In documents prepared by these companies, the Special Rapporteur has read studies of the comparative costs of various peacekeeping operations and what these companies would charge, with the difference, according to them, of greater efficiency for breaking up pockets of resistance, extinguishing hardline opposition or opening up avenues for humanitarian assistance. Advertising for these companies and the services and jobs they offer can be found on the Internet and leave no doubt as to what is being offered and the connection with mercenary agents.

65. In contrast, national States are showing no sign of a reaction that focuses on these companies' international expansion and the dangers it entails for State sovereignty and objectives. In his correspondence with Member States, the Special Rapporteur has asked for their opinion on this matter, but the replies received have not dealt with it. This silence is alarming inasmuch as there are situations in which a country's press reports in abundant detail on the presence of companies involved in matters of national security and public safety without regard for human rights and in open contradiction with constitutional provisions that categorically state that internal order and security are the exclusive responsibility of the State.

66. South Africa is the country that has adopted the clearest position on these companies and their supply of military assistance services abroad.

A South African law enacted in 1998 regulates military assistance abroad and defines the jurisdiction of private companies in this regard. It introduces a sentence of no more than 10 years' imprisonment and a fine of no more than 1 million rand for nationals or foreigners resident in South Africa who participate in military missions outside South African territory unauthorized by the State. It also imposes limits on the freedom of security companies to deal with military matters. It regulates, but does not prohibit, the existence of these companies which employ mercenaries. Meanwhile, Executive Outcomes, a company which was established in 1989 on the initiative of Eeben Barlow and has broad experience in military assistance in African countries, has complied with the law and registered with the Department of Defence as an organization providing military assistance, mainly training, abroad. In any event, it will be necessary to see the practical consequences of the law in order to form a definite opinion on its usefulness and efficiency.

67. During his mission to South Africa in October 1996 at the invitation of President Mandela's Government, the Special Rapporteur was told by Barlow that his company had no problems with the promulgation of that law, provided that it did not represent an added administrative burden that could reduce his organization's operational efficiency. As to the rest, he said, his company did business only with legitimate Governments and not with rebels or insurgents and made every effort to be scrupulous in its respect for the human rights of the inhabitants of the countries in which it intervened and for the rules of international humanitarian law. However, a recent Channel 4 documentary on 'The War Business' shown on British television in May 1998 charged that Executive Outcomes had exploded a napalm bomb in a market in an African village, killing 500 people, including civilians, of course, in a single day.

68. Regardless of the declarations of good intentions and respect for human rights and humanitarian law made by the directors of these companies, the Special Rapporteur emphasizes that a matter of principle is the basis of the analysis and of the stance that the Commission on Human Rights should adopt. National security and public safety and action to combat rebels, traffickers and terrorists are not merchandise that can be freely sold. They are, instead, matters related to a State's very existence and raison d'être; providing security and maintaining law and order is solely the State's responsibility.

69. One argument is that, at a time of the globalization of the economy, information and communications, the privatization of security and law enforcement is also permissible. This argument holds that, when transnationals arrive in an underdeveloped country, they have to have their own experienced security guards to police the perimeter of their

facilities, since the security forces of the country in question do not offer any guarantees. If the area is prone to insurgency or organized crime, why, they ask, not allow these private guards to intervene and clean up the region, fighting even rebels or traffickers? The answer is clear: who guarantees the human rights of the inhabitants? Who guarantees that the provisions of the 1949 Geneva Conventions and their Additional Protocols will be respected in such fighting? Who guarantees that these companies, whose sole purpose is gain, will not artificially intensify or prolong the conflicts and situations of insecurity for the sole purpose of extending their stay and earning more money? It is not acceptable for private companies, which are perfectly legitimate in civil or commercial activities and whose sole aim is money, to replace a country's army and police force in providing national security and public safety and protecting the exercise of civil rights.

70. If Governments accept this substitution, they forsake their peoples and expose them to the risks of private protection that can discriminate among population groups for reasons of race or ideology; make use of military offensive and combat weapons normally reserved for forces that are the expression of the authority of the State; and, in that context, commit every kind of human rights abuse and violation.

71. If, as indicated above, private companies providing security services and military assistance and advice resort to the use of mercenaries, it must be made clear that the explanation given, i.e. that they act in the service of a constitutional or legitimate Government or to restore one to power, is not acceptable. The distinction between using mercenaries for good or evil ends is no more admissible than is the distinction between good and bad mercenaries. A State's weakness, a State's impoverishment and disintegration, the breakdown of the constitutional system, internal armed conflicts and anything that might constitute a grave risk for public order and peace in a country must be resolved on the basis of the multi-lateral security agreements that exist in all regions and on all continents, calling on international cooperation and strengthening all the peace-building and peacekeeping operations which are, according to the San Francisco Charter, to be carried out by the United Nations.

72. Specifically, it is neither lawful nor advisable, no matter how often short-term or emergency reasons are invoked, to entrust a country's security and the speedy settlement of armed conflicts to private companies which hire mercenaries to achieve those objectives and will earn substantial economic profits for their participation. Consider also that countries facing a situation in which they call on these companies are usually in a poor economic and financial position and lack the funds to pay for their services. They are therefore obliged to do so by granting

concessions for resources that are part of the national heritage. The companies are prepared for this highly lucrative eventuality and have set up various branches and subsidiaries. Thus, Strategic Resources Corporation (SRC), the holding company for Executive Outcomes, has other independent companies such as the Branch Energy mining and oil company; Heritage Oil and Gas; Diamond Works, established in Vancouver in 1996 and today Canada's largest diamond producer; and airlines such as Ibis Air and other transport, logistics and service enterprises.

73. Of the companies offering security services and military assistance and advice on the globalized world market, mention should be made of Defence Systems Limited (DSL), which provides security to several mining and oil companies, Saladin Security; Control Risks Group; Braddock, Dunn and McDonald Inc. (BDM) headed by a former United States Secretary of Defense and headquartered in McLean, Virginia; Integrated Security Systems; Booz Allen and Hamilton, which trains the Saudi military forces; Vinnell Corporation, with Vietnam war experience; O'Gara Protective Services; Science Applications International Corporation (SAIC); and Military Professional Resources Incorporated (MPRI), with headquarters in Alexandria, Virginia, which trained the Croatian army in 1995 and is currently training the armed forces of Bosnia and Herzegovina. These companies may affect the exercise of sovereignty in the countries in which they operate, create resentment among the inhabitants and contribute to impunity for the crimes committed by the mercenaries they hire.

74. Who is responsible for the human rights violations committed? The company will say it is the mercenaries acting individually or abusing their powers. The State will say that responsibility lies with the company and not with its own officials or forces. It is time for Member States and for the Commission on Human Rights to make a detailed study of this matter. More and more Governments are hiring these companies to settle the military conflicts that are destabilizing them, knowing full well that a mercenary component is included in the offer. Other bodies, including international organizations, sign contracts with them for security and logistical support. It is predictable that some Governments may also resort to this type of company to undertake unilateral action in another country, on the pretext of establishing order or ensuring peace in a given region. All of this is tantamount to officially tolerated mercenary intervention, even if it is with 'the best of intentions'.

75. Must the Commission on Human Rights accept such a situation? Is it not obvious to the international community that the trend is to replace traditional peacekeeping forces, which, under international law, are the

responsibility of the United Nations and regional organizations, by operations carried out by these companies? It is rather strange that, in parallel with the growth of these companies' activities, a smear campaign has been waged against peacekeeping operations and that reference is being made to the unfortunate events in Somalia. The Commission should pay priority attention to the fact that a kind of privatization of war is being promoted through unilateral positions implemented by these private companies, with unforeseeable implications for the exercise of human rights. The Special Rapporteur's point of view has already been stated: without undermining the principles on which its very existence is based, the international community cannot allow the free and globalized market to function as well for operations for the sale of military assistance and peacekeeping and peace-building operations that are the province of the international organizations. To do otherwise would mean, in practice, allowing paramilitary forces with a mercenary component to interfere in internal affairs. In view of its special importance, however, the Special Rapporteur is, in accordance with his terms of reference, continuing to study this question in depth.

76. When this report is published, the Special Rapporteur will already have visited the United Kingdom, a country where some of the companies working in this line of military assistance and advisory services are registered. The results of the visit will form the basis of the Special Rapporteur's next report to the General Assembly. The Special Rapporteur would nevertheless consider it advisable for the Commission on Human Rights and other United Nations bodies to look into the possibility of convening an international conference of experts to study and take specific decisions on this matter, paying particular attention to the fact that these companies would like to be regarded as an alternative to the conduct of United Nations peace operations; that the hiring of such companies may lend legitimacy to the use of mercenaries; and, in particular, that a system should be devised for the protection of international human rights and the rules of international humanitarian law in a world of privatized wars and private combatants and police.

V. CURRENT STATUS OF THE INTERNATIONAL CONVENTION AGAINST THE RECRUITMENT, USE, FINANCING AND TRAINING OF MERCENARIES

77. The Special Rapporteur has shown in his preliminary reports to the Commission on Human Rights that the International Convention against the Recruitment, Use, Financing and Training of Mercenaries, adopted by the General Assembly on 4 December 1989 in its resolution 44/34, expands the international regulation of the question and

confirms the legal nature of the resolutions and declarations of United Nations bodies condemning mercenary activities. Its entry into force will contribute to preventive cooperation among States, better identification of situations involving mercenaries and the clear determination of jurisdiction in each case and will facilitate procedures for the extradition of mercenaries and the effective prosecution and punishment of offenders.

78. Unfortunately, only 16 States have completed the process of expressing their willingness to be bound by the International Convention, while a further 22 are required for its entry into force. Those States are: Azerbaijan, Barbados, Belarus, Cameroon, Cyprus, Georgia, Italy, Maldives, Mauritania, Saudi Arabia, Seychelles, Suriname, Togo, Turkmenistan, Ukraine and Uzbekistan. Ten other States have signed the International Convention, but have not yet ratified it. They are: Angola, Congo, Democratic Republic of the Congo, Germany, Morocco, Nigeria, Poland, Romania, Uruguay and Yugoslavia. Pursuant to article 19, the International Convention is to enter into force on the thirtieth day following the date of deposit of the twenty-second instrument of ratification or accession with the Secretary-General. Its failure to enter into force nine years after its adoption means that international legislation on mercenaries continues to be limited to article 47 of the 1977 Protocol I Additional to the Geneva Conventions of 1949 and the 1977 OAU Convention on the Elimination of Mercenarism in Africa.

VI. CONCLUSIONS

79. Mercenary activities continue to exist in many parts of the world and to take on new forms. The recruitment and hiring of mercenaries by private companies providing security services and military assistance and advice and, in turn, the hiring of these companies by Governments which entrust them with responsibility for security, maintaining public order and safety and even armed combat against rebel forces and organized crime are a serious challenge to the international human rights protection system currently in force.

80. Although mercenaries pose as technicians or military experts hired as such by private companies providing security services and military assistance and advice, or by Governments, this changes neither the nature nor the status of those who hire themselves out to meddle and cause destruction and death in foreign conflicts and countries.

81. Since the nature of the act and the classification of a mercenary have not changed, although the forms and operational methods have, the condemnation of mercenary activities and the use of mercenaries by the

Commission on Human Rights and other United Nations bodies is still valid. Mercenary activities impede the exercise of the right of peoples to self-determination and jeopardize the sovereignty of States, the principle of non-interference in internal affairs, the stability of constitutional Governments and the enjoyment of the human rights of the peoples concerned. They are, by definition, wrongful and unlawful activities.

82. The diversification and modernization of operational methods do not mean that mercenary activities are disappearing; rather, the conclusion is that, as mercenaries have become better organized and their pay has increased, their numbers have grown and more persons are prepared to do the job, although, for reasons of self-esteem, they sometimes prefer to pose and think of themselves as military experts or soldiers of peace.

83. The information received by the Special Rapporteur suggests that, even though Africa continues to be the continent most affected by mercenary activities, these activities have spread to other continents, although the methods of operation vary according to the situation in the country in which their services are hired.

84. Mercenaries have been particularly active in Sierra Leone, where they were taken first by Executive Outcomes, the security and military advisory and assistance company registered in South Africa, and then by Sandline International, which is registered in the Bahamas and has offices in London. These two played an important role in the overthrow of the military junta of the alliance formed by the Armed Forces Revolutionary Council (AFRC) and the Revolutionary United Front (RUF) and in restoring democratically elected President Tejan Kabbah to power. They also trained the Kamajor paramilitary troops who are fighting the rebel forces of the overthrown Government with the acquiescence of the current Government and are responsible for grave human rights violations against prisoners and the civilian population. As a result of this intervention, a number of affiliates of Executive Outcomes and Sandline International are now exploiting Sierra Leone's mineral resources and the Diamond Works company has become Canada's largest diamond producer thanks to its operations in this West African country.

85. The Special Rapporteur has observed that the current situation is marked by the inadequacy of the international rules which deal with and punish mercenarism. The legal gaps and ambiguities detected suggest that the existing set of rules is by no means effective for successfully combating mercenary activities.

86. Mercenarism is not classified as a separate crime in the criminal legislation of most States, a situation that prevents legal action from being taken against mercenaries, except when they have committed related offences and are charged therefor.

87. Terrorist acts may be committed by mercenaries. Further consideration should be given to studies and the adoption of anti-terrorist policies; mercenarism should be seen as an aggravating factor in the crime of terrorism when there is proof that a mercenary planned, took part in or committed the terrorist act.

88. The increase in the number of companies providing security services and military assistance and advice on the international market and their recruitment and hiring of mercenaries raise serious questions about responsibility for human rights violations and encourage impunity. Governments are therefore tempted to transfer their responsibility for the violations committed to these companies and they, in turn, transfer it to the mercenaries they have recruited. The Commission on Human Rights must give priority to this matter because it will affect the entire international system for the protection of human rights.

89. The hiring of private companies providing security services and military assistance and advice is also an indicator of vested interests, more particularly those of third Powers, which see the use of this type of company as a way of effectively intervening in another country's internal affairs, without having responsibility for such intervention being attributed to them directly, without their own military forces risking casualties and without having to bear military costs.

90. The growth of this type of company is indicative of little, if any, international reaction to their activities. Even some international organizations have given in to the temptation of using their services to obtain logistical support to or open up avenues of humanitarian assistance. There is also a chance that traditional Government peace-building and peacekeeping forces may be replaced by these companies, in conjunction with a smear campaign against the Government forces and reminders of serious incidents that have occurred, such as those in Somalia. The possibility of an artfully devised campaign and intentionally promoted tolerance of this type of company warrants serious investigation.

91. The legal gaps, defects and ambiguities that currently facilitate mercenary operations by these polyvalent companies should be remedied through explicit rules that regulate and clearly limit what these private companies may and may not do internationally, while clearly defining their responsibility for human rights violations and abuses and other

crimes and offences, as well as that of the States that hire them and that of the individuals who recruit them. The United Nations mandate and its work in peace-building and peacekeeping operations must be strengthened at the same time.

92. As the tenth anniversary of the General Assembly's adoption of the International Convention against the Recruitment, Use, Financing and Training of Mercenaries approaches, only 16 States have agreed to be bound by it. The fact that it has not entered into force continues to contribute to the increase in the criminal activities of mercenaries.

VII. RECOMMENDATIONS

93. The Commission on Human Rights should pay priority and urgent attention to the challenge to the system for the international protection of human rights created by the growth and development of companies providing security services and military assistance and advice, their recruitment and use of mercenaries and the increase in the use of such companies by Governments facing internal armed conflicts or the aggression of organized crime or terrorism. Because of the difficult situation they are in, these Governments do not have the funds to pay for the services of these companies and have to grant them major concessions of mineral and oil resources that account for a valuable share of their national heritage. The blurred lines of responsibility for human rights violations and the resulting impunity must be carefully studied by the Commission.

94. The Commission on Human Rights must continue to uphold its explicit condemnation of mercenary activities, regardless of the form they take, requesting the Member States of the United Nations to classify mercenarism as a crime in their internal criminal law and to make the fact of being a mercenary an aggravating factor in the commission of other wrongful criminal acts, especially acts of terrorism.

95. The Commission on Human Rights must also communicate once more with all Member States of the Organization to recommend that they should explicitly prohibit the use of their territory for the recruitment, training, assembly, transit, financing and use of mercenaries.

96. Given the legal ambiguities and gaps that currently facilitate the use of mercenaries and the increase in their numbers, it is recommended that the Commission on Human Rights should invite Member States to ratify or accede to the International Convention against the Recruitment, Use, Financing and Training of Mercenaries.

97. The Commission on Human Rights also has to consider the serious risk of joint action by terrorists and mercenaries in carrying out attacks in which some are motivated by ideological, political or religious beliefs, or simply by hatred, while others just want to earn money. It should therefore be recommended that the studies, plans and action which the Commission considers in order to prevent and punish human rights violations and abuses, particularly terrorism, should also take account of the dimension of participation by mercenaries.

98. The Commission should recommend to the Economic and Social Council that it allocate the necessary financial and budget resources to the Office of the United Nations High Commissioner for Human Rights to enable it to disseminate information, in the bulletins it is publishing, on the adverse effects of mercenary activities on the enjoyment of human rights and the exercise of the right of peoples to self-determination. The Office of the High Commissioner must also be allocated the resources it needs to provide technical assistance services to countries which have suffered the consequences of mercenary activities, if they so request.

99. The Commission on Human Rights must remind all States and international organizations of the need for constant vigilance in monitoring companies that employ mercenaries, particularly those that offer security services and military assistance and advice on the globalized international market.

100. The Commission on Human Rights must carefully monitor the human rights situation and the exercise of the right of peoples to self-determination in the countries where these companies operate.

101. The Commission must also remember that mercenaries base their comparative advantage and greater efficiency on the fact that they do not regard themselves as being bound to respect human rights or the rules of international humanitarian law. Greater disdain for human dignity and greater cruelty are considered efficient instruments for winning the fight. The participation of mercenaries in armed conflicts and in any other situation in which their services are unlawful may jeopardize the self-determination of peoples and always hampers the enjoyment of the human rights of those on whom their presence is inflicted.

Notes on Contributors

J. 'Kayode Fayemi is the Director of the Centre for Democracy & Development. A civil-military relations scholar, Dr Fayemi studied at the Universities of Lagos, Ife and London where he received his doctorate in War Studies from King's College. He has written extensively on the military in politics, democratic control of the military, defence planning and democratic transitions in Africa.

Abdel-Fatau Musah is the Research and Publications Coordinator at the Centre for Democracy & Development. Dr Musah studied at the Universities of Ghana, Legon and the Moscow State University, where he received his doctorate degree in International Political History. Prior to joining CDD, he worked with the British-American Security and Information Council (BASIC). His research interests include globalisation, regional security and light weapons proliferation.

Kevin A. O'Brien is a research associate at the Centre for Defence Studies, King's College, University of London. Formerly a doctoral fellow in Security Studies at the University of Hull, Dr O'Brien is the Executive Director of Hussar International Research Group based in South Africa.

Khareen Pech is an independent journalist specialising in private security. She has written for a number of publications including the *Mail & Guardian* (South Africa) and the Independent. She is currently writing a book on the defunct South African based private military company Executive Outcomes. She lives in Johannesburg.

Johan Peleman is the Director of the International Peace Information Service in Antwerp, Belgium. He specialises in, and has written extensively on, private security research, weapons transfers and civil-military relations.

Kofi Oteng Kufuor is lecturer in International Law at the University of East London, England.

Eboe Hutchful is Professor at the Department of African Studies, Wayne State University, Detroit, USA and Executive Director, Defence Research and Advocacy Centre (DEPRAC) in Accra, Ghana. Dr Hutchful has written and taught extensively on security sector reform, structural adjustment and privatisation in Africa, and civil-military relations.

'Funmi Olonisakin is a MacArthur Post-Doctoral Research Fellow in the Department of War Studies, King's College, University of London. Dr Olonisakin has written on regional security, conflict management and peace support operations. She is the author of *Reinventing Peacekeeping* (Almqvist & Wiksell), forthcoming.

Alex Vines works for the Arms and Africa Divisions of Human Rights Watch. He is the author of several monographs on the human rights, light weapons and landmines situation in Africa. His chapter in this book was written in his individual capacity as a Research Associate of Queen Elizabeth House, University of Oxford, UK.

Index

Compiled by Auriol Griffith-Jones

Note: Italic page numbers refer to Tables; bold refers to Appendices

Abacha, Gen Sani, president of
   Nigeria, 95, 101, 102, 103–4
Abdallah, Ahmed (Comoros Is.), 6,
   17
Abidjan Peace Accord, 90, 97
Abubakar, Gen A.A., Nigerian CDS,
   93
accountability, 14, 217; of PMCs,
   189, 204–5, 236–7
Acker, Gary Martin, 22
ACRI, 224, 242–4
ADFL, 125, 126, 131, 135–6;
   counter-offensive (1997), 143–5;
   victory, 146
Adson Holdings, 61
Advance Systems Communications,
   66
advocacy, 38
Afghanistan, mercenaries from, 35
Africa: Cold War conflicts, 48,
   120–1, 257; Cold War superpower
   rivalry in, 17, 19, 246, 247;
   colonialist nationalism, 19–20;
   concept of security in, 26, 30–4,
   37–8, 211–12, 225, 245–7;
   external interests in, 25–6; illiberal
   democracies, 226, 257; independence movement (1960s), 5–6,
   245; inter-state border wars, 17,
   223, 257; regional alliances,
   147–8, 218; rise of internal
   conflicts, 16, 17, 26; root causes of

conflicts, 29, 37, 109, 217,
   237–8, 245–7, 262–3; trade and
   mercenaries, 17, 121, 124, 212
African Centre for Security Studies,
   224
African High Command, Nkrumah's
   call for, 215, 245
African leaders: authoritarian
   dictators, 31, 37, 226, 257;
   impunity of, 262; 'new
   Metternichs', 223; and perceived
   threats, 30–1, 225; and rise of civil
   society, 213, 217, 257; ruler's
   dilemma, 211–12, 213
African nationalism, 118
African Peoples and Human Rights
   Court, 39
Africans: in colonial armies, 19–20;
   mercenaries, 123, 148
Afro Mineiro, 66
aircraft: in Angola, 172, 188; in
   Congo, 138, 143; Russian
   helicopters, 60; Sandline's, 105
AirScan, 57, 62–3, 188
Alba Marine, 68
Alberts, Carl, 105
Alerta, in Cabinda, 188
Algeria, 6, 20, 51, 61, 221; political
   Islam in, 217
alienation, popular, 220
Alpha-5, 53, 173, 185–6
America Diamond Buyers, 156

America Mineral Fields, 69, 162, 164–5, 166; Kabila and, 56, 57, 155–6, 164
American mercenaries, Angola, 21
ANC, and EO connection, 50
Anglo-American mining, 164–5
Angola, 6, 17, 21–2, 48, 218, 226; Boulle in, 162–3; Cabinda enclave, 56, 57, 188; Cafunfo operation, 172, 173; and Central African crisis, 56–7; and Congo, 58, 120, 128, 130, 147; diamond fields, 23–4, 52; DSL contract, 53–4, 185; EO in, 43, 51–4, 172–5, 200; IDAS contract, 163–5; oil industry, 51, 53, 57; PMCs in, 35, 51–4, 59, 185; SADF unit in, 49, see also UNITA
Annan, Kofi, UN Secretary General, 25–6, 62, 71, 102
anti-poaching operations, 47
Applied Electronics Services, 66
APR, 125
AquaNova, 66
Arakis Energy, 63
armies: democratic principles within, 229; rebellions and coups, 223; retrenchment, 3–4, 17, 217, 225; weakness of national, 220, 242
Armor Holdings, 184
arms procurement: by EO, 176; for Congo, 128–9; for Sierra Leone, 99, 100–1, 158
arms proliferation, 7, 16, 217; attempts to control, 226, 259; in Sierra Leone, 86, 108–9; small arms, 26, 224–5
arms trade, 7, 26, 213, 217, 225; PMC links with, 26–7, 48, 236; Sandline and, 177, 179–80; South African, 49–50
ARMSCOR, 49
assassinations, 6, 47
Australia, 162, 170

Babangida, Gen Ibrahim, Nigeria, 103
Balkan states, 1, 54, 181, 184
Ballesteros, Enrique see UN Special Rapporteur
Banks, John, 47

Baramoto Kata, Gen Kpama, 59, 129, 132–3, 144, 145; and South African mercenaries, 141, 142, 182
Barlow, Eeben, 6, 17, 23, 173; and De Beers, 50, 66; on EO in Sierra Leone, 28; and EO–Branch Heritage links, 67–9; founded EO (1989), 48–50; in Kenya, 61; and Mobutu, 130–1; and Sandline, 60, 100; and Stabilco, 143; and Strategic Resources, 66, 67–8; subsidiary companies, 52
Barrick Gold Corporation, 132
Barril, Paul, 50
BDM International, 18, 56
Belgium, 118; mercenaries, 40n, 137, 181
Bell, Alan, 58
Benin, 1970s war, 6, 17, 22
Berewa, Solomon, Sierra Leone, 85, 110; policy document, 93–5
Berg, Nick van den, 6, 60–1
Betac Corporation, 56
Biafran War, 29, 36
Bio, Steven, 89
Biya, Paul, president of Cameroon, 50
Bizimungu, Gen Augustin, 136
Bockarie, Sam, Sierra Leone, 108
Botswana, 61, 62
Boulle, Jean-Raymond, 157–60, 161, 162, 163–4, 165–6; and America Mineral Fields, 156, 157, 165; dispute with Friedland, 69, 157–8, 161, 162; in Liberia, 60, 105; rivalry with EO, 27, 57, 105
Boulle, Max, 162
boundaries, 30; artificial nature of African, 33, 37, 216, 245
Bowen, Rupert, Sandline, 60, 98, 100
BP (British Petroleum), in Colombia, 186–7
Branch Energy, 23, 24, 64, 100, 158; bought out by Carson Gold, 53, 68–9; and DiamondWorks, 180; and EO, 51, 148; in Sierra Leone, 65, 92, 161; subsidiaries, 25, 65–6
Branch Energy (Uganda), 61
Branch Mining Ltd, 52, 66, 158, 172

Branch-Heritage, 62, 105; links with
    EO, 63–4, 67–70
Brazza, Pierre de, 121
Brenco Trading, 139
Bridge International, 52, 66
Bridge Resources, 53, 63
British South Africa Company, 43
Buckingham, Anthony, 51, 105; and
    Barlow, 67; Branch Energy, 23,
    64–5; DiamondWorks, 92;
    director of Sandline, 23, 25, 60,
    100, 162; and EO, 50, 66, 67;
    Heritage Oil, 24, 64
Buckley, Charles, 106
Bulgaria, arms from, 60, 99, 180
Bunia airport (Congo), 127, 132,
    144
Burkina Faso, 86, 108
Burundi, 96, 126, 217, 218, 221;
    arms supplies for, 56; army revolt,
    223

Cabinda Gulf Oil Company, 57
Cafunfo (Angola), 52
Callan Georgiou, 'Colonel', 6, 22
CAMA (Campaign against
    Mercenaries in Africa), 25, 263–4
Cameroon, 50, 62
Canada, 58
Capricorn Africa, 20
Capricorn Air Systems, 68
Capricorn Systems Ltd, 52
Carlucci, Frank, 18, 40n
Carr-Smith, Maj-Gen Stephen, 18
Carson Gold, 23, 25, 52–3, 68–9
CCB, 49
Cekovic, Jovan, 140
Central African Republic, 223
CFAO, 17
Chad, 147, 246
Chan, Sir Julius, Papua New Guinea,
    27
Chanas, Jean-Louis, 50
Charlie, 'Captain', 22
Chevron Oil (US), 57, 64, 188
child soldiers, 7, 86, 109, 219
China, 107
CIA, 21, 128
CIAS, French subsidiary of DSL, 62
civil society, 213, 217, 230, 263–4
Cleary, Sean, 59

Cline, Ray, CIA director, 159
Clinton, W., US President, 54, 157,
    165
COCOM (Co-ordinating Committee on
    Multilateral Exports), 3
COIN Security, in Angola, 53
Cold War, 3, 14, 31, 48; effect of
    ending on African state security,
    210, 216–18, 224, 247, 257
Colombia, 106, 186–7
Combat Force, 63
Comoros Islands, 6, 17, 22, 134
conflict management: long-term
    strategies, 244–5; mechanisms, 7,
    33–4; PMC standard package, 28;
    tools of, 14–15
conflict resolution: need for new
    approach, 31, 37, 190–1, 238–9;
    realist policy on, 14–15, 38; trends
    in PMC involvement, 46–9; use of
    force, 239, 243, see also peacekeep-
    ing
conflicts: lack of political objectives,
    220; new forms of, 218–25;
    savagery of, 217, 219, see also
    internal conflicts
Congo, Belgian, 117–19; 1960s war,
    6, 17, 20, 43, see also Congo,
    Democratic Republic (DRC); Zaire
Congo, Democratic Republic (DRC),
    27, 29, 217, 218; and Central
    African wars, 122–3; continuing
    destabilisation, 145–6; EO in,
    128–9, 130–1, 141, 143;
    government employment of
    mercenaries, 35, 120–1; mineral
    resources, 58, 118; mining
    concessions, 155–6, 157–8; PMCs
    in, 55, 58–9, 130–1, 133,
    134–45, 148; reaction to Sierra
    Leone coup (1997), 96; Rwandan
    Tutsi-led incursion, 124–6,
    127–8; Rwandan and Ugandan
    troops in, 35–6, 127, 136, 146,
    147; war in eastern Zaire, 124–6;
    White Legion in, 55, 181–2
Congo-Brazzaville (Republic of
    Congo), 56, 96, 182–3, 221;
    Central African war in, 57–8,
    145–6; mercenaries in, 182–3
Conte, Lansana, Guinea, 101

Control Risks Group, 47, 88
Cook, Robin, UK Foreign Secretary, 102
Corporate Trading International, 53, 63
Corps of Commissionaires, 47
corruption, 83, 226
Cote d'Ivoire, 226
criminality, 221, 226
Crooke, Ian, 47
Cuban mercenaries, 35, 183
Cuban troops, 36, 48, 173

De Beers, 156, 157; contract for EO, 50, 51; in Sierra Leone, 77, 82; use of mercenaries, 46, 66
de Clerq, Antoine, 135
De Kock, Eugene, 63
de Matos, Joao, Angola, 51, 53, 131
demilitarisation, 26, 226–8; role of civil society, 230
demining specialists, 45, 52, 163
democracy, 26, 210, 226–7; as conflict management tool, 14; and militarisation, 220–1; and powers of parliament, 228–9
Denard, Bob (Gilbert Bourgeaud), 6, 17, 22, 134–5, 137
development: nature of, 216; and security, 33, 211
DGSE, 55
Diamond Field Resources, 69, 158
diamond smuggling, 46, 51; Congo, 129, 132
DiamondWorks, 23, 25, 68–9, 92, 157, 190; in Angola, 180; in Sierra Leone, 99, 100, 161, 180–1
Dixon, Professor W.H., 131
Dos Santos, José, president of Angola, 51, 53
Double-A-Design, 66
DRC see Congo, Democratic Republic (DRC)
drugs trade, 7, 262
DSL (Defence Systems Limited), 47–8, 62, 184–6, 263; in Angola, 53–4, 185; in Colombia, 186–7; network links, 58, 69; in Sierra Leone, 88, 105; UK military links, 18

East African Community, revival of, 34, 218
ECOMOG, 32, 33, 41n, 218; in Liberia, 240, 241, 249; Nigerian forces, 93, 102, 103–4, 240, 242; in Sierra Leone, 60, 93, 97–8, 104–6, 107–8, 109
economic structural adjustment, 210, 217, 257
ECOWAS, 218, 240–1, 248–50, 258; compared with SADC, 253, 254; in Liberia, 249; and Sierra Leone, 89–90, 97, 105, 249; structural weaknesses, 249–50
Egypt, 41n, 181
Ehlers, Willem ('Ters'), 56
Elf Aquitaine, 96, 139
Ellis, Neil, 53, 105, 141–2, 143
Eluki onga Aundu, Gen, Congo, 127
Endiama, Angolan parastatal, 53, 162, 164
EO (Executive Outcomes), 13, 24, 49–51, 158; in Angola, 43, 51–4, 130, 172–5, 200; and Branch-Heritage links, 63–4, 67–70; in Burundi, 56; in Congo, 128–9, 130–1, 141, 143; in Congo-Brazzaville, 57–8; diamond concessions, 23–4, 91–2; and Mobutu, 55–6; network connections, 25, 57, 260; operations in Africa, 61–2; operations terminated, 61, 62, 67, 189; origins, 48–9, 200; role in Sierra Leone, 36, 88–9, 94; in Sierra Leone, 28, 60, 65, 91–3, 175–7, 234–5; and Sierra Rutile mine, 161; and Stabilco, 143; unbundling of, 123, 148; and UNITA, 27, 52, 57, 172, 177, 260
Erasmus, J.C., 53
Eric SA (French PMC), 50
Eritrean War of Independence, 29
Ethiopia, 229, 246
ethnic identity, 227
ethnic pogroms, 210, 217, 219, 223
ethnicity, politicisation of, 32, 220, 221
EU (European Union), 101
Executive Research Associates, 59

FAA (Forcas Armadas Angolanas), 52

Falconer Systems, 52

Falconstar, 47

Fanon, Frantz, 121

FAR (Forces Armeés Rwandaise), 124, 126, 136

FAZ (Zairean Armed Forces), 124, 126, 127, 133, 135, 137–8, 145–6; counter-offensive, 144; mercenary training, 182

FLEC (Cabinda), 57

FNLA, 21–2, 128

Foccart, Jacques, 20, 135

Foday Sankoh, Alfred, 84, 95, 106, 108

France, 27, 50, 224; African colonies, 17, 20; and Congo, 118, 127, 131, 137; links with Serbs, 138–9; use of intelligence services, 70, 138, 139

Franklin, Marc, 162

French mercenaries, 123; Congo, 130, 134–5, 137, 181

Friedland, Eric, 158

Friedland, Robert, 69, 161; rivalry with Boulle, 157–8, 161, 162

Front Line States (1974), 251

Frontline Security Forces, 77

Gambia, 223

Gearhart, Daniel, 22

Gecamines (Congo), 155

Gemini Video Productions, 66

Geneva Convention, Article 47, 21, 170

GeoLink, 55, 139

Ghana, 101, 109, 112–13n, 243

Ghanda, Isa, Zairean diplomat, 129, 142

globalisation, 32, 39

GMR Pty Ltd, 56

Golan, Amos, 58

Goma (Congo), 124, 125–6

governance: concept of, 211; as conflict management tool, 14; need to address issues of, 226–7; OAU and, 215

Gray Security, 53, 186

Grillo, Gustavo Marcelo, 22

Group 4, 88

Grunberg, Michael, 28, 67; and DiamondWorks, 25, 181; and Sandline, 60, 64–5, 100

Guevara, Che, 35, 125

Guinea, 6, 20, 101, 218, 223

Guinea-Bissau, 218, 223, 249

Gulf War, 163

Gurkha Security Guards (GSG), 77, 163, 164, 183; in Sierra Leone, 87–8, 158, 159–60, 183

Hansard Management Services, 69

Hansard Trust Company, 69

Harms Commission investigation, 49

Hayes, John, NZ official, 179

Heritage Oil & Gas, 24, 51, 64–5, 190; and Sandline, 61, 100

Hinga-Norman, Sam, 81, 95, 107

Hoare, 'Mad Mike', 6, 17, 20, 22

Hogg, Douglas, 106

Honey Badger Arms and Ammunition, 63

Hood, David, 106

Hooper, Jim, 174

Howe, Herb, 13, 27, 174

Hugo, Renier, 141, 142

human rights, 7, 146, 172; record of PMCs, 6, 172, 174, 187, 234

Human Rights Watch: in Angola, 173, 174; Colombia, 186–7; in Congo, 126; on EO in Burundi, 56; in Sierra Leone, 175

humanitarian aid, by PMCs, 31–2

Hutus, in Congo, 126, 146, 222

Ibis Air, 24, 52, 66, 68, 99, 109; in Angola, 172; EO and, 176; in Kenya, 61

ICI (International Chartered Incorporated), 105

IDAS, 57, 162–3, 166; in Angola, 163–5; demining, 163; and Kabila, 56; network connections, 57, 69

ideology, as mercenary motive, 35

IMF, 4; and Sierra Leone, 91, 166

Indigo Sky Gems, Namibia, 65

Interahamwe (Hutu militia), 126, 146, 222

internal conflicts: changing nature of, 15, 200, 202; exacerbated by end

of Cold War, 6, 210, 257–8;
nature of African, 15, 26; pattern
in West Africa, 85; and proliferation of light weapons, 26; revolt
against authoritarianism, 31; root
causes, 29, 37, 39, 217, 237–8,
262–3; use of extreme violence,
219–20; use of light weapons, 26
International Alert, 88, 89
international community: attitude to
PMCs, 44; intervention fatigue, 2;
morality of, 261; reluctance to
help with conflict resolution, 31,
38, 45, 239; responsibilities of,
6–7
International Consulting Services, 58
International Convention against the
Use, Financing and Training of
Mercenaries, 21
International Criminal Court, 38–9,
236
international law, 7, 29, 203; and
mercenarism, 34–7, 38–9
Intersec, 47
intervention, multilateral, 2, 199,
244; by African states, 204,
215–16, 241, 243; and Cold War,
3; duty of, 218; unilateral, 223
Investment Surveys, 63
IRIS, in Angola, 59
Israel, legal controls on PMCs, 70
Israeli mercenaries, 26, 106, 173,
183
Ivanhoe Capital Investments, 158

J&S Franklin, 87, 158–60, 162, 163,
178, 183
J & P Security Ltd, in Sierra Leone, 88
Janssen, General, 118, 120
Jean-Pierre, Michel, IDAS, 162, 163
Johnson, Jim, KMS, 47
Joubert, Joup, 105, 143
Jupiter Mining Company, 98, 99,
101, 105

Kabba, Allie, MAP leader, 84, 85
Kabbah, Tejan, president of Sierra
Leone, 28, 85, 106; and 1996
elections, 91; continuing
opposition to, 107–8, 110;
counter-coup, 97–8, 176,

179–80; and EO, 91–2, 100; and
Sandline, 60, 95, 98–9; strategic
policy, 92–6, 104; UK support for,
101–2, 107
Kabila, Laurent, Congo (DRC), 35,
55–6, 120; and America Mineral
Fields, 56, 57, 155–6, 157, 164;
government of Congo, 146–7,
155; PMC contracts, 56, 58–9; rise
of, 122, 125, 126, 145
Kagame, Paul, Rwanda, 123, 126
Kamajors (Sierra Leone militia), 80,
89, 91, 94, 95, 99; new role for,
104, 107
Kanu, Abu, RUF leader, 84
Kardan Investment, 57
Karefa-Smart, John, SLPP, 79
KAS Enterprises, 47
Kasavubu, Joseph, Congo, 119, 120
Kaunda, Kenneth, 18
Kengo wa Dondo, Leon, Congo, 127,
132, 139, 142, 156
Kenya, 20, 61, 62
Kerekou, Matthew, Benin, 22
Khobe, Brig-Gen Maxwell, ECOMOG,
104
Kisangani (Congo), 144–5
Klein, Lt Col Yair, 106
KMS company, 47
Koroma, Johnny Paul, Sierra Leone,
60, 96, 98, 175
Kosovo, 32, 34, 138, 180
Kulinda Security Ltd, 46

landmines, 38, 108
Le Carro, Alain, White Legion, 55,
130, 181
Le Roux, Mauritz, 53, 59, 141, 142,
182
legislation, to ban mercenaries, 7, 38,
170–1, 236
Leibenberg, Ian, EO, 57–8
Leopold II, King of the Belgians, 117
Lesotho, 218, 223
Levdan (Israeli PMC), 57
Liberia, 86, 101, 108, 218, 219;
mercenary activity in, 60, 62
Libya, 84, 108, 147
LifeGuard Management, 24, 53, 64,
66, 68

LifeGuard Systems, in Sierra Leone, 92, 100, 105, 175
Likulia Bolongo, Congo, 142
Lintveldt, Johannes, 142
Lissouba, Henry, president of Congo-Brazzaville, 26, 57–8, 146, 180, 183, 188
Longreach Pty Ltd, 47, 53, 63
Lonrho, mine security, 18, 47, 183
Luitingh, Lafras, 66, 130–1, 143; and EO, 50, 57, 67–8
Lumumba, Patrice, Congo, 6, 17, 118, 119, 120

Maada Bio, Brig Julius, Sierra Leone, 89, 90
Mackenzie, Robert, 87, 88, 158, 159
McKerron, Allan, 161–2
Mackinlay, A., MP, 24
McKinney, Cynthia, 155
Madagascar, 61
Madima, Gen Mavua, Congo, 134
Mahele Liego Bokungu, General Marc, 142, 145
Malacrino, Capt Dominique, 135
Malawi, 61, 62
Mali, 226, 230, 243
Mallants, Col Willy, 164
Mandela Doctrine (duty of intervention), 218
Mandela, Nelson, 122, 218
Mann, Simon, 50, 51, 60, 100
Mansaray, Rashid, RUF leader, 84
MAP (Mass Awareness and Participation), Sierra Leone, 84
Marchiani, Jean-Charles, 138, 139
Margai, Albert, Sierra Leone, 79
Margai, Charles, 107
Margai, Sir Milton, SLPP, 79
Marine Protection Services, 77
market capitalism, 3
masculinity, and militarism, 227
Maynard, James, with GSG, 87
Mbeki, Thabo, 122, 174
Mechem, 52, 66
mercenaries, 5, 16, 45–6, **265–74**; in 1960s wars, 6, 17–18, 20–1, 43, 120–1; in Angola, 21–2; colonial use of, 1, 5, 16; financial motive, 23, 35, 199, 235–6, 259–60; impact on conflicts, 1,

199–200, 235–8, 257, 263; individual, 261; Mafia-backed, 261; modern firms see PMCs; pay, 40n, 71, 137, 142, 200; supply of, 4, 40n, 47–8, 70–1; support for policies of home governments, 22, 35, see also PMCs
mercenarism: definitions, 16–19, 170; future for, 28–9, 259–60, 263; legislation against, 7, 38, 170–1; long-term elimination of, 244–5, 254; OAU definition, 21, 34–6, 201–2; regulation of, 34–8, 190, 233–4, 260; UN definition, 36–7, 208–9n
Middle East, 18, 46
migrations, forced, 210
MIL Investments, 157, 161–2
militarism, 26, 220–1, 226–8; and masculine identity, 227
Miller, Harold, Stabilco, 141–2
mineral resources: Branch-Heritage companies, 65; control as factor in conflicts, 219; diamond concessions, 23–4, 91–2; exploitation, 4, 234–5; illegal trade in, 7; PMC interests in, 123–4
mining companies, transnational, 1, 26, 45, 77, 123, 201
mining concessions, 23–4, 65, 190, 261–2; Congo, 155–6, 157–8; Sandline, 98–9, 105, 178; Sierra Leone, 24, 91–2, 160
Mirow, Jean, GeoLink, 139
Mitterand, François, 130
Mobuto Sese Seko, President of Zaire, 22, 31, 35, 120; escape to Chad, 143–4; fall of, 122, 144–5; illness, 126–7, 132, 133–4; personal rule and wealth, 127, 129; rise of, 118, 119; use of mercenaries, 55–6, 120–1, 127, 128–9, 181–2
Moi, Raymond, Kenya, 61
MOJA, in Liberia, 83
Momoh, Maj-Gen Joseph, Sierra Leone, 83, 160
Montoya, Robert, White Legion, 55, 130, 181
Morafono, Fred, 105
Morgenthau, Hans, 30
Morrison, Alastair, DSL, 47, 184

Mozambique, 6, 17, 47, 49, 159, 230; PMCs in, 61, 62; UN monitoring, 238
MPLA, 21–2, 27
MPRI (Military Professional Resources Inc), 18, 54, 105, 172, 263; and Kabila, 55, 58; in Liberia, 62
Mugabe, Robert, 252
Muhammad, Gen. Murtala, Nigeria, 103
Mulele, Pierre, Congo, 119
multilateral intervention, 2, 199, 244
multilateralism, 258–9, 262
Museveni, Yoweri, Uganda, 122–3, 126, 145
Myres, Andrew, with GSG, 87

Namibia, 49, 147, 238
Nammock, John, 22
National Islamic Front, 35
NATO, in Kosovo, 32, 34
negotiation, failure of, 2
Neimöller, Johan, 58–9
neoliberalism, 3
Netherlands, VOC trading company, 43
Neto, Dr Agostinho, MPLA (Angola), 21
NGOs (humanitarian), contracts with DSL, 48, 62
Nguesso, Denis, Congo-Brazzaville, 146, 183, 188, 221
Niger, army coup, 223
Nigeria, 101, 221; civil war (1967-9), 20, 36; EO security for gold fields, 61–2; foreign policy, 102–3; instability, 101–2, 249–50; regional hegemony, 32–3, 34, 218, 240, 250, 253–4; role in ECOMOG, 32, 93, 103, 104, 249–50; role in Sierra Leone civil war, 93, 95, 97–8, 99, 110
Nkrumah, Kwame, president of Nigeria, and regional security, 215–16, 245
non-interference, principle of, 36, 202–4, 216, 218
Nord Resources, 158, 161, 166
Norman, S.H., Sierra Leone, 93

NPRC (Sierra Leone), 24
Nyasaland, 20
Nzimbi Ngbale, Gen, 129, 133–4, 142, 144, 145–6

OAS, 20, 137
OAU, 17, 31, 206n, 258; Convention for Elimination of Mercenaries (1972), 286–8; Convention on Elimination of Mercenarism (CEMA), 21, 34, 44, 198–205, 246, 275–80; declaration on mercenaries (1971), 20–1, 283–5; definition of mercenary, 34–6, 201–2; errors in founding Charter, 215–16, 245–7; and issue of governance, 215, 246; need for review of charter, 37, 204, 205, 247; Ouagadougou summit (1998), 215–16, 218; principle of non-interference, 36, 202–4, 216, 245–6; Resolution on Activities of Mercenaries (1967), 281–2; review of conflict resolution, 36, 247; and Sierra Leone, 89–90, 96–7; view of Tshombe, 41–2n
Obasanjo, Gen Olusegun, Nigeria, 103
Occidental air company, 109
O'Connell, D.C., EO, 173
Omega Support Ltd, 53, 56, 59, 63
OPM Support Systems, 66
Oppenheimer, Harry, de Beers, 46
Opperman, Daan, Panasec, 63

Pakenham, Thomas, 118
Palemis, Milorad, 55, 138
Palm, Nicolaas, EO, 66
Palmer, Philip, Sierra Leone, 95
Pan Africanism, 215, 216, 246
PANAFU, in Sierra Leone, 83, 85
Panasec Corporate Dynamics, 53, 63
Papua New Guinea, 25, 177, 184; Sandline operation, 24, 26–7, 60–1, 68, 162, 177–9
Pasqua, Charles, 138–9
Patel, Samir, 98
peacekeeping: ACRI proposals, 224, 242–4; multilateral, 2, 260, 262–3; and need for standing

army, 32–3, 243, 250; OAU view of, 246–7; regional, 32–3, 249–50, 251–3, 262–3, *see also* ECOMOG; UN

Penfold, Peter, British High Commissioner, 60, 98, 99, 101

Perette, Philippe, GeoLink, 139

Plaza 107 Ltd, 24, 25, 64, 67; and Sandline–EO links, 68–9, 100

PMCs (private military companies), 23, 25, 45, 188–91, 221–2; achievement of short-term stability, 28, 32, 45, 175, 204–5; advantages of, 2, 6, 13, 27; in Angola (1993-97), 50–4; and arms trade, 26–7, 262; in Central Africa, 54–9, 148–9; competition between, 27, 189; in Congo, 130–1, 133, 134–45, 148; connections with TNCs, 16, 44, 45; contacts with Western governments, 44, 45, 46, 189; cost to contractors, 199, 234–5; demining specialists, 45; effectiveness of, 46, 233–4; human rights record, 6, 172, 174, 187, 234; links with mineral companies, 4–5, 59–60, 63–70, 109, 261–2; links to African governments, 26, 44, 189; mining concessions, 23–4; networks, 7, 23–5, 65, 66; payment of, 51, 65, 143, 175; payment problems, 101, 137, 189–90; pre-1990 firms, 46–9; shifting loyalties, 27, 123, 260; strategic alliances between, 123; training contracts, 31, *see also especially* DiamondWorks; DSL; EO; mercenaries; Sandline; Stabilco

poverty, 6, 39, 82–3, 149

private security companies, 43–4, 169, 183–4, 214–15, 222, 261; local, 184, 192; power of, 261; regulation, 193

public relations, PMC use of, 23

Punic Wars, 5

Ranger Oil West Africa Ltd, 51

Rapport Research & Analysis Ltd, 88

Ratte, Willem, 49

realist school of conflict management, 14–15, 38

recruitment: by PMCs, 47–8, 70–1, 169, 200; illegal, 170–1

refugees, from Liberia, 249

regional hegemonies, 218, 253, 254; capacity of, 240–1; role in conflict resolution, 240–2, 262; structural weaknesses within, 249, 258

regionalism: and intervention, 241, 243, 247, 249; mistrust within, 147–8; Museveni's proposals for, 122–3; Nkrumah's ideas for security, 215–16; and prospects for security mechanisms, 32–4, 111, 218, 225, 251–2, *see also* ECOMOG; ECOWAS; SADC

religious conflict, 210

RENAMO, 159

Rhodesia, 20, *see also* Zimbabwe

Rio Tinto Zinc, 177

Ripley, Tim, 87–8

Roberto, Holden, FNLA, 21–2

Ronco, in Rwanda, 54

RPF, 54, 56, 124

RUF (Sierra Leone), 77, 84–7, 91, 93–6, 108, 159–60; violence of, 84, 85–6

Rundels, Col Fred, 106

Russia, 63, 261; helicopters from, 60, 89, 134

Russian pilots, 105

Rwanda, 6, 217, 218, 221; and Congo, 35, 122, 123; and PMCs, 54–5, 62; Tutsi-led rebellion into Congo, 124–6

SADC, 218; compared with ECOWAS, 253, 254; ISDSC, 251, 253; peace operations, 34, 251; regional security, 251–2; structural weaknesses, 252–3

SADF, 47, 128, 129; redundant personnel, 48–9, 71, 141, 200

SafeNet, 59, 141

Saladin Security, 47

Saleh, Maj-Gen Salim, Uganda, 61

Sanders, Robin, US NSC, 155

Sandline International, 13, 19, 100, 158; aircraft for Nigerian forces, 105; mineral concessions, 98–9,

105, 178; other operations in
Africa, 60–1; ownership of, 24–5;
and Papua New Guinea, 24, 26–7,
60–1, 68, 162, 177–9; payment
problems, 189–90; relationship
with EO, 60, 64, 68; in Sierra
Leone, 19, 60, 92, 98–9, 100–1,
102–3, 104–5, 179–80; as
successor to EO, 61, 67; UK parliamentary inquiry, 14, 19, 24–5,
45; value of contracts, 179, 191,
192
Sapelli SARL, 58
Saracen International, 52, 56, 66,
173
Saracen Uganda, 61, 66
Savimbi, Dr Jonas, UNITA, 21, 139
Saxena, Rakesh, 98, 101
Schramme, Jacques, 6, 17, 20, 135
Schultz, Kobus, 63
Secrets, in Cameroon, 50
security, 29, 211; and demilitarisa-
tion, 226–8; and demonisation,
214; human, 227–8; notion of
African, 26, 30–4, 37–8; privatisa-
tion of, 221–2; as racket, 212–15;
ruler's dilemma, 211–12, 213;
selective, 214
Security Advisory Services (SAS) Ltd,
47
security vacuums, 31, 217
Senegal, 41n, 101, 218, 226, 249
Serb mercenaries: in Angola, 173; in
Congo, 123, 134, 137, 138–40,
181–2
Serbia, 140
Sexwale, Tokyo, 50
Seychelles, 6, 22, 47
Shafer, Nico, 106
Shearer, David, 13, 65, 174
Shibata Security, 53, 66, 172–3
Shield Security, 63
Sieromco, 175
Sierra Leone, 27, 29, 78, 219; Armed
Forces Revolutionary Council, 96,
98, 106; army revolt, 223; civil
war, 76–7, 83–9; colonial history,
78–9; cost of PMCs, 199–200,
234–5; early private security,
77–8; ECOMOG forces in, 104–6,
218; economic decline, 78, 82–3,

86–7; elections (1996), 90–1; EO
in, 28, 60, 65, 91, 175–7, 200;
EO's role in, 36, 88–9, 94; ethnic
identities, 79–80; Kabbah's
counter-coup, 97–8, 99; military
coup (May 1997), 95–7, 110;
military and defence structures,
79–81, 86–7; mineral resources,
77, 78, 82, 104; mineral transnationals, 158–63; mining
concessions, 24, 91–2, 160; pact
with Nigeria, 93; peace agreement,
89–90; PMCs in, 35, 59–60, 87–9,
91–4, 104–6, 110–11; political
collapse, 79, 81, 82; prospects for
peace and democracy in, 106–8,
109–11; rise of RUF, 77, 84–7;
and Sandline, 19, 60, 92, 98–9,
100–1, 102–3, 179–80; UN
embargo, 97, 99, 108–9;
Watchguard operation, 47;
weapons proliferation in, 108–9,
110, see also Kabbah, Tejan
Sierra Leone All People's Congress,
79
Sierra Leone People's Party (SLPP),
79, 91, 106, 107
Sierra Rutile mine, 158, 161, 166,
175, 180
Silver Shadow PMC, 58, 187
Simpson, Dan, US ambassador, 136
Singirok, Brig-Gen, PNG commander,
162, 179
Sky Air, 109
Slovak Republic, arms from, 109
Smith, John, 53
social relations, breakdown of, 220
Société Commerciale Ouest Africaine,
17
SOCIR oil refinery, Congo, 58
Somalia, 226, 246; UN peacekeeping
mission, 6, 32, 238, 239
Sonangol, Angolan parastatal, 51,
53, 64, 172
SorussAir, 89, 109
South Africa, 122, 226; arms
embargo, 48–50; Defence Review,
227, 228–9; definition of security,
227–8; Foreign Military
Assistance Law, 38, 62, 70, 171;
Inkatha Freedom Party, 221; links

with UNITA, 59; military
  reductions, 17, 48–9; and origins
  of EO, 48–50; political transition,
  48, 202, 217; private security
  firms, 169; regional hegemony,
  218, 240, 254; and SADC, 252–3
South African mercenaries, 123,
  137; in Angola, 173; in Congo,
  140–4, 146, 181; in Congo-
  Brazzaville, 183
South African PMCs, 63, 169; in
  Angola, 59; in Congo (DRC), 58,
  134; origins of, 44, 47, 48–50, 63
sovereignty: based on community
  consensus, 37, 258; and market
  capitalism, 3; of nation state, 36,
  203, 246; and national security,
  29; and regional institutions,
  248–9, 253; as responsibility, 218
Soviet Union, and Congo, 119
Spaarwater, Maritz, 63
Special Projects Services Ltd, 77
Specialist Services International
  (German), 77
Spicer, Tim, 6, 17–18, 181; and
  Papua New Guinea operation, 27,
  68, 179; Sandline, 19, 24–5, 67,
  98, 99; in Sierra Leone, 60, 98
SRC, 52, 64, 66
Stabilco, 59, 63, 182; in Angola, 53;
  in Congo, 55, 134, 137, 141–3;
  and EO, 143
Stanisic, Jovica, 140
state, 15; failures, 217, 224;
  legitimacy of, 260; regime security
  and survival of, 29, 212–15; and
  responsibility for security, 228;
  and sovereignty, 36, 37, 203, see
  also boundaries
Stevens, Siaka, Sierra Leone, 79, 81,
  82–3, 84
Stewart, Robert, 164
Steyl, Crause, 52
Stilltoe, Sir Percy, 46, 77
Stirling, Col David, 20, 47;
  Watchguard, 18, 19, 46–7
Strasser, Valentine, Sierra Leone
  NPRC, 86, 88–9, 159–60, 161
Strategic Concepts Pty Ltd, 59, 63
Stringham, Joe (US Brigadier retd.),
  57

Stuart Mills International, 66, 68,
  173
student movements, Sierra Leone,
  83, 84
Sudan, 29, 35, 62, 226, 246; arms
  for, 62, 188
Sunshine Boulle, 77, 160–1
Sunshine Mining, 69, 161

Tanganyika/Tanzania, 20, 215, 226,
  251
Tarawali, Major Abou, 87
Tavernier, Christian: in Algeria, 20;
  in Congo/Zaire, 20, 22, 55, 133,
  134–8, 145, 181
Taylor, Charles, Liberian NPF, 84,
  86, 106, 160
Teleservices, in Angola, 53, 185–6
Tensae, Tsadkan Gebre, 229
terror, use of, 219
Thompson-Armement, 139
Tindemans, Leo, Belgian prime
  minister, 136
Togo, 217
Touré, Ahmed, 101
trading companies, private security
  forces, 17, 43, 200
training, 26, 43–4
TransAfrica Logistics, 66
transnational corporations (TNCs):
  mining, 1, 26, 45, 77, 123, 201;
  oil, 51, 57, 64; in Sierra Leone,
  158–63; use of PMCs, 44, 45, 193,
  261
transparency, 228–9
Tshombe, Moise, 20, 118, 119–20
Tunisia, 243
Turle, Arish, 47
Tutsi: Rwandan incursion into
  Congo, 124–6; in Zaire, 122, 125,
  147

Uganda, 122, 185, 218, 226; and
  ACRI, 243; EO and, 61, 131
Ugandan troops, in Congo, 35
UK, 224; and African independence
  movements, 20; attitude to PMCs,
  44, 70; Foreign Enlistment Act
  (1870), 170–1; Labour government's ethical foreign policy, 99,
  101–3; links with J&S Franklin,

158–9, 165; Military Assistance Training Teams, 43; origin of early PMCs, 46–7; parliamentary inquiry into Sandline, 14, 19, 24–5, 45, 102; policy on Nigeria, 101–2; provision of arms by, 27; Secret Intelligence Service (SIS), 46; and Sierra Leone, 107, 159–60; tacit support for Sierra Leone operations, 88, 99, 102–3, 180

Ukrainian mercenaries, 173, 183
Ukrainian pilots: in Congo, 134, 137–8, 181; for Nigerian forces, 105, 106
Umkhonto weSizwe, trained by EO, 50
UN: contracts with DSL, 48, 62; definition of mercenaries, 36–7, 208–9n, 246; overall responsibility, 34, 263; Resolution 1132 (1997), 97, 108; Resolution (1989), 44; and Sierra Leone peace talks, 89–90, 97
UN Commission on Human Rights, role for, 191–2
UN peacekeeping: in Congo (1961), 20, 119, 215; cost of, 235; Somalia, 6, 32; traditional monitoring role, 238–9
UN Special Rapporteur, 24, 179, 181–2, 184; report, 170, 171–2, 176, **289–320**; role of Commission on Human Rights, 191–2
UNHCR (UN High Commission for Refugees), 31
Unison, 18, 19
UNITA, 21, 172, 173–5; and Congo, 128, 130, 143, 144, 147; control of oil industry, 51, 64, 163–4; and EO, 27, 52, 57, 172, 260; PMC assistance for, 59
United African Company, 17
United States: and ACRI, 224, 242–4; attitude to PMCs, 18, 44, 70; and Belgian Congo, 118; Defence Intelligence Agency, 165; DSL contracts, 185; military training in Rwanda, 56, 125;

MPRI licences refused, 54, 55; Neutrality Act (1937), 170; reluctance to commit forces, 2–3; and Sierra Leone, 107, 180; view of Africa, 39n
USDS, subsidiary of DSL, 58

van Heerden, Roelf, 141–2
Vietnam, 2–3, 18, 70
vigilantism, 222
Vinnell Corporation (US), 18, 70
Vuono, Carl (US General), 18

Walker, David, KMS, 47
Walker, Gen Sir Walter, UNISON, 18, 19
Walsham, Bruce, 181
Watchguard International, 18, 19, 46–7
weapons: inhumane and illegal, 7, 174–5; and masculinity, 227; rising costs of, 213; Sandline contracts, 191, 192, see also arms proliferation; arms trade
Western governments: military advisors in Africa, 48; new interventions, 224; policy towards PMC activity, 45; security of interests, 64, 217, 247
White Legion, in Zaire, 55, 181–2
Wibaux, Fernand, 135
Williamson, Craig, Longreach, 47, 56
women: and masculinism, 226; role in peace movements, 230
World Bank (IBRD), 4; contracts with DSL, 48, 62
Wrangal Medical, 66, 68

Yale, Jean Seti, 139
Yerine, Col Zev, Levdan, 57
Yugo, 'Col Dominic', 134, 138

Zaire, 120, 215; Katanga rebellion, 22, 118, 119; PMCs in, 55, 120–1, see also Congo, Democratic Republic (DRC)
Zambia, 47, 61, 184, 223, 251
Zimbabwe, 147, 226; and SADC, 252–3, 254

Printed and bound by CPI Group (UK) Ltd, Croydon, CR0 4YY

27/10/2024

14580224-0005